# Electrical and Electronic Engineering Principles

# Electrical and Electronic Engineering Principles

**Noel M Morris**
Consultant
and formerly Principal Lecturer,
Staffordshire University

 LONGMAN

**Pearson Education Limited**
Edinburgh Gate, Harlow
Essex, CM20 2JE, England
*and Associated Companies throughout the world.*

© Longman Group UK Limited 1994

First published 1994
Second impression 1995
Third impression 1997
Fourth impression 1998
Fifth impression 2000

**British Library Cataloguing in Publication Data**
A CIP record for this book is available from the British Library

ISBN 0 582 098157

Set by 6 in Monotype Times 10/12 pt
Printed in Malaysia, VVP

# Contents

# Preface

Electrical and Electronic Principles at the National Certificate and Diploma level cover an impressively wide range of topics. The reader will find in this book information about each of these topics, and will provide more than adequate support for any additional electrical or electronic subjects studied.

This book is unique in providing coverage of computer software for the solution of circuits and applications in both Certificate and Diploma courses. The software described in the text is widely available at colleges, and can be obtained for use at home from addresses given later in the book. The software described includes SPICE (Simulation Program with Integrated Circuit Emphasis), CODAS (Control System Design And Simulation) and PCS (Process Control Simulation).

The purpose of the first of these is to solve almost any electrical circuit problem from a simple d.c. circuit, through to polyphase and transient problems (and much more!). The latter two items of software can be used to solve control system and process control problems, respectively. The software can be used in support of lectures, projects and practical work.

The advice and assistance of Mr M. Cole and Mr P. Goss, respectively Managing Director and Technical Manager of ARS Microsystems, Dr J. Golten and Mr A. Verwer of Golten and Verwer Partners, and Mr A. Lewis, Head of Management Services, Longdean, Hemel Hempstead are gratefully acknowledged.

Finally, I would like to thank my wife for her assistance and encouragement during the writing and preparation of the book.

Noel M. Morris

# 1 Units and the electrical circuit

## 1.1 Basic units

The SI system of units (Système International d'Unités) has been adopted as an international unified system; this defines basic units for length, mass, time, electric current, etc., the unit abbreviations being listed in Table 1.1.

**Table 1.1** Fundamental units

| Quantity | Symbol | Unit | Abbreviation |
|---|---|---|---|
| Length | $L$ or $l$ | metre | m |
| Mass | $M$ or $m$ | kilogram | kg |
| Time | $T$ or $t$ | second | s |
| Electric current | $I$ or $i$ | ampere | A |
| Absolute temperature | $\theta$ | kelvin | K |
| Luminous intensity | $I$ | candela | cd |

Where a quantity, such as current, varies with time so that, at one moment it has a value, say, of 10 A, and a short time later it may be 12 A; in this case we say that the quantity has an **instantaneous value** at a particular time. Such a quantity is assigned a lower-case alphabetical symbol ($i$ in this case); thus we may say that $i = 10\,\text{A}$ when $t = 20\,\text{s}$, and $i = 12\,\text{A}$ when $t = 22\,\text{s}$.

When the quantity reaches its **steady-state value**, it is assigned an upper-case (capital) character ($I$ in this case). For example, if the steady current in the circuit is 20 A, then we say that $I = 20\,\text{A}$.

## 1.2 SI multiples

To satisfy the wide variety of values involved in engineering, the SI system has a range of multiples to accommodate them. For example, the unit of electrical power, the watt, is too small to describe the power produced by a power station, and is too large to describe the power dissipated in a transistor. Selected multiples are listed in Table 1.2. Examples of the use of multiples include:

$$1.5\,\text{kA} = 1500\,\text{A}$$

$$2.3\,\text{mW} = 0.0023\,\text{W}$$

## 1.3 Atomic structure

**Table 1.2** SI multiples

| Prefix | Abbreviation | Multiple |
|--------|--------------|----------|
| tera- | T | $10^{12}$ |
| giga- | G | $10^{9}$ |
| mega- | M | $10^{6}$ |
| kilo- | k | $10^{3}$ |
| centi- | c | $10^{-2}$ |
| milli- | m | $10^{-3}$ |
| micro- | $\mu$ | $10^{-6}$ |
| nano- | n | $10^{-9}$ |
| pico- | p | $10^{-12}$ |
| femto- | f | $10^{-15}$ |
| atto- | a | $10^{-18}$ |

Every **element** consists of smaller parts known as **atoms**, and each of these is built up from even smaller particles, the principal particles being **protons, electrons** and **neutrons**. So far as we are concerned in electrical engineering, the main difference between them is their relative size and the electrical charge they carry.

A proton is 1840 times more 'massive' than an electron, and it carries a positive charge; an electron carries an equal but opposite (negative) charge. A neutron has the same mass as the proton, but carries no electrical charge.

*An atom has no overall electrical charge*, and carries an equal number of protons and electrons. The relatively massive protons (and neutrons) are concentrated at the centre or **nucleus** of the atom, and the less massive electrons orbit around the nucleus in **layers, energy bands** or **shells**. It is the electrons in the outermost shell (known as the *valence shell*) which interest electrical engineers, for these are the electrons which are most easily 'liberated' from the atom, and enable current to 'flow' in a circuit.

A **conductor** (usually a metal) offers very little resistance to current flow because the outer shells overlap, and the transfer of electrons between them is easy.

An **insulator** (usually a non-metal such as glass or plastic) has a very high resistance to current flow. One reason for this is that the valence shells of the atoms do not overlap, making transfer of electrons difficult.

A **semiconductor** has a resistance which is midway between that of a good conductor and a good insulator. Many semiconductors have properties which make them specially applicable to electronics.

## 1.4 Direct current and alternating current

There are two main types of electrical current, namely **direct current** (d.c.) and **alternating current** (a.c.). In a direct current circuit, the current flows in one direction only, that is the current is **unidirectional**; the current from a battery is of this kind. In an alternating current circuit, the current flows in one direction at one instant of time and, a few moments later, the direction of the current reverses; that is, the current *alternates*. For the moment we concentrate on d.c. circuits.

## 1.5 Electrical charge ($Q$) and current ($I$)

The unit of **charge** (symbol $Q$) is the **coulomb** (unit symbol C); the charge carried by an electron is $-1.602 \times 10^{-19}$ C (a very small charge indeed!)

Electrical **current** (symbol $I$) whose unit is the **ampere** (unit symbol A), is the *rate of movement of charge through a circuit*; a current of 1 A represents the movement of over six million billion electrons per second! The relationship between electrical charge, current and time is given by

$$Q = It \qquad\qquad [1.1]$$

If $I$ is in amperes and $t$ in seconds, then $Q$ is in coulombs.

Current can be measured by any one of several methods, the simplest being by measuring the mechanical force on a current-carrying conductor on a magnetic field. From this point of view, the ampere is defined as follows:

> **When a current of one ampere flows in each of two infinitely long parallel conductors placed 1 m apart in a vacuum, the force between the conductors is $2 \times 10^{-7}$ N per metre length (or $0.2\,\mu N/m$).**

*Worked example 1.1*   If a current of 2.3 A flows through a circuit for 2 min 10 s, calculate the charge which has passed through the circuit.

*Solution*   The total time is

$$t = 2\,\text{m} \; 10\,\text{s} = (2 \times 60) + 10\,\text{s} = 120\,\text{s}$$

From eqn [1.1] we see that

$$Q = It = 2.3 \times 130 = 299\,\text{C}$$

## 1.6 Voltage, potential difference and e.m.f.

**Voltage** is the potential to produce current flow in a circuit. The **potential difference** (p.d.) is the difference in electrical potential between two points in a circuit, and the **electromotive force** (e.m.f.) is the 'force' producing an electric current in a circuit. An e.m.f. can arise from many effects including chemical, magnetic, thermal, etc. E.m.f. and p.d. are measured in **volts** (unit symbol V).

## 1.7 Ohm's law

The relationship between the e.m.f. applied to a resistive circuit and the current in the circuit is given by Ohm's law as follows:

$$E = IR \qquad\qquad [1.2]$$

If $E$ is in volts and $I$ is in amperes, then $R$ is the **electrical resistance** of the circuit in **ohms** (unit symbol $\Omega$). Alternatively, if a p.d. of $V_1$ volts appears between the terminals of resistor $R$, Ohm's law can be written

$$V_1 = IR \qquad\qquad [1.3]$$

*Worked example 1.2*   If 1000 C of electricity passes through a circuit of resistance $20\,\Omega$ in 1 m 40 s, calculate the p.d. across the resistor.

*Solution*    The time involved is $t = 1\,\text{m}\,40\,\text{s} = 100\,\text{s}$ hence

$$I = Q/t = 1000/100 = 10\,\text{A}$$

Ohm's law tells us that

$$\text{Potential difference} = IR = 10 \times 20 = 200\,\text{V}$$

## 1.8 Conductance

**Conductance**, $G$, is the *reciprocal of resistance*, that is

$$G = 1/R \qquad\qquad [1.4]$$

If the resistance is in ohms, the conductance is in **siemens** (unit symbol S). A high value of resistance corresponds to a low value of conductance, and vice versa.

For example, a resistance of $0.01\,\Omega$ has a conductance of $1/0.01 = 100\,\text{S}$, and a resistance of $1\,\text{M}\Omega$ has a conductance of $1/10^6 = 10^{-6}\,\text{S}$.

## 1.9 Resistivity and conductivity

Consider a conductor of resistance of $R$ ohms, which has a length of $l$ metres, a cross-sectional area of $a$ square metres, and which carries a current of $I$ amperes when $V_1$ volts is applied to the ends of the conductor. If two of these conductors are connected end to end, i.e. they are in series with one another, then, to force the same current through them, twice the p.d. must be applied to the circuit. That is, doubling the length of the conductor doubles the resistance, or

**the resistance of the conductor is proportional to the length of the conductor.**

That is

$$R \propto l \qquad\qquad [1.5]$$

If the two conductors are connected in parallel with one another, then each conductor carries one-half of the total current. Connecting the two conductors in parallel has the same effect as doubling the cross-sectional area so, effectively, *doubling the area of the conductor halves the resistance of the circuit*, or

**the resistance of a conductor is inversely proportional to the area of the conductor.**

That is

$$R \propto 1/a \qquad\qquad [1.6]$$

Combining eqns [1.5] and [1.6] gives

$$R \propto l/a \qquad\qquad [1.7]$$

We can remove the proportional sign simply by inserting a constant of proportionality known as the **resistivity** (symbol $\rho$)

of the conductor material, as follows:

$$R = \rho \, \frac{l}{a} \qquad [1.8]$$

From this equation we see that $\rho = Ra/l$, and the dimensions of resistivity are

$$\frac{\text{dimensions of } R \times \text{dimensions of } a}{\text{dimensions of } l}$$

$$= \frac{\text{ohms} \times (\text{length} \times \text{length})}{\text{length}}$$

$$= \text{ohms} \times \text{length or } \Omega \, m$$

The resistivity of a number of materials is given in Table 1.3. The reader should bear in mind that the value quoted in Table 1.3 depends on a number of factors including the method of production of the material (i.e. is it annealed or hard-drawn?), the operating temperature, the purity of the material, etc. The materials manganin, constantan and nichrome have a higher resistivity, and are frequently used in wire-wound resistors. Insulators, such as wood, plastic, mica, etc. have a much higher resistivity than conductors.

**Conductivity** (symbol $\sigma$) is the reciprocal of resistivity, and has dimensions of siemens per metre (S/m), that is

$$\sigma = 1/\rho$$

**Table 1.3**  Resistivity at $0\,^{\circ}C$

| Material | Resistivity ($\Omega \, m$) |
| --- | --- |
| Silver | $1.47 \times 10^{-8}$ |
| Copper | $1.55 \times 10^{-8}$ |
| Aluminium | $2.5 \ \times 10^{-8}$ |
| Zinc | $5.5 \ \times 10^{-8}$ |
| Nickel | $6.2 \ \times 10^{-8}$ |
| Iron | $8.9 \ \times 10^{-8}$ |
| Manganin | $41.5 \ \times 10^{-8}$ |
| Constantan | $49.0 \ \times 10^{-8}$ |
| Nichrome | $108.3 \ \times 10^{-8}$ |
| Plastic | $10^{7} - 10^{9}$ |
| Wood | $10^{8} - 10^{11}$ |
| Glass | $10^{9} - 10^{12}$ |
| Mica and mineral oil | $10^{11} - 10^{15}$ |

***Worked example 1.3***   If the resistivity of annealed copper wire referred to $20\,^{\circ}C$ is $0.0173 \, \mu\Omega \, m$, calculate the resistance at $20\,^{\circ}C$ of a length of wire (a) of length $10 \, m$ and cross-sectional area $1 \, mm^2$ and (b) of length $200 \, m$ and diameter $2 \, mm$.

*Solution*   The resistivity of the copper is $0.0173 \, \mu\Omega \, m$ or $0.0173 \times 10^{-6} \, \Omega \, m$.

(a) The area of the conductor is $1 \times 10^{-6} \, m^2$, hence from eqn [1.8]

$$R = \rho \times l/a = 0.0173 \times 10^{-6} \times 10/1 \times 10^{-6}$$

$$= 0.173 \, \Omega$$

(b) The area of the conductor is

$$\frac{\pi d^2}{4} = \frac{\pi \times (2 \times 10^{-3})^2}{4} = 3.142 \times 10^{-6} \, m^2$$

and

$$R = \rho \times l/a = 0.0173 \times 10^{-6} \times 200/3.142 \times 10^{-6}$$

$$= 1.1 \, \Omega$$

**1.10 Effect of temperature change on resistance**

When the temperature of a conductor rises, the atomic nuclei gain energy and the atoms 'vibrate', *resulting in an increase in resistance to current flow*. Conversely, a reduction in temperature results in a reduction in resistance. Generally speaking, over a wide range of temperature, *the change in resistance is proportional to the temperature change* (in some cases there is a variation to this rule, and this is referred to later).

The effect is shown in the graph in Fig. 1.1. The resistance of the conductor at $0\,°C$ is $R_0$, at $\theta_1\,°C$ it is $R_1$, and at $\theta_2\,°C$ it is $R_2$.

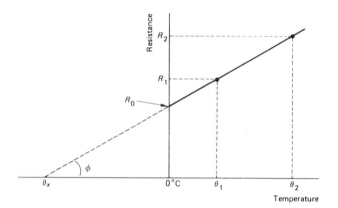

**Fig. 1.1** Effect of temperature change on the resistance of a conductor

The **temperature coefficient of resistance**, $\alpha$, of a conductor is the change of resistance from an original temperature (say $0\,°C$) to some other resistance, say $R_1$, at a new temperature ($\theta_1\,°C$), expressed as a fraction of its original temperature. Referring to Fig. 1.1, the temperature coefficient, $\alpha_0$, referred to $0\,°C$ can be defined as

$$\alpha_0 = \frac{R_1 - R_0}{(\theta_1 - 0)R_0} = \frac{R_1 - R_0}{\theta_1 R_0} = \frac{\tan\phi}{R_0} \qquad [1.9]$$

where $\phi$ is the angle of the graph in Fig. 1.1. Hence

$$\alpha_0 \theta_1 R_0 = R_1 - R_0$$

where $\alpha$ has the dimensions of $(°C)^{-1}$ or 'per $°C$'. Hence

$$R_1 = R_0(1 + \alpha_0 \theta_1) \qquad [1.10]$$

Alternatively, we may calculate the value of $\alpha$ referred to some other temperature, say $\theta_1$, from

$$\alpha_1 = \frac{\tan\phi}{R_1} = \frac{R_2 - R_1}{(\theta_2 - \theta_1)R_1} \qquad [1.11]$$

which is the resistance–temperature coefficient referred to $\theta_1$.

If the graph for a copper conductor is extended until it intersects the zero axis, intersection occurs at $\theta_3 = -234.5\,°C$. This temperature is the *inferred absolute zero temperature* of copper.

Different conducting materials have different inferred absolute zero temperatures, and a selection are listed in Table 1.4.

**Table 1.4**  Inferred absolute zero temperature

| Material | Inferred absolute zero (°C) |
|---|---|
| Gold | −274 |
| Aluminium | −236 |
| Copper | −234.5 |
| Silver | −243 |
| Iron | −162 |

We have assumed that the graph in Fig. 1.1 is linear but, in fact, at its lower end it departs from a straight line and reaches zero resistance at absolute zero of temperature ($-273.15\,°C$).

Manipulation of eqn [1.11] shows that, for a reference temperature $\theta_n$, then

$$\alpha_n = \frac{1}{\theta_n + |\theta_3|} \qquad [1.12]$$

where $|\theta_3|$ is the *modulus* or *magnitude* of the inferred absolute zero temperature $\theta_3$. That is, $\alpha_0$ for copper is

$$\alpha_0 = 1/(0 + 234.5) = 0.004\,264 \text{ per } °C$$

and at a temperature of $20\,°C$

$$\alpha_{20} = 1/(20 + 234.5) = 0.003\,93 \text{ per } °C$$

Further manipulation of the equation gives the relationship between $\alpha_n$ and $\alpha_0$ as

$$\alpha_n = \frac{1}{\theta_n + 1/\alpha_0} \qquad [1.13]$$

It is sometimes useful to relate the value of $R_1$ (see Fig. 1.1) to $R_2$ as follows. From eqn [1.11]

$$R_1 = R_0(1 + \alpha_0\theta_1) \quad \text{and} \quad R_2 = R_0(1 + \alpha_0\theta_2)$$

then

$$\frac{R_2}{R_1} = \frac{R_0(1 + \alpha_0\theta_2)}{R_0(1 + \alpha_0\theta_1)} = \frac{1 + \alpha_0\theta_2}{1 + \alpha_0\theta_1} \qquad [1.14]$$

At very high values of temperature, the graph in Fig. 1.1 curves away from linearity, and the resistance $R_n$ at temperature $\theta_n$ is given by

$$R_n = R_0(1 + \alpha\theta + \beta\theta^2) \qquad [1.15]$$

where $\alpha$ is the *linear temperature coefficient* referred to above, and $\beta$ is the *quadratic temperature coefficient*. For copper the value of $\beta$ is $1.12 \times 10^{-6}$; at normal operating temperature, the quadratic term may be neglected.

Certain resistive materials have a *negative temperature coefficient of resistance*. These employ oxides of copper, cobalt, manganese and nickel, and are made into beads, discs and other shapes and are known as **thermistors**. The equation for the resistance of these elements is given by

$$R = a \, e^{b/T}$$

where $a$ and $b$ are constants (whose value depends on the material), $e = 2.71828$ is the base of Napierian logarithms, and $T = \theta \, °C + 273.15$ is the absolute temperature. Certain thermistors give a relatively large change in resistance with temperature variation, and are widely used for temperature measurement.

*Worked example 1.4*  The resistance of a length of wire is found to be $56.36\,\Omega$ at $30\,°C$, and $62.72\,\Omega$ at $60\,°C$. Calculate the resistance–temperature coefficient referred to $0\,°C$ for the material.

*Solution*  Referring to Fig. 1.1, we have sufficient data to calculate the tangent of the slope angle, $\phi$, of the temperature–resistance graph as follows:

$$\tan \phi = \frac{R_2 - R_1}{\theta_2 - \theta_1} = \frac{62.72 - 56.36}{60 - 30} = 0.212$$

Since we know both $\theta_1$ and $\theta_2$, we can calculate the resistance–temperature coefficient referred to these values as follows:

$$\alpha_{\theta_1} = \tan \phi / R_1 = 0.212/56.36 = 0.003762$$

and

$$\alpha_{\theta_2} = \tan \phi / R_2 = 0.212/62.72 = 0.00338$$

From eqn [1.13] we have

$$\alpha_n = \frac{1}{\theta_n + 1/\alpha_0} \qquad\qquad [1.16]$$

or

$$\alpha_0 = \frac{1}{(1/\alpha_n) - \theta_n}$$

Using $\alpha_{\theta_1}$ and $\theta_1$ we have

$$\alpha_0 = \frac{1}{(1/0.003762) - 30} = 0.00424 \text{ per } °C$$

A similar result is obtained using $\alpha_{\theta_2}$ and $\theta_2$.

**1.11 Electrical energy, W**

When current flows in a resistive circuit, energy is consumed, and is given by

$$W = I^2Rt \qquad [1.17]$$

If $I$ is in amperes, $R$ in ohms and $t$ in seconds, the energy is in *joules* (unit symbol J) or *watt seconds*. Applying Ohm's law to the above equation gives

$$W = EIt = \frac{E^2}{R}t = I^2RT \qquad [1.18]$$

The joule is a small amount of energy (a 60 W lamp consumes 216 000 J in 1 h!). For industrial and commercial purposes the kilowatt hour (kWh) or 1000 watt hour is more useful, where

$$1\,kWh = 1\,kW \times 1\,h = 1000\,W \times (60 \times 60)s$$

$$= 3\,600\,000\,J = 3.6\,MJ$$

That is, a 5 kW heater running for 8 h consumes 40 kWh or 144 MJ of energy.

*Worked example 1.5*  Calculate the amount of electrical energy dissipated in 15 min by a 2 kΩ resistor which carries 120 mA.

*Solution*  The current in amperes is $120 \times 10^{-3} = 0.12$ A. From eqn [1.18] we see that

$$W = I^2Rt = 0.12^2 \times 2000 \times (15 \times 60)$$

$$= 25\,920\,J \text{ or } 7.2\,Wh$$

**1.12 Electrical power, P**

*Power is the rate of doing work*, that is

$$P = \frac{\text{energy}}{\text{time}} = \frac{I^2Rt}{t} = I^2R \text{ watts (W)} \qquad [1.19]$$

Since $I = E/R$, then

$$P = \left[\frac{E}{R}\right]^2 R = \frac{E^2}{R} \qquad [1.20]$$

and

$$P = IR \times I = EI \qquad [1.21]$$

The most popular multiples of power in heavy current work are the kilowatt (kW or $10^3$ W) and the megawatt (MW or $10^6$ W), and in light current work are the milliwatt (mW or $10^{-3}$ W) and the microwatt ($\mu$W or $10^{-6}$ W).

***Worked example 1.6***     If a circuit consumes 5.8 kW when carrying a current of 900 A, calculate (a) the resistance of the circuit and (b) the voltage across the circuit.

*Solution*     (a) From eqn [1.19] $P = I^2R$, hence

$$R = P/I^2 = 5800/900^2 = 7.16 \times 10^{-3}\,\Omega \text{ or } 7.16 \text{ m}\Omega$$

(b) From Ohm's law

$$V = IR = 900 \times 7.16 \times 10^{-3} = 6.44 \text{ V}$$

**1.13   Electrical circuit definitions**

An **electrical network** is a set of interconnected elements such as supply sources, resistors, etc. and an **electrical circuit** is a network containing at least one **closed path**; for our purposes, networks and circuits are equivalent terms.

A **node** is a point in a circuit which is common to two or more elements in a circuit, and a **junction** or **principal node** has three or more elements connected to it. For example, the points labelled A to F in Fig. 1.2 are nodes, and nodes A and C are principal nodes. In the majority of networks, we select one node as a **reference node** or **datum node**, and the voltage at this node is usually taken to be zero.

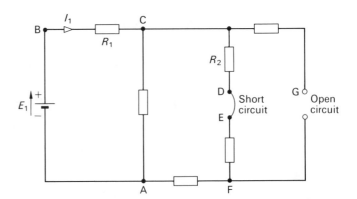

**Fig. 1.2**   An electrical network

A **branch** is a path in a circuit containing one circuit element; for example, in Fig. 1.2, AB is a branch, AF is a branch, CD is a branch, etc. A **path** in a circuit is a set of connected elements that is traversed without passing twice through the same node. Elements $E_1$, $R_1$ and $R_2$ are in the path connecting node A to node D, i.e. the path ABCD.

An **open circuit** exists between points which are not electrically connected; an open circuit exists between points F and G in Fig. 1.2. A **short circuit** exists between two points linked by a conductor of zero resistance; a short circuit exists between points D and E in Fig. 1.2.

A **loop** is a closed path in a network; Fig. 1.2 contains loops ABCA, ACDEFA and ABCDEFA. The reader will note that each loop begins and ends at the same node.

## 1.14 Voltage and current notation

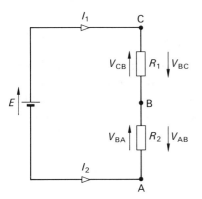

**Fig. 1.3** Voltage and current arrows

When analysing electrical circuits, engineers adopt a notation for voltage and current which tries to avoid many pitfalls, the general notation being illustrated in Fig. 1.3.

A **current arrow** is drawn *on the branch in which it flows* in the *assumed direction* of current flow (*note*: current flows in a circuit). Two current arrows are shown in Fig. 1.3; all we are saying here is that we can assume the current to flow in either direction. Since both currents flow in the same wire, we are merely saying that $I_1 = -I_2$. If $I_1 = 10\,\text{A}$, then $I_2 = -10\,\text{A}$; that is, we can either have a current of $+10\,\text{A}$ flowing in the direction of $I_1$, or $-10\,\text{A}$ flowing in the direction of $I_2$.

A **voltage arrow** is drawn *by the side of the circuit between a pair of nodes*. The arrowhead points towards the node which is assumed to have the more positive potential. We may also use a **double-suffix notation** to describe the voltage as follows. The voltage $V_{CB}$ (drawn to the left of $R_1$ in Fig. 1.3) between nodes C and B in Fig. 1.3 is

$$V_{CB} = \text{potential of node C with respect to node B}$$

$$= V_C - V_B$$

We must ask here 'what do we mean by $V_C$?' The voltage $V_C$ is the potential of node C *relative to the reference node*. We have not, as yet, nominated the reference node, so let it be node A. We may therefore write

$$V_C = V_{CA} \quad \text{and} \quad V_B = V_{BA}$$

By definition, the reference node is the zero voltage node and, for this reason, we can omit the reference node from the definition of $V_B$ or $V_C$; none the less, we need to know which node is the reference node.

Also, since current flows from a node of higher potential to one of lower potential, current $I_1$ enters node C and leaves node B, that is

$$V_{CB} = I_1 R_1$$

and

$$V_{BA} = V_B - V_A = I_1 R_2$$

Alternatively, we may draw the potential arrows in the reverse direction, as shown by $V_{BC}$ and $V_{AB}$. In the case of $V_{BC}$, we have *assumed* that node B is positive with respect to node C, that is

$$V_{BC} = V_B - V_C$$

In this case, we have *assumed* that the positive direction of current is into B and out of C, or

$$V_{BC} = I_2 R_1$$

But, as we have already seen, $I_2 = -I_1$ hence

$$V_{BC} = I_2 R_1 - I_1 R_1$$

Also

$$V_{AB} = V_A - V_B = I_2 R_2 = -I_1 R_2$$

We can also see from the potential arrow associated with the battery, $E$, that node B is positive with respect to node A.

We use common sense when selecting the direction of voltage and current arrows and, for Fig. 1.3, we would normally take the current to be $I_1$, and would assume node C to be positive with respect to node B which, in turn, is positive with respect to node A.

## 1.15 Resistors in series

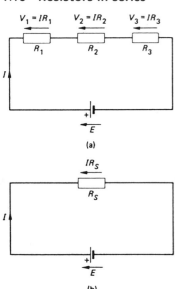

Fig. 1.4 Resistors in series

Resistors are connected in series *when each carries the same current*, as shown in Fig. 1.4; series-connected resistors are often referred to as a *string of resistors*. The total voltage applied to the circuit is the sum of the p.d.s. across the individual resistors; that is, for three resistors in series

$$E = IR_1 + IR_2 + IR_3$$

If $R_E$ is the **equivalent resistance** of the complete circuit, then

$$E = IR_E = IR_1 + IR_2 + IR_3 = I(R_1 + R_2 + R_3)$$

that is

$$R_E = R_1 + R_2 + R_3$$

If there are $n$ resistors in series, the equivalent resistance of the complete circuit is

$$R_E = R_1 + R_2 + \cdots + R_n \qquad [1.22]$$

**Clearly, the equivalent resistance of a series circuit is greater than the greatest individual value of resistance in the circuit.**

## 1.16 Resistors in parallel

Resistors are connected in parallel with one another *when each resistor has the same voltage across it*, as shown in Fig. 1.5. In this case, the sum of the currents in the circuit is equal to the current drawn from the supply, that is

$$I = I_1 + I_2 + \cdots + I_n$$

From Ohm's law $I_1 = E/R_1$, $I_2 = E/R_2$, etc. hence

$$I = \frac{E}{R_1} + \frac{E}{R_2} + \cdots + \frac{E}{R_n}$$

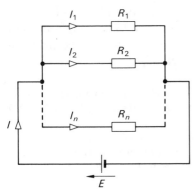

**Fig. 1.5**  Resistors in parallel

If $R_E$ is the *equivalent resistance of the parallel circuit*, then

$$I = \frac{E}{R_E} = \frac{E}{R_1} + \frac{E}{R_2} + \cdots + \frac{E}{R_n}$$

That is to say, the equivalent resistance of the parallel circuits is

$$\frac{1}{R_E} = \frac{1}{R_1} + \frac{1}{R_2} + \cdots + \frac{1}{R_n} \qquad [1.23]$$

**The equivalent resistance of a parallel circuit is lower in value than the lowest individual resistance in the circuit.**

In the special case of two resistors in parallel, the equivalent resistance of the circuit is

$$\frac{1}{R_E} = \frac{1}{R_1} + \frac{1}{R_2} = \frac{R_1 + R_2}{R_1 R_2}$$

or

$$R_E = \frac{R_1 R_2}{R_1 + R_2} \qquad [1.24]$$

In terms of conductance, *the equivalent conductance, $G_E$*, of the parallel circuit is

$$G_E = G_1 + G_2 + \cdots + G_n \qquad [1.25]$$

*Worked example 1.7*  Calculate the equivalent resistance of three resistors of 25, 35 and 40 $\Omega$ connected (a) in series, (b) in parallel. If the supply voltage is 10 V, calculate the power consumed in each case.

*Solution*  (a) For the series connection

$$R_E = R_1 + R_2 + R_3 = 25 + 35 + 40 = 100\,\Omega$$

which is greater than the largest individual resistance in the circuit.
   The power consumed in this circuit is

$$P = V_S^2/R_E = 10^2/100 = 1\,\text{W}$$

(b) For the parallel circuit

$$\frac{1}{R_E} = \frac{1}{R_1} + \frac{1}{R_2} + \frac{1}{R_3} = \frac{1}{25} + \frac{1}{35} + \frac{1}{40}$$

$$= 0.04 + 0.0286 + 0.025 = 0.0936\,S$$

and

$$R_E = 1/0.0936 = 10.68\,\Omega$$

which is less than the smallest value of resistance in the circuit.
   The power consumed is

$$P = V_S^2/R_E = 10^2/10.68 = 9.36\,\text{W}$$

## 1.17 Series–parallel resistor combinations

Each circuit needs to be treated on its merits and, where necessary, groups of resistors (parallel or series) need to be combined before including them in the remainder of the circuit. The following example illustrates the general method.

*Worked example 1.8*

Determine the equivalent resistance of the circuit in Fig. 1.6(a) given that $R_1 = 50\,\Omega, R_2 = 25\,\Omega, R_3 = 10\,\Omega, R_4 = 20\,\Omega, R_5 = 5\,\Omega$ and $R_6 = 4\,\Omega$.

*Solution*

The overall approach to the solution is shown in Fig. 1.6(b) and (c). Initially, we calculate the value of $R_{P1}$ (which is equivalent to $R_1$, $R_2$ and $R_3$ in parallel), and $R_{P2}$ (equivalent to $R_4$ and $R_5$ in parallel).

Next $R_{P1}$ and $R_{P2}$ are grouped in series to form the series resistance $R_S$. Finally, the parallel resistance of $R_6$ and $R_S$ is calculated.

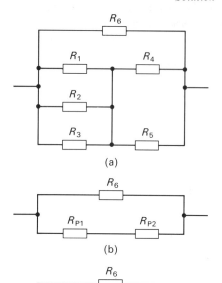

(a)

(b)

(c)

**Fig. 1.6** Figure for Worked Example 1.8

*Calculation of $R_{P1}$*

$$\frac{1}{R_{P1}} = \frac{1}{R_1} + \frac{1}{R_2} + \frac{1}{R_3} = \frac{1}{50} + \frac{1}{25} + \frac{1}{10}$$

$$= 0.02 + 0.04 + 0.1 = 0.16\,\text{S}$$

or $R_{P1} = 1/0.16 = 6.25\,\Omega$.

*Calculation of $R_{P2}$*

$$R_{P2} = R_4 R_5/(R_4 + R_5) = 20 \times 5/(20 + 5) = 4\,\Omega$$

*Calculation of $R_S$*

$$R_S = R_{P1} + R_{P2} = 6.25 + 4 = 10.25\,\Omega$$

*Equivalent resistance of the complete circuit*

$$R_E = R_6 R_S/(R_6 + R_S) = 4 \times 10.25/(4 + 10.25)$$

$$= 2.877\,\Omega$$

## 1.18 Important mechanical units

Many items of electrical plant produce a mechanical output, and an electrical engineer must have a working knowledge of the units involved. In the following we define the most important of these.

$$\text{Linear velocity, } v = \frac{\text{change in linear movement}}{\text{time}}\ \text{m/s}$$

$$\text{Linear acceleration, } a = \frac{\text{change in linear velocity}}{\text{time}}\ \text{m/s}^2$$

Deceleration is a reduction in velocity with time and is, effectively, a negative acceleration.

$$\text{Density} = \frac{\text{mass of a certain volume of substance}}{\text{volume of substance}}\ \text{kg/m}^3$$

$$\text{Relative density} = \frac{\text{mass of a certain volume of substance}}{\text{mass of an equal volume of water}}$$

Force, $F$ = mass × acceleration = $ma$ newtons (N)

The *newton* is that force which gives a mass of 1 kg an acceleration of $1 \, \text{m/s}^2$.

Energy or work, $W$ = force × distance moved

$$= Fl \text{ joules (J)}$$

The *joule* is the work done when 1 N acts through a distance of 1 m in the direction of the force.

Heat gained or lost, $Q$ = mass × specific heat capacity

$$\times \text{ temperature change}$$

$$= mc\theta \text{ joules (J)}$$

The dimensions of specific heat capacity are J/g K or kJ/kg K, and the dimensions of temperature are K. Typical values of specific heat capacity are

| Substance | Specific heat capacity ( J/kg K ) |
|-----------|-----------------------------------|
| Water     | 4190 |
| Aluminium | 950  |
| Iron      | 500  |
| Copper    | 390  |

Torque, $T$ = force × radius = $Fr$ newton metres (N m)

Torque is the *turning moment* produced by a force about an axis or centre of rotation.

Linear power, $P$ = force × velocity

$$= Fv \text{ watts or joules per second}$$

Rotary power, $P$ = angular velocity × torque

$$= \omega T = 2\pi nT \text{ watts or joules per second}$$

where $\omega$ is the *angular velocity* in rad/s, $n$ is the rotational speed in rev/s, and $T$ is the torque in N m.

$$\text{Efficiency, } \eta = \frac{\text{output power}}{\text{input power}} = \frac{\text{output energy}}{\text{input energy}}$$

*Worked example 1.9*    A kettle heats 2 litres of water from 10 to 100 °C. Given that 1 litre of water has a mass of 1 kg, the specific heat capacity of water is 4190 J/kg K, the cost of electricity is 10p per unit, and the efficiency of the kettle is 80 per cent, calculate (a) the electrical energy consumed in (i) megajoules, (ii) kWh when boiling the water, and (b) the cost of electricity involved.

*Solution*    The mass of water in the kettle is $2 \times 1 = 2$ kg, and the temperature rise is $100 - 10 = 90\,°C$ or 90 K. From the reference to specific heat capacity in section 1.18, we see that:

(a) (i)

$$\text{Heat} = mc\theta = 2 \times 4190 \times 90$$

$$= 754\,200 \text{ J or } 0.7542 \text{ MJ}$$

The energy taken from the supply to boil the water is

$$\text{Input energy} = \frac{\text{output energy}}{\text{efficiency}} = \frac{0.7542}{0.8} = 0.9428 \text{ MJ}$$

and from section 1.12 we have

$$\text{Input energy in kWh} = \frac{\text{input energy in MJ}}{3.6\,(\text{MJ/kWh})}$$

$$= 0.9428/3.6 = 0.262 \text{ kWh}$$

(b) The cost of electricity is

$$\text{Cost} = \text{number of electrical units} \times \text{cost per unit}$$

$$= 0.262 \times 10 = 2.62\text{p}$$

### Problems

**1.1**    Calculate the force required to accelerate a mass of 150 kg to $0.5 \text{ m/s}^2$.

$$[75\,\text{N}]$$

**1.2**    (a) A *rod* (also called a *perch* or *pole*, and dates from 1450) is equal to 16.5 ft (imperial). An electrical cable has a length of 2.5 rods, what is its length in metres? (b) A UK *gallon* [which has a mass of 10 lb (imperial) of water at 62 °F] has a volume of $0.1605 \text{ ft}^3$. Calculate the volume of 1 gallon in $\text{m}^3$.

$$[(\text{a})\ 12.57\,\text{m};\ (\text{b})\ 4.546 \times 10^{-3}\,\text{m}^3]$$

**1.3**    The rotational speed of a motor is 2900 rev/min; calculate the speed in rad/s. If the power output is 80 kW, the overall efficiency is 80 per cent, and the supply voltage 500 V, calculate (a) the input power and current and (b) the shaft torque in N m.

$$[303.7 \text{ rad/s}; (\text{a})\ 100\ \text{kW},\ 200\ \text{A}; (\text{b})\ 263.4\ \text{N m}]$$

**1.4**    An electrical machine has been exposed to a water spray, and two 1 kW electric heaters are used to dry the machine out. The heaters are used for 70 h on a 250 V supply, the cost of electricity being 10p per kWh. Calculate (a) the current consumed by the heaters, (b) the energy consumed in kWh, and (c) the cost of the energy.

$$[(\text{a})\ 8\ \text{A}; (\text{b})\ 140\ \text{kWh}; (\text{c})\ £14]$$

**1.5** An electric kettle containing 1.5 litres of water at 15 °C boils the water in 10 min, the supply voltage being 250 V. The efficiency of the kettle is 80 per cent; assume that 1 kg calorie = 4.2 kJ. Calculate (a) the current drawn by the kettle and (b) the cost of electricity consumed at 10p per kWh.

[(a) 0.744 A; (b) 1.86p]

**1.6** A battery discharges 3.9 A for 3 h. What quantity of electricity does it supply in (a) coulombs, (b) ampere-hours?

[(a) 42 120 C; (b) 11.7 A h]

**1.7** Two voltmeters A and B, having a resistance of 7 and 14 kΩ, respectively, are connected in series to a 250 V supply. Calculate the reading of each voltmeter.

[A: 83.33 V; B: 166.67 V]

**1.8** In problem 1.7, what power is absorbed by each voltmeter, and what energy is consumed in an 8 h period?

[A: 0.992 W, 7.936 Wh; B: 1.984 W, 15.873 Wh]

**1.9** A supply source of e.m.f. $E$ volts having an internal resistance $r$ ohms, supplies current to a water heater. Calculate the resistance $R$ of the heater if (a) the water is to be heated the most rapidly, and (b) 80 per cent of the total energy is absorbed by the water.

[(a) $R = r$ (see also the maximum power transfer theorem in Chapter 12); (b) $R = 4r$]

**1.10** A two-branch parallel circuit contains two series-connected resistors in each branch. Branch A contains a 3 and a 7 Ω resistor, and branch B contains a 2 and a 6 Ω resistor. If the parallel circuit takes a current of 2.25 A, calculate (a) the equivalent resistance of the parallel circuit, (b) the voltage across the parallel circuit, (c) the current in each branch, and (d) the power consumed by the 7 Ω resistor.

[(a) 4.44 Ω; (b) 10 V; (c) A: 1 A, B: 1.25 A; (d) 7 W]

**1.11** A two-branch parallel circuit has a resistance of 10 Ω in one branch and 20 Ω in the other. The parallel combination is in series with a 7 Ω resistor, the complete circuit being energized by a 50 V d.c. supply. Calculate (a) the equivalent resistance of the parallel combination, (b) the equivalent resistance of the complete circuit, (c) the current drawn by the circuit, (d) the voltage across the 7 Ω resistor, and (e) the power dissipated in each resistor.

[(a) 6.667 Ω; (b) 13.667 Ω; (c) 3.66 A; (d) 25.62 V; (e) 7 Ω, 93.77 W; 10 Ω, 59.44 W; 20 Ω, 29.72 W]

**1.12** A conductance of 0.1 S is connected in series with two parallel-connected conductances of 0.0667 S. What conductance must be connected in parallel with this combination if the current drawn from a 40 V supply is 3 A?

[0.0178 S]

**1.13** A two-branch parallel circuit containing resistances of 20 and 10 Ω, respectively, is connected in series with an unknown resistance. The total power dissipated by the circuit is 50 W when the applied voltage is 20 V. Calculate the value of the unknown resistance.

[1.333 Ω]

**1.14**  Determine the resistance of a 2 km length of copper cable whose resistivity at the operating temperature is $1.7 \times 10^{-8}\,\Omega\,m$. The cross-sectional area of the wire is $8.6\,mm^2$.

[$3.95\,\Omega$]

**1.15**  The resistance of a copper wire of cross-sectional area $1\,mm^2$ and length 1 m is $0.02\,\Omega$. Determine the resistance of a wire of similar material having a length of 2 km and cross-sectional area $5\,mm^2$.

[$400\,\Omega$]

**1.16**  The current in a circuit is limited by a 'liquid' resistance comprising parallel-plate electrodes immersed in a liquid of resistivity $0.22\,\Omega\,m$. If the current density is $2500\,A/m^2$, and the liquid resistance absorbs 50 kW when the voltage between the electrodes is 450 V, calculate the area of the electrodes and the distance between them.

[$0.0444\,m^2$; $0.817\,m$]

**1.17**  The resistance–temperature coefficient of phosphor bronze is $39.4 \times 10^{-4}$ per °C referred to 0 °C. Determine the value of the coefficient referred to (a) 20 °C and (b) 100 °C.

[(a) $36.5 \times 10^{-4}$; (b) $28.3 \times 10^{-4}$]

**1.18**  Show that if $R_1$ is the resistance of a conductor at $\theta_1$ °C, $\alpha_1$ is the resistance–temperature coefficient referred to $\theta_1$, and $R_2$ is the resistance of the conductor at $\theta_2$ °C, then

$$R_2 = R_1(1 + \alpha_1(\theta_2 - \theta_1))$$

**1.19**  The field coils of a generator have a total resistance of $500\,\Omega$ at 15 °C. After a test run, the resistance increases to $520\,\Omega$. If the temperature coefficient of resistance of the conductor material is 0.004 per °C referred to 0 °C, calculate the final temperature of the coils.

[25.6 °C]

**1.20**  If the ambient temperature is 15 °C, calculate the current at switch-on of a 60 W, 250 V lamp given that the linear resistance–temperature coefficient of the filament referred to 15 °C is 0.005. The operating temperature of the lamp is 2000 °C; the effect of the quadratic resistance–temperature coefficient, $\beta$, may be neglected.

[2.62 A]

# 2 Basic circuit theory

## 2.1 Introduction

In this section we deal with a number of topics with which we need to be familiar throughout our engineering career, and to which we shall refer many times later in the book.

## 2.2 Division of voltage between series-connected resistors

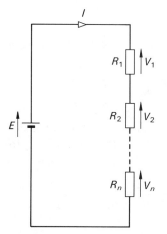

**Fig. 2.1** Division of voltage in a series circuit

Many electrical circuits contain series-connected resistors, and engineers need to be able to calculate the voltage across each resistor in the circuit. We will now look at a method of doing this.

The series circuit in Fig. 2.1 contains $n$ series-connected resistors, and the applied voltage across the resistors is

$$E = V_1 + V_2 + \cdots + V_n = IR_1 + IR_2 + \cdots + IR_n$$
$$= IR_E$$

where $R_E$ is the equivalent resistance of the series circuit (see also Chapter 1). The ratio of the voltage across the $n$th resistor to the supply voltage is

$$\frac{\text{Voltage across } n\text{th resistor, } V_n}{\text{Supply voltage, } E} = \frac{IR_n}{IR_E} = \frac{R_n}{R_E}$$

That is, the voltage across the $n$th resistor is

$$V_n = E \frac{R_n}{R_E} \qquad [2.1]$$

Since $V_n$ is proportional to $R_n$, the greater the value of the resistance the higher the voltage across it.

For resistor $R_1$

$$V_1 = ER_1/(R_1 + R_2 + \cdots + R_n)$$

and for $R_2$

$$V_2 = ER_2/(R_1 + R_2 + \cdots + R_n)$$

etc.

*Worked example 2.1*  Resistors of 10, 5 and 25 $\Omega$ are connected in series to a 100 V supply. Calculate the voltage across each resistor, and the power consumed by the 5 $\Omega$ resistor.

*Solution*   The effective resistance of the resistor chain is

$$R_E = R_1 + R_2 + R_3 = 10 + 5 + 25 = 40\,\Omega$$

From eqn [2.1]

$$V_1 = ER_1/R_E = 100 \times 10/40 = 25\,V$$

$$V_2 = ER_2/R_E = 100 \times 5/40 = 12.5\,V$$

and

$$V_3 = ER_3/R_E = 100 \times 25/40 = 62.5\,V$$

*Note*: the voltage across the circuit is

$$V_1 + V_2 + V_3 = 100\,V = E$$

The power consumed in the $5\,\Omega$ resistor is

$$\frac{V_2^2}{R_2} = \frac{12.5^2}{5} = 31.25\,W$$

## 2.3  Division of current in a parallel circuit

The total current, $I$, drawn by a parallel circuit containing $n$ branches is (see Fig. 2.2)

$$I = I_1 + I_2 + \cdots + I_n = \frac{E}{R_1} + \frac{E}{R_2} + \cdots + \frac{E}{R_n} = \frac{E}{R_E}$$

where $I_n$ is the current in the $n$th branch, and $R_E$ is the equivalent resistance of the parallel circuit. The ratio of the current in $R_n$ to the supply current is given by

$$\frac{\text{Current in } n\text{th resistor, } I_n}{\text{Current drawn from supply, } I} = \frac{E/R_n}{E/R_E} = \frac{R_E}{R_n}$$

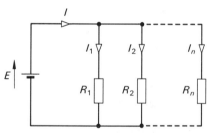

**Fig. 2.2**   Division of current in a parallel circuit

hence

$$I_n = I\frac{R_E}{R_n} \tag{2.2}$$

That is, the current in $R_1$ is

$$I_1 = IR_E/R_1$$

the current in $R_2$ is

$$I_2 = IR_E/R_2$$

etc. For the *special case of a two-branch parallel circuit*

$$R_E = R_1R_2/(R_1 + R_2)$$

hence

$$I_1 = I\frac{R_E}{R_1} = \frac{IR_2}{R_1 + R_2} \tag{2.3}$$

and

$$I_2 = I \frac{R_E}{R_2} = \frac{IR_1}{R_1 + R_2}$$ [2.4]

***Worked example 2.2*** A three-branch parallel circuit contains a $20\,\Omega$ resistor in one branch, a $25\,\Omega$ resistor in the second branch, and a $40\,\Omega$ resistor in the third branch. If the circuit draws a current of 10 A, calculate the current in each branch, the voltage across the circuit, and the total power consumed.

*Solution* The equivalent resistance of the parallel circuit is calculated from

$$\frac{1}{R_E} = \frac{1}{R_1} + \frac{1}{R_2} + \frac{1}{R_3} = \frac{1}{20} + \frac{1}{25} + \frac{1}{40}$$

$$= 0.05 + 0.04 + 0.025 = 0.115\,S$$

hence

$$R_E = 1/0.115 = 8.696\,\Omega$$

Equation [2.2] tells us that

$$I_1 = IR_E/R_1 = 10 \times 8.696/20 = 4.348\,A$$

$$I_2 = IR_E/R_2 = 10 \times 8.696/25 = 3.478\,A$$

$$I_3 = IR_E/R_3 = 10 \times 8.696/40 = 2.174\,A$$

*Note*: $I_1 + I_2 + I_3 = 10\,A$.
    The voltage across the circuit is

$$E = IR_E = 10 \times 8.696 = 86.96\,V$$

and the total power consumed is

$$P = I^2 R_E = 10^2 \times 8.696 = 869.6\,W$$

## 2.4 Kirchhoff's laws

Gustav Robert Kirchhoff (1824–87) was the first person to write down two basic rules for the presentation of electrical circuit equations, and are as follows.

**Kirchhoff's current law (or KCL)**

> **The total current flowing towards a junction, *J*, in a circuit is equal to the total current leaving it.**

or, alternatively

> **The sum of the current flowing towards a junction is zero, i.e. $\sum I = 0$.**

Referring to Fig. 2.3, the first statement of KCL says

$$I_A + I_B + I_D = I_C$$

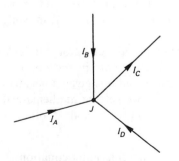

**Fig. 2.3** Kirchhoff's current law (KCL)

The second statement of the law says

$$I_A + I_B - I_C + I_D = 0$$

**Kirchhoff's voltage law (or KVL)**

Once again, there are two statements of the law as follows:
> **In any closed circuit or mesh, the algebraic sum of the potential drops and e.m.f.s is zero.**

or, alternatively
> **In any closed circuit or mesh, the algebraic sum of the e.m.f.s is equal to the algebraic sum of the p.d.s.**

Before we can write down KVL for the closed loop in Fig. 2.4, it is necessary to make some assumptions. Firstly, we will assume that the current, $I$, flows in a clockwise direction around the loop (this also assumes that $E_1 > E_2$). The assumptions could, of course, be incorrect; this does not matter since we can sort out the actual direction of $I$ once we have completed the calculation.

Having made this assumption, it implies that node B is at a higher potential than node C, so that the potential 'arrow' associated with $R_1$ points from node C to node B (see also section 1.14). Similarly, the potential arrow associated with $R_2$ points from node A to node D (the potential arrows would have been reversed had we assumed that the current flowed in the opposite direction).

The first statement of KVL states that if we travel around the loop and return to the starting node, the net gain in voltage is zero. We write down the voltage equation for the loop ABCDA as follows:

$$E_1 - IR_1 - E_2 - IR_2 = 0$$

It is important at this point to fully appreciate the way in which the equation is obtained, and we look at it in some detail. Commencing at node A and moving to node B, we see that an e.m.f. arrow points from A to B, so we give the associated e.m.f. a positive sign. That is, *if we move in the direction of an e.m.f. (or p.d.) arrow, we give it a positive sign.* If the e.m.f. (or p.d.) arrow points in the opposite direction to the way in which we are moving, we give it a negative sign.

Continuing to move around the loop in the direction ABCDA, the potential arrow associated with $R_1$ points in the opposite direction to the way in which we move, and it is given a negative sign. Thus, after moving from A to C we get the mathematical expression for the voltage of node C relative to node A of

$$V_{CA} = E_1 - IR_1$$

Applying these rules to the complete loop results in the equation

$$E_1 - IR_1 - E_2 - IR_2 = 0$$

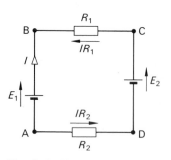

**Fig. 2.4**  Kirchhoff's voltage law (KVL)

KVL also applies even if we commence at a different node and move around the loop in a different direction. For example, the equation for the loop CBADC is

$$IR_1 - E_1 + IR_2 + E_2 = 0$$

which gives the same general result as the equation above.

The second statement of KCL gives the same general form of equation but, in the experience of the author, can occasionally result in errors in the circuit equation unless care is taken. The following suggestion may be useful. Travel around the loop in one direction (say clockwise) when adding the e.m.f.s, and in the opposite direction (anticlockwise) when adding the p.d.s. Then write down the circuit equation in the form

Sum of e.m.f.s = sum of p.d.s

Taking the loop ABCDA, and moving around the loop in the direction ABCDA when adding the e.m.f.s we get

Sum of e.m.f.s = $E_1 - E_2$

and moving around the loop in the *opposite direction* when adding the p.d.s we get

Sum of p.d.s = $IR_1 + IR_2$

Hence

$$E_1 - E_2 = IR_1 + IR_2$$

Alternatively, commencing at node C we can add e.m.f.s in the direction CBADC, and add p.d.s in the direction CDABC to give

$$-E_1 + E_2 = -IR_2 - IR_1$$

Applications of Kirchhoff's laws are illustrated below.

*Worked example 2.3*  Determine $I_1$, $I_2$ and the voltage of node B with respect to node D ($V_{BD}$) in Fig. 2.5.

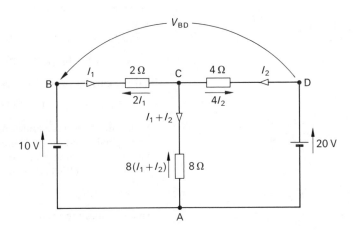

**Fig. 2.5**  Figure for Worked Example 2.3

*Solution*  We see that $I_1$ and $I_2$ approach node C, and $(I_1 + I_2)$ leaves it, that is there are two unknown values of current. That means we need two loop equations to solve for them and, once we have their value, the voltage $V_{BD}$ can be calculated.

Initially, we draw labelled *current arrows on the circuit*, and the *potential arrows by the side of the batteries and resistors.*

There are three loops in the circuit, namely ABCA, ADCA and ABCDA. We need to choose two of these to give the circuit equations; let us select the first two.

*Loop ABCA*
Moving around the loop in the direction ABCA gives

$$10 - 2I_1 - 8(I_1 + I_2) = 0$$

or

$$10 = 10I_1 + 8I_2 \qquad\qquad [2.5]$$

*Loop ADCA*
Moving around the loop in the direction ADCA gives

$$20 - 4I_2 - 8(I_1 + I_2) = 0$$

or

$$20 = 8I_1 + 12I_2 \qquad\qquad [2.6]$$

If we multiply eqn [2.5] by 12/8 and subtract eqn [2.6] from it, we eliminate $I_2$ as follows:

$$
\begin{aligned}
15 &= 15I_1 + 12I_2 \quad ([2.5] \times 12/8) \\
20 &= \phantom{1}8I_1 + 12I_2 \\
\hline
\text{Subtract} \quad -5 &= \phantom{1}7I_1
\end{aligned}
$$

that is

$$I_1 = -5/7 = -0.714\,\text{A}$$

The negative sign associated with $I_1$ simply tells us that we chose, in the first instance, the 'wrong' direction for $I_1$. That is to say, the 10 V battery is charged by a current of 0.714 A! However, at this stage, *we must not alter the direction of $I_1$*: consequently, whenever we see $I_1$ in an equation, we must insert the value $-0.714$ A in the equation (see below).

Substituting the value of $I_1$ into eqn [2.5] gives

$$I_2 = (10 - 10I_1)/8 = (10 + 7.14)/8 = 2.143\,\text{A}$$

and the current in the 8 Ω resistor is

$$I_1 + I_2 = -0.714 + 2.143 = 1.429\,\text{A}$$

At this stage we have some answers, but we do not know if they satisfy the original equations. A simple way of checking the solution is to insert the value of $I_1$ and $I_2$ in any one of the original equations (say eqn [2.6]), and check whether the numerical value is correct as follows:

$$8I_1 + 12I_2 = (8 \times (-0.714)) + (12 \times 2.143) = 20\,\text{V}$$

which verifies the results.

The voltage $V_{BD}$ (see also section 1.14) is $V_{BD}$ = potential of node B with respect to node D. To obtain this voltage, we move from node D (assumed to be at zero volts for our purpose) to node B, adding the voltages associated with potential arrows which 'point' from D to B, and subtract those which 'point' in the reverse direction. If we take the path through the 4 and 2 Ω resistors we get

$$V_{BD} = -4I_2 + 2I_1 = (-4 \times 2.143) + (2 \times (-0.714))$$

$$= -10\,V$$

Alternatively, we could determine $V_{BD}$ via the two batteries as follows:

$$V_{BD} = -20 + 10 = -10\,V$$

***Worked example 2.4***    Calculate $I_1$ and $I_2$ in Worked Example 2.3 using determinants.

*Solution*    The solution of simultaneous equations is fraught with difficulties, and all engineering students will have experienced difficulties in this area. However, help is at hand in the form of **determinants**.

To understand the reason behind the solution of simultaneous equations by determinants requires a deal of mathematical knowledge. However, to drive a complex machine like a car does not mean that we need to be a skilled car mechanic! So it is with determinants; we merely need to follow simple rules to quickly and accurately solve simultaneous equations, as shown below.

In the first instance we need to write down the circuit equations in the manner described in Worked Example 2.3. These are

$$10 = 10I_1 + 8I_2 \qquad\qquad [2.5]$$

$$20 = 8I_1 + 12I_2 \qquad\qquad [2.6]$$

Next we write the coefficients or multipliers of $I_1$ and $I_2$ on the right-hand side of the equations in 'determinant' form as follows, and call it **det**:

$$\mathbf{det} = \begin{vmatrix} 10 & 8 \\ 8 & 12 \end{vmatrix}$$

The two vertical lines tell us that this is a determinant, and its numerical value is calculated as follows:

$$\mathbf{det} = \begin{vmatrix} 10 & 8 \\ 8 & 12 \end{vmatrix} = (10 \times 12) - (8 \times 8) = 56$$

That is, we multiply the coefficients linked by the downward-pointing arrow, and subtract the product of the coefficients linked by the upward-pointing arrow.

Next, we replace the coefficients of $I_1$ in **det** by the values on the left-hand side of the simultaneous equations, and call this **det $I_1$**. The value of **det $I_1$** is evaluated in the manner outlined above, as follows:

$$\mathbf{det}\,I_1 = \begin{vmatrix} 10 & 8 \\ 20 & 12 \end{vmatrix} = (10 \times 12) - (8 \times 20) = 120 - 160$$

$$= -40$$

We then replace the coefficients of $I_2$ in **det** by the values on the left-hand side of the simultaneous equations and call it **det** $I_2$, and evaluate it as described above.

$$\textbf{det } I_2 = \begin{vmatrix} 10 & 10 \\ 8 & 20 \end{vmatrix} = (10 \times 20) - (10 \times 8) = 120$$

These values are related by the equation

$$\frac{1}{\textbf{det}} = \frac{I_1}{\textbf{det } I_1} = \frac{I_2}{\textbf{det } I_2}$$

hence

$$I_1 = \frac{\textbf{det } I_1}{\textbf{det}} = \frac{-40}{56} = -0.714 \text{ A}$$

$$I_2 = \frac{\textbf{det } I_2}{\textbf{det}} = \frac{120}{56} = 2.143 \text{ A}$$

By this means we have solved the simultaneous equations quickly and without errors! While this does not test the mathematical skill of the reader, most of us are not in the business of mathematics (which is simply one of the tools the reader must use), but are in the business of solving electrical problems.

*Worked example 2.5*    Calculate the value of $I_1$, $I_2$ and $I_3$ in Fig. 2.6, together with the voltage $V_{AC}$.

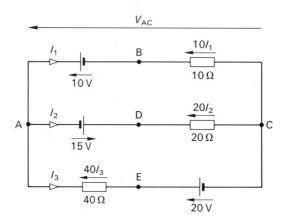

**Fig. 2.6**  Figure for Worked Example 2.5

*Solution*    The three-branch parallel circuit has a battery and a resistance in each branch, and we have assumed that all currents flow from node A. Clearly, this is not possible since, according to KCL, whatever current flows away from node A must return to it! That is, at least one current must return to node A.

Next, we assign potential arrows to the resistors, which reflect the assumed direction of current flow in each branch. Applying KCL to node A (and node C) gives

$$I_1 + I_2 + I_3 = 0$$

or

$$I_3 = -(I_1 + I_2) \qquad [2.7]$$

That is, we only need to solve for two currents, since the third current is defined in terms of the other two. This means that we need two simultaneous equations, which are obtained below. We select loops ABCDA and ADCEA; the loop equations are deduced using the methods outlined earlier.

*Loop ABCDA*
The loop equation is

$$-10 - 10I_1 + 20I_2 - 15 = 0$$

or

$$-25 = 10I_1 - 20I_2 \qquad [2.8]$$

*Loop ADCEA*
The equation for this loop is

$$15 - 20I_2 + 20 + 40I_3 = 0$$

or

$$35 = 20I_2 - 40I_3$$

From eqn [2.7] $I_3 = -(I_2 + I_3)$. Substituting this in the above equation gives

$$35 = 40I_1 + 60I_2 \qquad [2.9]$$

We solve eqns [2.8] and [2.9] for $I_2$ and $I_2$ as follows:

$$-75 = 30I_1 - 60I_2 \quad ([2.8] \times 3)$$
$$\underline{35 = 40I_1 + 60I_2} \qquad [2.9]$$
$$\text{Add} \quad -40 = 70I_1$$

or

$$I_1 = -40/70 = -0.571 \, \text{A}$$

Substituting this value into eqn [2.9] yields

$$35 = (40 \times (-0.571)) + 60I_2 = -22.84 + 60I_2$$

hence

$$I_2 = (35 + 22.84)/60 = 0.964 \, \text{A}$$

Therefore

$$I_3 = -(I_1 + I_2) = -(-0.571 + 0.964)$$

$$= -0.393 \, \text{A}$$

Alternatively, we can solve the equations using determinants as follows:

$$I_1 = \frac{\begin{vmatrix} -25 & -20 \\ 35 & 60 \end{vmatrix}}{\begin{vmatrix} 10 & -20 \\ 40 & 60 \end{vmatrix}} = \frac{(-25 \times 60) - (-20 \times 35)}{(10 \times 60) - (-20 \times 40)} = \frac{-800}{1400}$$

$$= -0.571\,\text{A}$$

$$I_2 = \frac{\begin{vmatrix} 10 & -25 \\ 40 & 35 \end{vmatrix}}{\begin{vmatrix} 10 & -20 \\ 40 & 60 \end{vmatrix}} = \frac{(10 \times 35) - (-25 \times 40)}{(10 \times 60) - (-20 \times 40)} = \frac{1350}{1400}$$

$$= 0.964\,\text{A}$$

From eqn [2.6]

$$I_3 = -(I_1 + I_2) = -(-0.571 + 0.964) = -0.393\,\text{A}$$

Since $I_1$ and $I_3$ have negative values, they indicate that we initially nominated the wrong direction to these two currents! That is, current flows out of the positive terminal of each battery, i.e. they each discharge power into the circuit. The power supplied by each battery is

10 V battery, $P = 10 \times 0.571 = \phantom{0}5.71\,\text{W}$

15 V battery, $P = 15 \times 0.964 = \phantom{0}14.46\,\text{W}$

20 V battery, $P = 20 \times 0.393 = \phantom{0}\underline{7.86\,\text{W}}$

Total $\phantom{15 V battery, P = 20 \times 0.393 = } \underline{28.03\,\text{W}}$

The reader will find it an interesting exercise to verify that the total power consumed by the resistors in the circuit comes to this value.

*Worked example 2.6*  Determine the voltage $V_{FB}$ in Fig. 2.7, i.e. the voltage of node F with respect to node B.

*Solution*  This problem is an interesting departure from the circuits described hitherto, for the reason that there is no return path for the current in the 3 Ω resistor connecting node A to node D. The net result is that *there*

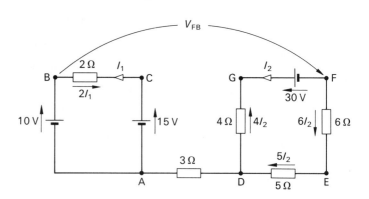

**Fig. 2.7**  Figure for Worked Example 2.6

*is no current in this resistor*, and that nodes A and D have the same electrical potential, i.e. they are *equipotential points*!

In effect, we can calculate the value of $I_1$ and $I_2$ as though they were in separate circuits. As with other circuits, we nominate the direction of $I_1$ and $I_2$, and draw potential arrows by the sides of the e.m.f.s and resistors as before. The loop equations are given below.

*Loop ABCA*

$$10 + 2I_1 - 15 = 0$$

or

$$I_1 = 5/2 = 2.5\,\text{A}$$

*Loop DEFGD*

$$-5I_2 - 6I_2 + 30 - 4I_2 = 0$$

or

$$I_2 = 30/15 = 2\,\text{A}$$

Since we are asked to evaluate $V_{FB}$, we start at node B and make our way to node F *via any path*.

*Path BCADEF*
The equation for this path is

$$V_{FB} = 2I_1 - 15 + \text{p.d. across } 3\,\Omega \text{ resistor} - 5I_2 - 6I_2$$

$$= (2 \times 2.5) - 15 + 0 - (5 \times 2) - (6 \times 2) = -32\,\text{V}$$

That is, node F is $-32\,\text{V}$ with respect to node B. Let us take another path, as shown below.

*Path BADGF*

$$V_{FB} = -10 + \text{p.d. across } 3\,\Omega \text{ resistor} + 4I_2 - 30$$

$$= -10 + 0 + (4 \times 2) - 30 = -32\,\text{A}$$

## 2.5 Ideal voltage and current sources

An *ideal source* is one whose output (either a voltage or a current) is independent of the load which is connected to it; these sources are known as *independent sources*. You must remember that we are talking about ideal sources which, strictly speaking, do not exist!

An *ideal voltage source* is one that will maintain a *constant voltage* across a load, *no matter how high or how low the resistance of the load*. Obviously, there are limits beyond which we cannot take this discussion, because if we connect a resistance of zero ohms (or thereabouts) to the source, the current in the load is infinitely high! Our views on ideal sources will be tempered by the discussion in section 2.6 on practical sources.

An *ideal current source* is one that will maintain a *constant current* into a load, no matter what the resistance of the load. If, in this case, we try to connect a load of infinite resistance (an open circuit) to a current source, the terminal voltage is infinity! As with the ideal voltage source, we must temper our thoughts with a little common sense.

## 2.6 Practical voltage and current sources

An example of a *practical voltage source* is a battery, which can be thought of as an ideal voltage source that has an *internal resistance*, R (also known as the *source resistance*, or *output resistance*), which is connected in series with the ideal voltage source, as shown in Fig. 2.8(a) and (b).

In a practical voltage source, the terminal voltage ($V_T$) is equal to the ideal source value ($E$) when no current is drawn from the source. When a load is connected, current flows in the load, and the p.d. in the internal resistance causes the terminal voltage to fall below $E$. This is illustrated in the load characteristic in Fig. 2.8(c). Everyone has experienced this in the home because, when a large item of electrical equipment is switched on, the electric lights momentarily dim. The equation for the terminal voltage is

$$V_T = I_L R_L$$

where $I_L$ is the load current whose value is

$$I_L = \frac{E}{R + R_L}$$

that is

$$V_T = \frac{E R_L}{R + R_L}$$

An example of a *practical current source* is the output circuit of a bipolar junction transistor. It comprises an ideal current source (symbolized by a circle with an arrow inside it representing the direction of the current) shunted by its output resistance, $r$, as shown in Fig. 2.9(a).

The voltage–current characteristic of a practical current source is shown as the full line in Fig. 2.9(b). When the load resistance is zero, the internal resistance is shorted out and the terminal voltage is zero; at the same time, all of the current, $I$, produced by the ideal source flows in the short circuit. As the load resistance increases in value, so the terminal voltage rises, reaching $V_T = Ir$ when the terminals are open-circuited.

Clearly, we can select to use a practical voltage source or a practical current source to represent any type of power supply. That is, there is a relationship between the two types of source, which we will show below.

For the practical voltage source in Fig. 2.8, the terminal voltage is

$$V_T = E - I_L R$$

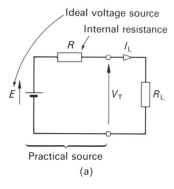

Ideal voltage source

Internal resistance

Practical source

(a)

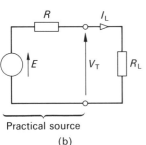

Practical source

(b)

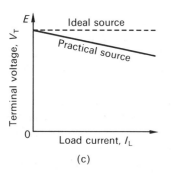

(c)

**Fig. 2.8** (a) and (b) Two versions of practical voltage source. (c) Typical *V–I* characteristic for a practical voltage source

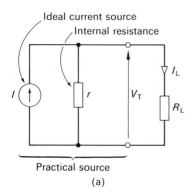

Practical source

(a)

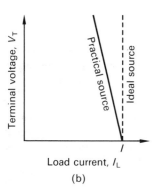

Load current, $I_L$

(b)

**Fig. 2.9** (a) Equivalent circuit of a practical current source and (b) its V–I characteristic

hence

$$I_L = \frac{E}{R} - \frac{V_T}{R}$$

or

$$\frac{E}{R} = I_L + \frac{V_T}{R} \qquad [2.10]$$

For the practical current source in Fig. 2.9, current $I$ enters the parallel circuit containing $r$ and $R_L$, and is given by

$$I = I_L + \frac{V_T}{r} \qquad [2.11]$$

For the practical voltage and current sources to be equivalent to one another, eqns [2.10] and [2.11] must be equivalent. That is

$$I = \frac{E}{R} \qquad [2.12]$$

and

$$\frac{V_T}{R} = \frac{V_T}{r}$$

hence

$$R = r \qquad [2.13]$$

*Worked example 2.7*   A 12 V battery has an internal resistance of 0.1 Ω. What values are associated with its equivalent current source? If a load of resistance 1.1 Ω is connected to the terminals of the source calculate, for both sources, the load current and the terminal voltage.

*Solution*   From the data provided (see Fig. 2.8)

$$E = 12\,\text{V} \qquad R = 0.1\,\Omega$$

From eqn [2.12], the 'ideal' current source is

$$I = E/R = 12/0.1 = 120\,\text{A}$$

and from eqn [2.13], the internal resistance of the current source is

$$r = R = 0.1\,\Omega$$

That is, the practical current source comprises a 120 A ideal current source shunted by a 0.1 Ω resistance (see also Fig. 2.9).

If $R_L = 1.1\,\Omega$, then in the case of the practical voltage source (see Fig. 2.8)

$$I_L = \frac{E}{R + R_L} = \frac{12}{0.1 + 1.1} = 10\,\text{A}$$

32 ELECTRICAL AND ELECTRONIC ENGINEERING PRINCIPLES

and

$$V_T = I_L R_L = 10 \times 1.1 = 11 \, \text{V}$$

In the case of the practical current source, 120 A divides between the $0.1 \, \Omega$ internal resistance and the $1.1 \, \Omega$ load resistance. From the work in section 2.3, the load current is

$$I_L = I \, \frac{r}{r + R_L} = 120 \, \frac{0.1}{0.1 + 1.1} = 10 \, \text{A}$$

and, once more, the terminal voltage is

$$V_T = I_L R_L = 10 \times 1.1 = 11 \, \text{V}$$

### Problems

**2.1** A $10 \, \text{k}\Omega$ resistor and a voltmeter of resistance $30 \, \text{k}\Omega$ are connected in series. If the circuit is supplied at 100 V, determine the voltage across the $10 \, \text{k}\Omega$ resistor, and the voltage indicated by the voltmeter.

[25 V; 75 V]

**2.2** A resistor $R$ is connected in series with a two-branch parallel circuit which contains a resistor of $8 \, \Omega$ in one branch and $12 \, \Omega$ in the second branch, the voltage applied to the circuit being 30 V. If the voltage across the parallel circuit is 9.6 V, calculate the value of $R$.

[$10.2 \, \Omega$]

**2.3** A three-branch parallel circuit has a resistance of $5 \, \Omega$ in one branch, $10 \, \Omega$ in the second branch, and $20 \, \Omega$ in the third branch. The circuit is connected in series with a $4 \, \Omega$ resistor. If 2 A flows in the $10 \, \Omega$ resistor, calculate the current in, and the voltage across, each resistor.

[$5 \, \Omega$: 4 A, 20 V; $10 \, \Omega$: 2 A, 20 V; $20 \, \Omega$: 1 A, 20 V; $4 \, \Omega$: 7 A, 28 V]

**2.4** A battery of e.m.f. 25 V and internal resistance $3 \, \Omega$ is connected in parallel with a battery of e.m.f. 20 V and internal resistance $2 \, \Omega$. What is the terminal voltage of the two batteries, and what current circulates between them? If a load resistance of $4 \, \Omega$ is connected to the terminals of the combination, determine the current supplied by each battery, the current in the load and the p.d. across the load.

[22 V, 1 A; 2.7 A, 1.54 A, 4.24 A, 16.92 V]

**2.5** Calculate the current flowing out of the positive pole of each battery in Fig. 2.10.

[24 V: 1.474 A; 20 V: $-1.04$ A; 18 V: $-0.43$ A]

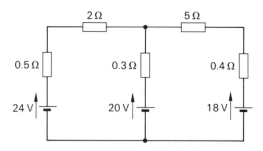

**Fig. 2.10**

**2.6** Calculate $V_{AB}$ in Fig. 2.11.

$$[-1.2\,V]$$

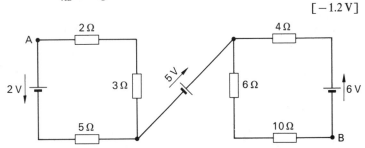

**Fig. 2.11**

**2.7** When a $30\,\Omega$ load is connected to a cell, the current is $41.2\,mA$. When the $30\,\Omega$ load is replaced by a $50\,\Omega$ load, the current is $25.93\,mA$. Calculate the e.m.f. and the internal resistance of the cell. What is the terminal voltage when the connected load is $10\,\Omega$?

$$[1.4\,V;\ 4\,\Omega;\ 1\,V]$$

**2.8** Calculate $I_1$ and $I_2$ in Fig. 2.12 if (a) $R = 2.5\,\Omega$, (b) $R = 0.5\,\Omega$. (c) What is the value of $R$ when $I_1$ is zero?

$$[(a)\ I_1 = 49\,A,\ I_2 = 51.9\,A;\ (b)\ I_1 = -37.5\,A,\ I_2 = 225\,A;\ (c)\ 0.8\,\Omega]$$

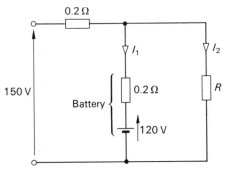

**Fig. 2.12**

**2.9** When a power source supplies $4\,A$ its terminal voltage is $6\,V$, and when it supplies $2\,A$ its terminal voltage is $8\,V$. If the source is (a) a voltage source, (b) a current source, determine the parameters of each.

$$[(a)\ 10\,V,\ 1\,\Omega;\ (b)\ 10\,A,\ 1\,\Omega]$$

**2.10** Calculate the value of $I$ in Fig. 2.13.

$$[19\,A]$$

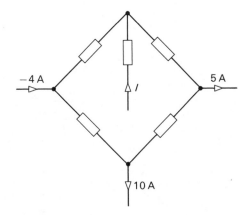

**Fig. 2.13**

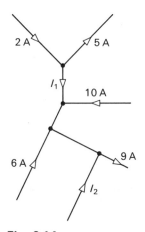

**Fig. 2.14**

**2.11**  Determine $I_1$ and $I_2$ in Fig. 2.14.

$$[-3\,\text{A}; \ -4\,\text{A}]$$

**2.12**  If 80 W of power is consumed by the 5 Ω resistor in Fig. 2.15, calculate the value of $R$. Determine also $V_{\text{BA}}$, $V_{\text{DC}}$ and $V_{\text{AC}}$.

$$[2\,\Omega; \ -20\,\text{V}; \ 28\,\text{V}; \ 40\,\text{V}]$$

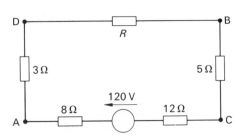

**Fig. 2.15**

# 3 Electric fields and capacitance

## 3.1 Introduction

The topic of electric fields is as old as electricity, and was one of the first properties of electricity discovered by man. We look at the relationships which exist in capacitors that are of interest to us as engineers, and the capacitance of interconnected capacitors. Since capacitors are one of the few devices which can store electricity, they are of importance to us not only in the related fields of electricity and electronics, but also in many other fields including control systems.

## 3.2 Static electricity

As outlined in Chapter 1, atoms contain electrons, protons and neutrons. Each electron carries a negative charge, and each proton carries an equal but positive charge. Moreover, every atom has as many electrons as it has protons, so that its net charge is zero. The removal or addition of an electron causes the atom to assume an electrical charge; when this occurs, the atom is said to be *ionized*.

When two substances are rubbed together, a polythene rod and a woollen duster for example, electrons transfer from one substance to the other, upsetting the charge balance of the two substances. The substance which loses electrons (the woollen duster) acquires a positive charge, and the one gaining electrons (the polythene) acquires a negative charge.

## 3.3 Force between charged bodies

If two polythene rods are rubbed on a woollen duster, both rods become charged with the same polarity. If the two rods are supported on threads close to one another, they try to move away from one another. That is

**like charges repel.**

Similarly, if two rods charged with opposite polarity are suspended close to one another, they are attracted towards one another. That is

**unlike charges attract.**

Moreover, if an uncharged rod is suspended parallel to a charged rod, there is a force of attraction between them. That is

**charged objects attract uncharged objects.**

We have all experienced the latter, witness the attraction of dust to a television screen or to a computer monitor screen! The simple

Christmas trick of rubbing a blown-up balloon on your sleeve, at which point the balloon becomes electrically charged; if the balloon is pressed onto the wall (or ceiling), it remains for a very long period, and is an example of the force between charged bodies!

## 3.4 Electric flux and flux density

An electrostatically charged body is surrounded by an **electric field**, and a measure of the field strength is its **electric flux**, symbol $Q$, having units of the coulomb (C). The electric field is usually represented by a series of *imaginary lines* which radiate from the charged body; it is important to note that nothing actually 'moves' along these lines. The *direction of the electric field* at any point in space is given by the direction in which a 'free' positive charge would move if placed at that point. Since a positive charge is repelled by another positive charge, and is attracted by a negative charge, the 'lines of force' are generally assumed to emanate from a positive charge and terminate on a negative charge. Once again, it must be pointed out that nothing actually 'moves' along a line of electric force.

The unit of electric flux has been chosen so that unit charge produces unit electric flux, that is

Electric flux = $Q$ coulombs

A **dielectric** is a material which offers high resistance to the flow of electricity, but through which electric flux can pass; most insulating materials such as air, glass, mica, etc., are of this kind. If the electric flux passes through a dielectric of area $A$ (the area being measured perpendicular to the flux), the **electric flux density** (symbol $D$) in the dielectric is

$$D = \frac{Q}{A} \text{ coulombs per square metre} \qquad [3.1]$$

## 3.5 Electric field strength, *E*

When a p.d. of $V_C$ is applied between the two surfaces of a dielectric which are separated by $d$ metres, the **electric field strength**,[†] $E$, in the material is

$$E = \frac{V_C}{d} \text{ volts per metre} \qquad [3.2]$$

The electric field strength is also known as the *potential gradient*, or the *electric field intensity* or the *electric stress*. In many cases it can have a very high value; for example, if two parallel plates

---

[†] The reader should take care not to confuse the symbol $E$ meaning electric field strength with the symbol $E$ for e.m.f.

separated by 0.1 mm have a voltage of 200 V between them, the electric field strength in the dielectric is

$$E = 200/0.1 \times 10^{-3} = 2 \times 10^6 \text{ V/m} \quad \text{or} \quad 2\,\text{MV/m}$$

## 3.6 Permittivity

The permittivity (or *absolute permittivity*), symbol $\varepsilon$, of a material is a measure of its ability to allow electric flux to be established in it (and is the electrostatic equivalent of conductivity). It has the dimensions of *farads per metre* (F/m), and is given by

$$\varepsilon = \frac{\text{electrostatic flux density}}{\text{electric field strength}} = \frac{D}{E} \qquad [3.3]$$

The *permittivity of free space*, i.e. that of a vacuum, is given the special symbol $\varepsilon_0$, where

$$\varepsilon_0 = 8.85 \times 10^{-12} \text{ F/m} \quad \text{or} \quad 8.85 \text{ pF/m}$$

**Table 3.1**

| Material | Relative permittivity |
|---|---|
| Air | 1.0006 |
| Paper (dry) | 2–2.5 |
| Rubber | 2–3.5 |
| Mica | 3–7 |
| Bakelite | 4.5–5.5 |
| Glass | 5–10 |

The *relative permittivity*, $\varepsilon_r$, of a dielectric is the ratio of the absolute permittivity to $\varepsilon_0$ as follows:

$$\varepsilon_r = \frac{\varepsilon}{\varepsilon_0} \quad \text{or} \quad \varepsilon = \varepsilon_0 \varepsilon_r \qquad [3.4]$$

Since a ratio of two factors having the same dimensions is itself dimensionless, $\varepsilon_r$ is dimensionless and it has a value greater than unity. That is, any practical dielectric supports more electric flux for a given applied voltage than does free space (or air). Typical values of permittivity are listed in Table 3.1.

## 3.7 Capacitance

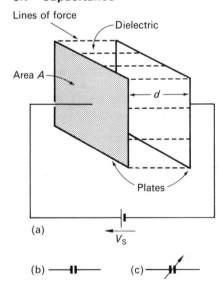

(a)

(b) ⊣⊢ (c) ⊣⊬

**Fig. 3.1** (a) Parallel-plate capacitor. Symbols for (b) fixed capacitor, (c) variable capacitor

A capacitor comprises two conducting surfaces separated by a dielectric.

A *parallel-plate capacitor* is shown in Fig. 3.1(a), and has two parallel plates separated by a dielectric. When a voltage is applied between the plates, an electric field is established in the dielectric and *energy is stored in the dielectric*. The energy is stored by producing a strain in the orbit of the electrons in the atoms of the dielectric, giving the electrons an elliptical orbit.

The **capacitance**, symbol $C$ (the unit is the farad, F), is a measure of the ability of the capacitor to store energy. Experiments show that the electric flux in the capacitor is related to the voltage between the plates and the capacitance of the capacitor as follows:

Electric flux, $Q$ = p.d. between the plates, $V_C$ × capacitance, $C$

that is

$$C = \frac{Q}{V_C} \qquad [3.5]$$

If $Q$ is in coulombs and $V_C$ in volts, then $C$ is in farads. It also follows from eqn [3.5] that

Stored charge, $Q = CV_C$ $\qquad [3.6]$

Capacitance can therefore be described as follows:

**A capacitance of one farad stores a charge of one coulomb when one volt is applied between its plates.**

In practical terms, a capacitance of 1 F is a very large value, and the values usually associated with practical capacitors are either a few microfarads ($\mu$F), where $1\,\mu F = 10^{-6}\,F$, or a few nanofarads (nF), where $1\,nF = 10^{-9}\,F$, or a few picofarads (pF), where $1\,pF = 10^{-12}\,F$.

Referring to eqns [3.1]–[3.3] and [3.6], we see that the permittivity of a capacitor is

$$\varepsilon = \frac{D}{E} = \frac{Q/A}{V_C/d} = \frac{Q}{V_C}\frac{d}{A} = C\frac{d}{A}$$

where $d$ is the distance between the plates of the capacitor (see Fig. 3.1), and $A$ is the area of each plate. That is

$$C = \varepsilon\frac{A}{d} = \varepsilon_0\varepsilon_r\frac{A}{d} \qquad [3.7]$$

From this equation, we see that the capacitor of a parallel-plate capacitor can be altered by altering the *area of the dielectric* (remember, energy is stored in the dielectric and not in the plates). If the capacitor has $n$ plates (as shown in Fig. 3.2), there are $(n-1)$ dielectrics, so that the area of the dielectric is increased $(n-1)$ times, to give the equation for an $n$-plate capacitor as

$$C = \varepsilon\frac{(n-1)A}{d} = \varepsilon_0\varepsilon_r\frac{(n-1)A}{d} \qquad [3.8]$$

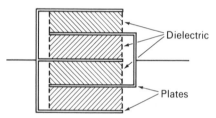

Dielectric

Plates

**Fig. 3.2** A multi-plate parallel-plate capacitor

*Worked example 3.1*

The area of each plate of a parallel-plate capacitor is 500 cm², and the thickness of the dielectric is 0.5 mm. Calculate (a) the capacitance of the capacitor with (i) air as the dielectric, (ii) a dielectric with a relative permittivity of 5. If the applied voltage is 400 V, determine (b) the charge stored in the capacitor and (c) the electric field strength in the dielectric.

*Solution*

From the data given

$$A = 500\ cm^2 = 500 \times (10^{-2})^2 = 0.05\ m^2$$

and $d = 0.5 \times 10^{-3}$ m.

(a) (i) From eqn [3.7]

$$C = \varepsilon_0 A/d = 8.85 \times 10^{-12} \times 0.05/0.5 \times 10^{-3}$$

$$= 0.885 \times 10^{-9}\ F \quad or \quad 0.885\ nF$$

(ii) $C = \varepsilon_0\varepsilon_r A/d$

$$= 8.85 \times 10^{-12} \times 5 \times 0.05/0.5 \times 10^{-3}$$

$$= 4.425\ nF$$

(b) (i) From eqn [3.6]

$$Q = CV_C = 0.885 \times 10^{-9} \times 400$$

$$= 3.54 \times 10^{-7}\,C \quad \text{or} \quad 0.345\,\mu C$$

(ii) $\quad Q = CV_C = 5 \times 0.345 = 1.77\,\mu C$

(c) From eqn [3.2]

$$E = V_C/d = 400/0.5 \times 10^{-3} = 800\,000\,\text{V/m} \quad \text{or} \quad 800\,\text{kV/m}$$

## 3.8 Methods of changing the capacitance of a capacitor

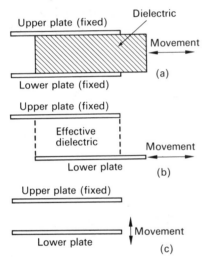

**Fig. 3.3** Variation of capacitance by altering (a) the effective permittivity of the dielectric, (b) the effective area of the dielectric and (c) the distance between the plates

## 3.9 Types of capacitor

For a parallel-plate capacitor, eqn [3.7] shows that

$$C = \varepsilon A/d$$

That is, the capacitance can be varied by altering:

1. the permittivity of the dielectric; or
2. the area of the dielectric; or
3. the distance between the plates.

The fact that the capacitance can be changed in a number of ways is particularly important in the field of instrumentation and measurement because, when we alter one of them, the change in capacitance is a measure of the effect which produces the change.

The way in which these effects change the capacitance of a simple parallel-plate capacitor is illustrated in Fig. 3.3. *Displacing the dielectric* to the right in Fig. 3.3(a) introduces more air between the plates and reduces the net permittivity, thereby reducing the capacitance. *Moving the lower plate* to the right in Fig. 3.3(b) reduces the net area of the dielectric and reduces the capacitance. *Moving the lower plate* downwards in Fig. 3.3(c) increases the distance between the plates, reducing the capacitance.

Capacitors in which one of the above changes can be made are known as *variable capacitors*.

Capacitors are classified according to the type of dielectric used, e.g. air, mica, paper, polystyrene, and according to whether the value of the capacitance is fixed or variable. In a fixed capacitor, the dimensions and type of dielectric cannot be altered, whereas they can be altered in a variable capacitor (see also section 3.7).

*Air capacitors* can either be fixed or variable, the fixed types generally being associated with laboratory standard capacitors.

*Paper dielectric capacitors* have a paper dielectric which has been impregnated with oil or wax, the plates being of metal foil, and the whole wrapped into a cylindrical shape. In *metallized paper capacitors* the paper is metallized on one side so that small gaps or voids between the plates and the dielectric are avoided.

*Plastic film dielectric capacitors* use plastic rather than paper as the dielectric. The plastics used include polystyrene, polyester, polycarbonate and polypropylene.

Mica is a material which can readily be split down into uniform thin sheets, and capacitors using mica as the dielectric are known as *mica capacitors*.

*Ceramic capacitors* have 'plates' of silver coating on opposite faces of ceramic cups, discs and tubes.

*Electrolytic capacitors* have the highest capacitance per unit volume of any type of capacitor, the dielectric in these capacitors being a thin coat of oxide formed either on one or both plates. The majority of capacitors are *polarized*, that is the p.d. between the terminals *must have the correct polarity*.

### 3.10 Parallel-connected capacitors

The combination of $n$ parallel-connected capacitors in Fig. 3.4(a) can be replaced by the equivalent capacitor, $C_E$, in Fig. 3.4(b).

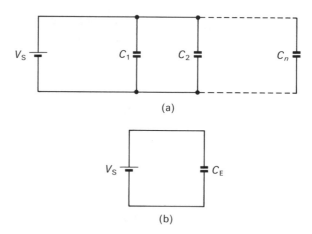

(a)

(b)

**Fig. 3.4** Parallel-connected capacitors

Before analysing the circuit we can, from previous work, get some idea how the parallel connection affects the effective capacitance of the circuit. From eqn [3.7], we see that an increase in the total area of the dielectric has the effect of increasing the capacitance of the capacitor. Connecting the capacitors in parallel with one another has the same effect of increasing the total area of the dielectric, so that we expect to find that the value of $C_E$ will be greater than any individual capacitor in the circuit.

The charge stored by $C_1$ in Fig. 3.4(a) is $Q_1 = C_1 V_S$, the charge stored by $C_2$ is $Q_2 = C_2 V_S$, and the charge stored by $C_n$ is $Q_n = C_n V_S$, hence the total stored charge is

$$C_1 V_S + C_2 V_S + \cdots + C_n V_S = (C_1 + C_2 + \cdots + C_n) C_S$$

If the parallel circuit in Fig. 3.4(a) is replaced by the equivalent capacitor $C_E$ in Fig. 3.4(b), the stored charge in this capacitor is $C_E V_S$. Since the circuits in Fig. 3.4 are equivalent, they both store the same charge so that

$$C_E V_S = (C_1 + C_2 + \cdots + C_n) V_S$$

or

$$C_E = C_1 + C_2 + \cdots + C_n \qquad [3.9]$$

That is

**the equivalent capacitance of parallel-connected capacitors is the sum of the individual capacitances in the circuit.**

Moreover, *the effective capacitance of parallel-connected capacitors is always greater than the largest individual value of capacitance in the circuit.*

*Worked example 3.2* Three capacitors are connected in parallel; one has a capacitance of $0.1\,\mu\text{F}$ and the second has a value of $200\,\text{nF}$. If the effective capacitance of the circuit is $0.5\,\mu\text{F}$, calculate the capacitance of the third capacitor. If $10\,\text{V}$ is applied to the circuit, determine the charge stored by each capacitor, and by the complete circuit.

*Solution* $C_1 = 0.1\,\mu\text{F}$ and $C_2 = 200\,\text{nF} = 200 \times 10^{-9}\,\text{F} = 0.2\,\mu\text{F}$. The effective capacitance of the circuit in $\mu\text{F}$ is

$$C_E = 0.5 = C_1 + C_2 + C_3 = 0.1 + 0.2 + C_3$$

Hence

$$C_3 = 0.5 - (0.1 + 0.2) = 0.2\,\mu\text{F}$$

The charge stored by each capacitor is

$$Q_1 = C_1 V_S = 0.1 \times 10^{-6} \times 10 = 1 \times 10^{-6}\,\text{C} \quad \text{or} \quad 1\,\mu\text{C}$$

$$Q_2 = C_2 V_S = 0.2 \times 10^{-6} \times 10 = 2 \times 10^{-6}\,\text{C} \quad \text{or} \quad 2\,\mu\text{C}$$

$$Q_3 = C_3 V_S = 0.2 \times 10^{-6} \times 10 = 2 \times 10^{-6}\,\text{C} \quad \text{or} \quad 2\,\mu\text{C}$$

and the total stored charge is

$$Q = Q_1 + Q_2 + Q_3 = 5\,\mu\text{C}$$

**3.11 Series-connected capacitors**

When capacitors are connected in series with one another, as shown in Fig. 3.5(a), the same charging current flows through each capacitor for the same length of time. That is, *each capacitor*

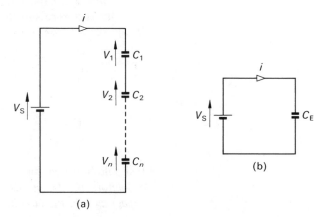

**Fig. 3.5** Series-connected capacitors

*receives the same electrical charge.* If $Q$ is the charge received by each capacitor then

$$Q = C_1 V_1 = C_2 V_2 = \cdots = C_n V_n$$

or

$$V_1 = \frac{Q}{C_1} \qquad V_2 = \frac{Q}{C_2}, \dots \qquad V_n = \frac{Q}{C_n} \qquad [3.10]$$

Since the capacitors are in series then, by KVL, the supply voltage is

$$V_S = V_1 + V_2 + \cdots + V_n = \frac{Q}{C_1} + \frac{Q}{C_2} + \cdots + \frac{Q}{C_n}$$

$$= Q\left(\frac{1}{C_1} + \frac{1}{C_2} + \cdots + \frac{1}{C_n}\right) \qquad [3.11]$$

If the series-connected capacitors are replaced by an equivalent capacitor, $C_E$, then

$$V_S = \frac{Q}{C_E} \qquad [3.12]$$

Since the circuits in Figs 3.5(a) and (b) are equivalent, then eqns [3.11] and [3.12] are equivalent. That is

$$V_S = \frac{Q}{C_E} = Q\left(\frac{1}{C_1} + \frac{1}{C_2} + \cdots + \frac{1}{C_n}\right)$$

or

$$\frac{1}{C_E} = \frac{1}{C_1} + \frac{1}{C_2} + \cdots + \frac{1}{C_n} \qquad [3.13]$$

That is

**when capacitors are connected in series, the reciprocal of the equivalent capacitance of the circuit is equal to the sum of the reciprocals of the individual capacitances in the circuit.**

In the special case of two capacitors in series

$$\frac{1}{C_E} = \frac{1}{C_1} + \frac{1}{C_2} = \frac{C_1 + C_2}{C_1 C_2}$$

or

$$C_E = \frac{C_1 C_2}{C_1 + C_2} \qquad [3.14]$$

It should be noted that *the effective capacitance of series-connected capacitors is always less than the smallest individual value of capacitance in the circuit.*

*Worked example 3.3*  Calculate the equivalent capacitance of a series circuit containing capacitors of capacitance 0.1, 0.8 and 0.01 $\mu F$.

*Solution*  From eqn [3.13]

$$\frac{1}{C_E} = \frac{1}{C_1} + \frac{1}{C_2} + \frac{1}{C_3} = \left[\frac{1}{0.1} + \frac{1}{0.8} + \frac{1}{0.01}\right] \times \frac{1}{10^{-6}}$$

$$= (10 + 1.25 + 100)/10^{-6} = 111.25/10^{-6} \, F^{-1}$$

or

$$C_E = 0.008\,99 \times 10^{-6} \, F \quad \text{or} \quad 8.99 \, nF$$

**3.12  Voltage division in a capacitor string**

As stated earlier, the charge stored by each capacitor in the string is equal to the total charge stored by the circuit. That is

$$Q = V_S C_E = V_n C_n$$

where $Q$ is the stored charge, $V_S$ is the supply voltage, $C_E$ is the equivalent capacitance of the circuit, $V_n$ is the voltage across the $n$th capacitor in the circuit, and $C_n$ is the capacitance of the $n$th capacitor. That is

$$V_n = V_S \frac{C_E}{C_n} \qquad [3.15]$$

In the special case of two capacitors connected in series, $C_E = C_1 C_2/(C_1 + C_2)$ and

$$V_1 = V_S \frac{C_1 C_2}{C_1(C_1 + C_2)} = V_S \frac{C_2}{C_1 + C_2} \qquad [3.16]$$

and

$$V_2 = V_S \frac{C_1}{C_1 + C_2} \qquad [3.17]$$

A careful study of eqn [3.15] reveals that *the smaller the value of capacitance in the series circuit, the larger the voltage it supports* (see also Worked Example 3.4).

*Worked example 3.4*  Three capacitors of 2, 4 and 5 $\mu F$ are connected in series to a 10 V supply. When the capacitors are fully charge, what is the voltage across each?

*Solution*  The equivalent capacitance of the circuit is

$$\frac{1}{C_E} = \frac{1}{2} + \frac{1}{4} + \frac{1}{5} = 0.95 \, \mu F^{-1}$$

or

$$C_E = 1.053 \, \mu F$$

From eqn [3.15]

$$V_1 = V_S C_E / C_1 = 10 \times 1.053/2 = 5.265\,\text{V}$$

$$V_2 = V_S C_E / C_2 = 10 \times 1.053/4 = 2.63\,\text{V}$$

$$V_3 = V_S C_E / C_3 = 10 \times 1.053/5 = 2.106\,\text{V}$$

*Note*: $V_1 + V_2 + V_3 = 10\,\text{V}$ (subject to small 'rounding' errors).

### 3.13 Series–parallel capacitor combinations

The effective capacitance of many apparently complex capacitor circuits can be obtained by reducing the circuit into groups of series and parallel capacitor combinations. Each problem is then dealt with on its merits, as indicated in Worked Example 3.5.

*Worked example 3.5*

Calculate the effective capacitance of the circuit in Fig. 3.6. If 10 V is applied to the circuit, determine the voltage across the parallel branch and across $C_4$.

*Solution*

Initially, we calculate the equivalent capacitance $C_{S1}$ of the series-connected pair of capacitors $C_1$ and $C_2$. This capacitance is

$$C_{S1} = C_1 C_2 / (C_1 + C_2) = 4 \times 6/(4 + 6) = 2.4\,\mu\text{F}$$

Next we determine the capacitance, $C_P$, of the parallel combination $C_3$ and $C_{S1}$ as follows:

$$C_P = C_3 + C_{S1} = 2 + 2.4 = 4.4\,\mu\text{F}$$

The capacitance of the complete circuit is equal to $C_P$ in series with $C_4$, giving an equivalent capacitance for the complete circuit of

$$C_E = C_P C_4 / (C_P + C_4) = 4.4 \times 5.6/(4.4 + 5.6)$$

$$= 2.464\,\mu\text{F}$$

Equation [3.10] tells us that the voltage $V_n$ across a capacitor $C_n$ is

$$V_n = Q/C_n$$

where $Q$ is the stored charge in $C_n$. However, in a series circuit, each capacitor stores the same charge as the complete circuit, which is

$$Q = C_E V_S = 2.464 \times 10^{-6} \times 10$$

$$= 24.64 \times 10^{-6}\,\text{C} \quad \text{or} \quad 24.64\,\mu\text{C}$$

The capacitance of the parallel branch is $C_P = 4.4\,\mu\text{F}$, hence the voltage across the parallel section of the circuit is

$$V_P = Q/C_P = 24.64 \times 10^{-6}/4.4 \times 10^{-6} = 5.6\,\text{V}$$

Similarly the voltage across $C_4$ is

$$V_4 = Q/C_4 = 24.64 \times 10^{-6}/5.6 \times 10^{-6} = 4.4\,\text{V}$$

*Note*: $V_P + V_4 = 5.6 + 4.4 = 10\,\text{V} = V_S$

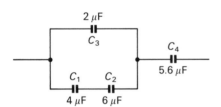

**Fig. 3.6** Figure for Worked Example 3.5.

**3.14 Energy stored in a capacitor**

When an uncharged capacitor is connected to an electricity supply, it begins to draw a *charging current* from the source, which results in energy being stored in the electric field of the capacitor. During the charging period, the voltage between the plates of the capacitor changes by d$v$ volts in d$t$ seconds. If the capacitance of the capacitor is $C$, and the change of stored charge in time d$t$ is d$q$, then the small change in charge is given by

$$dq = i \, dt = C \, dv$$

where $i$ is the *instantaneous current* which flows during time d$t$. That is, the charging current is

$$i = C \frac{dv}{dt}$$

The *instantaneous power*, $p$, supplied to the capacitor during time d$t$ is

$$p = vi = vC \frac{dv}{dt}$$

and the instantaneous energy, $w$, consumed in time d$t$ is

$$w = p \, dt = vi \, dt = vC \frac{dv}{dt} \, dt = vC \, dv$$

At the end of time d$t$, the p.d. between the capacitor plates has increased by d$v$, and the *total energy*, $W$, supplied when the capacitor is fully charged to $V_C$ is

$$W = \int_0^{V_C} vC \, dv = \tfrac{1}{2}C[v^2]_0^{V_C} = \tfrac{1}{2}CV_C^2 \qquad [3.18]$$

*Worked example 3.6*   A 10 $\mu$F capacitor is fully charged to 50 V. Calculate the energy it stores.

*Solution*   From eqn [3.18]

$$\text{Energy stored, } W = \tfrac{1}{2}CV_C^2 = \tfrac{1}{2} \times (10 \times 10^{-6}) \times 50^2$$

$$= 0.0125 \text{ J} \quad \text{or} \quad 12.5 \text{ mJ}$$

**Problems**

**3.1**   A p.d. of 250 V is maintained between two parallel plates separated by 0.04 mm. What is the electric field strength in the dielectric?

[6.25 MV/m]

**3.2**   What voltage must be applied to a 0.25 $\mu$F capacitor if the stored charge is 0.2 mC?

[800 V]

**3.3**   A 1.2 nF parallel-plate capacitor has an effective plate area of 500 cm$^2$, and the relative permittivity if the dielectric is 5. If the capacitor stores a charge of 0.4 $\mu$C, calculate (a) the electric flux in the dielectric, (b) the electric flux density, (c) the voltage between the plates and (d) the electric field strength in the dielectric.

[(a) 0.4 $\mu$C; (b) 8 $\mu$C/m$^2$; (c) 333.33 V; (d) 180.8 kV/m]

**3.4**   A capacitor is charged at a constant current of 0.1 A for 15 ms. If the voltage rises by 125 V, determine the capacitance of the capacitor.

[12 $\mu$F]

**3.5**   A capacitor is charged at a constant current of 0.1 A for 15 s, and then discharged at a constant rate of 0.7 A. How long does it take for the capacitor to become fully discharged?

[2.14 s]

**3.6**   A parallel-plate capacitor with an air dielectric has a plate area of 200 cm$^2$. Calculate the capacitance of the capacitor. When 250 V is applied to the terminals of the capacitor, calculate the total stored charge. If a dielectric of relative permittivity 4 is inserted between the plates, determine the new value of capacitance.

[0.354 nF; 88.5 nC; 1.416 nF]

**3.7**   A variable capacitor, adjusted to give its maximum capacitance, is connected to a 100 V supply. When fully charged, it is disconnected from the supply and the capacitor is adjusted so that its capacitance is one-quarter of its maximum value. Calculate the new voltage between the plates of the capacitor. If the energy stored at maximum capacitance is 5 $\mu$J, calculate the energy stored at one-quarter of its maximum capacitance.

[400 V; 20 $\mu$J]

**3.8**   Capacitors of 0.1, 0.2 and 0.4 $\mu$F are connected (a) in series then (b) in parallel. Calculate the capacitance in each case.

[(a) 0.571 $\mu$F; (b) 0.7 $\mu$F]

**3.9**   A 2 and a 4 $\mu$F capacitor are connected in series. What capacitance must be connected in parallel with them to give an equivalent overall capacitance of 5 $\mu$F?

[3.667 $\mu$F]

**3.10**   When two capacitors are connected (a) in series, (b) in parallel, the respective capacitance values are 6.667 and 30 $\mu$F. Determine the capacitance of each capacitor.

[10 $\mu$F; 20 $\mu$F]

**3.11**   A circuit of the type in Fig. 3.6 has an equivalent capacitance of 2.54 $\mu$F. If $C_1 = 6\,\mu$F, $C_3 = 1\,\mu$F and $C_4 = 10\,\mu$F, calculate the value of $C_2$.

[4 $\mu$F]

**3.12**   A network has three terminals A, B and C. Capacitors of 10 and 5 $\mu$F are connected between AB and BC, respectively. What value of capacitance must be connected between terminals A and C if the equivalent capacitance between A and C is 13.33 $\mu$F?

[10 $\mu$F]

**3.13**   If a 0.01 $\mu$F capacitor stores 0.001 J of energy, determine the capacitor voltage and the stored charge.

[447.2 V; 4.47 $\mu$C]

**3.14**   When the charge stored by a capacitor is 4 nC, the energy stored is 0.2 J. Evaluate the capacitor voltage and the capacitance of the capacitor.

[100 V; 40 $\mu$F]

**3.15**   When a 10 and a 100 $\mu$F capacitor are connected in series, the voltage across the 10 $\mu$F capacitor is 100 V. Calculate the supply voltage. Determine also (a) the charge stored by each capacitor and the total charge stored, (b) the energy stored by each capacitor and (c) the total energy stored.

[110 V; (a) the charge stored by each capacitor and the total circuit is 1 mC; (b) 10 $\mu$F, 50 mJ; 100 $\mu$F, 5 mJ; (c) 55 mJ]

# 4 Magnetic fields and circuits

## 4.1 Introduction

We look in this chapter at important aspects of magnetic fields and magnetic circuits so far as they concern electrical and electronic engineers. Commencing with a study of the production of a magnetic field by current, we move on to magnetomotive force (the electromagnetic equivalent of e.m.f.) and permeability, followed by the magnetization curve and the $B$–$H$ loop of ferromagnetic materials. Next, we turn our attention to magnetic circuits which, in many cases, have exact models of the electric circuit; so much so, that we can apply 'Ohm's law' and 'Kirchhoff's laws' to magnetic circuits. Finally, the attention of the reader is directed towards magnetic leakage and magnetic screening. A knowledge of these topics is vital to all students, for they form basic building blocks from which the subjects of electrical and electronic engineering are developed.

## 4.2 The magnetic field and flux density

It was discovered early in the nineteenth century that an electrical current produced a magnetic field. A magnetic field cannot be 'seen' by man, but its effects can be observed, for example, by the mechanical force it exerts on iron filings, or the mechanical force it produces on a current-carrying conductor. The presence of a magnetic field can also be detected by sensing the e.m.f. induced in a conductor or coil of wire when the magnetic field changes in value.

In fact, the magnetic field is the bedrock of electrical machine principles. The force acting on a current-carrying conductor is the basis of the motor, and the e.m.f. induced in a coil when a magnetic field changes is the principle of the electrical generator.

The name *magnetism* comes from *magnetite* (an iron oxide) whose magnetic properties were known before man understood the principle of electricity. The nature of magnetism can be understood from the make-up of the atom as follows. When electrons spin around the atomic nucleus, the moving charges are equivalent to a 'current', and the atomic 'current' gives rise to a tiny permanent magnet effect. Groups of electrons with like magnetic spins produce the same effect as a small magnet, and are described as **magnetic domains** or **dipole magnets**.

In a *demagnetized material*, the magnetic domains point in random directions, as shown in Fig. 4.1(a), and the net magnetic

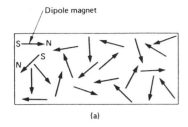

Dipole magnet

(a)

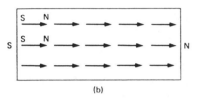

(b)

**Fig. 4.1** (a) Demagnetized iron and (b) magnetized iron

field produced by the iron is zero. When the iron is placed in a magnetic field, as shown in Fig. 4.1(b), the magnetic domains begin to align with the magnetic field. The stronger the applied magnetic field, the greater the proportion of domains which align with the field.

Finally, when all the magnetic domains are aligned, the iron is said to be magnetically **saturated** [see Fig. 4.1(b)]. Since no further domains can be aligned, any further increase in the external magnetic field does not produce any significant further increase in the magnetism in the iron.

If we could provide an isolated or 'free' magnetic *north-seeking pole* (an *N-pole*) and place it at a point in a magnetic field, *the direction of the magnetic field at that point is given by the direction of the force acting on the N-pole.* Experiments with a pair of permanent magnets show that:

1. **Like magnetic poles repel one another.**
2. **Unlike magnetic poles attract one another.**

Also, if the isolated N-pole was free to move, it would be repelled by another N-pole, and would be attracted to an S-pole. That is

**'lines' of magnetic force or flux are assumed to leave an N-pole and enter an S-pole.**

Strictly speaking, the concept of a 'line' of magnetic flux is man-made, and nothing actually 'moves' in the magnetic field; none-the-less, it is convenient to refer to the 'direction' of a magnetic field.

**Magnetic flux** is given the symbol $\Phi$, and is a measure of the magnetic field; its unit symbol is the *weber* (Wb). **Magnetic flux density**, symbol $B$, is the amount of flux passing through unit area perpendicular to the 'direction' of the flux. The unit of flux density is the *tesla* [unit symbol T $(=1\,\text{Wb}/\text{m}^2)$], and

$$B = \frac{\Phi}{A} \qquad\qquad [4.1]$$

where a flux of $\Phi$ passes through area $A$.

Materials which can be strongly magnetized in the direction of an external magnetic field are known as *ferromagnetic materials* and include iron, steel and a number of their alloys. Certain materials are 'non-magnetic' and can be classified either as *paramagnetic materials* or *diamagnetic materials*. Paramagnetic materials become weakly magnetized in the direction of the external magnetic field and include aluminium; diamagnetic materials become weakly magnetized in the opposite direction to the external magnetic field and include copper and gold.

*Worked example 4.1*    A magnetic field of 5 mWb exists in an iron core of circular cross-section of diameter 5 cm. Calculate the flux density in the iron.

*Solution*    The cross-sectional area of the iron is

$$A = \pi r^2 = \pi(2.5 \times 10^{-2})^2 = 1.96 \times 10^{-3}\,\text{m}^2$$

and from eqn [4.1]

$$B = \frac{\Phi}{A} = \frac{5 \times 10^{-3}}{1.96 \times 10^{-3}} = 2.55\,\text{T}$$

### 4.3  Magnetic field produced by a current-carrying conductor

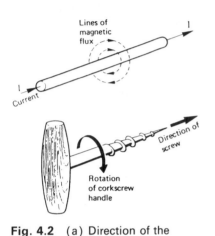

**Fig. 4.2**  (a) Direction of the magnetic flux around a current-carrying conductor and (b) the screw rule

We can determine the direction of the magnetic field produced by a current-carrying conductor in Fig. 4.2. Based on the results of experiments, engineers have proposed a simple *corkscrew rule*, depicted in Fig. 4.2, which enables us to predict the direction of the magnetic field as follows:

> **If we imagine a corkscrew pointing in the direction of current flow then, in order to propel the corkscrew in the direction of current flow, the handle must be turned in the direction of the magnetic field around the conductor.**

Alternatively, we can use a *right-hand grip rule* as follows:

> **If we grip the conductor in our right hand, with the thumb extended along the conductor in the direction of the current flow, then our fingers point in the direction of the magnetic field.**

### 4.4  Flux produced by a loop of wire

Using the information in section 4.3 on the flux produced by a current-carrying conductor, we can deduce the direction of the flux produced by a single loop of wire. If we look down from the top of the loop in Fig. 4.3(a), the section (X–X) is shown in Fig. 4.3(b). The current direction in the wire is assumed to be depicted by an 'arrow'. When the current approaches us, it is shown by a 'dot' which indicates the tip of the arrowhead [shown on the left-hand conductor in Fig. 4.3(b)]. If the current leaves us, it is shown by drawing a 'cross' which represents the crossed feathers of the arrow; this is shown on the conductor on the right-hand side of Fig. 4.3(b).

Applying the general rules outlined in section 4.3, we see that magnetic flux 'enters' the bottom of Fig. 4.3(b), and 'leaves' the top. Transferring these data to the loop in Fig. 4.3(a), it means that the side of the loop nearest to us is the S-pole, and the opposite side is an N-pole.

We can extend this to propose a simple diagrammatic method of determining the magnetic polarity produced by a single turn

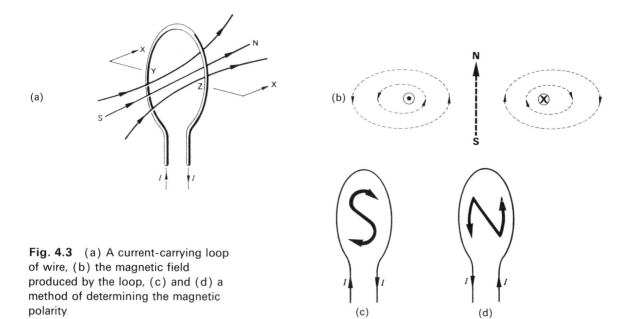

(a)

(b)

(c)          (d)

**Fig. 4.3** (a) A current-carrying loop of wire, (b) the magnetic field produced by the loop, (c) and (d) a method of determining the magnetic polarity

of wire, and is shown in Figs 4.3(c) and (d). Alternatively, we can use a version of the *right-hand grip rule*, as follows:

> **Grasp the loop of wire in the right hand, with the thumb extended at right angles to the fingers. If the fingers point in the direction of the current, then the thumb points in the direction of the magnetic field produced by the loop.**

### 4.5 Magnetic field produced by a current-carrying coil

The magnetic field produced by a current-carrying coil can be predicted by a simple extension to the work in the two previous sections.

Consider the coil in Fig. 4.4, with the current in the direction shown. Applying any of the rules in section 4.4, we see that the conductors both at the top and bottom of the figure cause magnetic flux to enter the left-hand end of the coil, and leave at the right-hand end. That is, the left-hand end is the S-pole, and the right-hand end is the N-pole.

Alternatively, applying the right-hand grip rule in section 4.4 to the complete coil, we come to the same conclusion.

A coil having an air core is known as a *solenoid*, and one having an iron core is an *electromagnet*. In either case, the current in the coil is described as the *excitation current*.

### 4.6 Magnetomotive force, *F*

The **magnetomotive force** (m.m.f.), symbol $F$, is the force which establishes magnetic flux in the magnetic circuit (its analogy in the electric circuit is e.m.f., which establishes a current in the

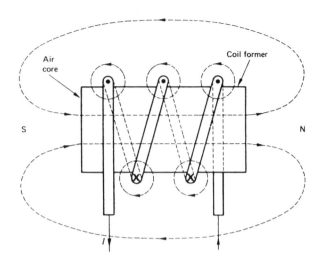

**Fig. 4.4** Magnetic field produced by a current-carrying coil of wire

electric circuit). An expression for m.m.f. is

m.m.f., $F$ = number of turns on the coil × current in the coil

$$= NI$$

Strictly speaking, the units of m.m.f. are (amperes × turns) or ampere turns. However, since the number of turns on the coil is a dimensionless quantity, the dimensions of m.m.f. are amperes (A). None the less, to avoid any confusion between the current in a coil and the m.m.f. produced by the coil, we will give m.m.f. the dimensions of ampere turns.

**4.7 Magnetic field intensity, _H_**

The **magnetic field intensity** (or **magnetic field strength** or **magnetizing force**), symbol $H$, is the m.m.f. per unit length of the magnetic circuit. That is

$$H = \frac{F}{l} = \frac{NI}{l} \qquad [4.2]$$

The dimensions of magnetic field intensity are (strictly speaking) amperes per metre, but can be expressed in ampere turns per metre.

*Worked example 4.2*

A coil of wire is wound with 4000 turns of wire, and carries a current of 1.2 A. If the coil is wound on a toroid, and the magnetic field intensity in the coil is 20 000 ampere turns per metre, calculate the mean diameter of the toroid.

*Solution*    m.m.f., $F = NI = 4000 \times 1.2 = 4800$ ampere turns or A

Since $H = F/l$, then

$$\text{Length of magnetic circuit}, l = F/H = 4800/20\,000$$

$$= 0.24 \text{ m}$$

If $d$ is the mean diameter of the toroid, then $l = \pi d$, or

$$d = l/\pi = 0.24/\pi = 0.0764 \text{ m} \quad \text{or} \quad 7.64 \text{ cm}$$

**4.8  Permeability, $\mu$**

When a coil is excited by current, the magnetic field intensity, $H$, in the magnetic circuit gives rise to a magnetic flux $B$. The relationship between the two is

$$B = \mu H \qquad\qquad\qquad [4.3]$$

where $\mu$ is the **absolute permeability** of the material, having dimensions of henrys per metre (H/m). The permeability is a measure of the ability of the material to concentrate magnetic flux. The **permeability of free space** (or of a vacuum) is given the special symbol $\mu_0$, where

$$\mu_0 = 4\pi \times 10^{-7} \text{ H/m}$$

The permeability of air has, to all intents and purposes, the same value as $\mu_0$. Clearly, if a coil has an air core, then the magnetic flux density inside the coil is

$$B = \mu_0 H$$

However, if the air core is replaced by an iron core, the magnetic flux density in the core increases significantly. In fact, the iron increases the permeability of the core by a factor $\mu_r$, where $\mu_r > 1$, hence the flux density in an iron core is

$$B = \mu_0 \mu_r H = \mu H$$

that is

$$\mu = \mu_0 \mu_r \qquad\qquad\qquad [4.4]$$

The factor $\mu_r$ is the *relative permeability* of the magnetic circuit, and can have a value up to about 7000, depending not only on the material but also on the operating flux density and temperature.

***Worked example 4.3***

A coil has a magnetic field intensity in its magnetic circuit of 600 ampere turns per metre. Calculate the flux density in the core if it has (a) an air core, (b) an iron core with $\mu_r = 500$ and (c) a transformer steel core with $\mu_r = 1600$.

*Solution*

In each case $H = 600$ ampere turns per metre.
(a) For air, $\mu = \mu_0 = 4\pi \times 10^{-7}$ H/m, hence

$$B = \mu_0 H = 4\pi \times 10^{-7} \times 600 = 0.754 \times 10^{-3} \text{ T} \quad \text{or} \quad 0.754 \text{ mT}$$

(b) For the iron core with $\mu_r = 500$

$$B = \mu_r \mu_0 H = 500 \times 0.754 \times 10^{-3} = 0.377 \text{ T}$$

(c) For the transformer steel when $\mu_r = 1600$

$$B = \mu_r \mu_0 H = 1600 \times 0.754 \times 10^{-3} = 1.206 \text{ T}$$

## 4.9   Magnetization curve of ferromagnetic materials

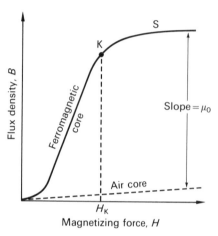

**Fig. 4.5**   Magnetization curve of a ferromagnetic material

In the case of a coil *with an air core*, an increase in the current in the voil results in a proportional increase in the flux (and the flux density) in the core, as is shown by the broken line in Fig. 4.5.

If the coil has an iron core, any increase in the current in the coil produces a fairly rapid increase in flux density, as shown in the graph by the full line in Fig. 4.5. The reason is that any increase in magnetizing force causes some of the magnetic domains in the core to align with the magnetic field, adding their own magnetic flux to that of the field. As the magnetizing force increases further, more and more domains align, producing a further increase in flux density in the core. Between the origin and the point marked K on the *B–H curve* or *magnetization curve* for a ferromagnetic material (see Fig. 4.5), the value of $\mu_r$ is very high.

However, beyond point K on the curve, any further increase in magnetizing force does not produce a significant increase in flux density. Point K is known as the 'knee' of the curve. By the time the knee has been reached, most of the magnetic domains in the core have aligned with the magnetic field produced by the current in the coil, and any further increase in excitation current only increases the magnetic flux by a small amount. The knee therefore marks the onset of *magnetic saturation*.

Finally, in the region of point S on the *B–H* curve, all domains in the iron core have aligned with the magnetic field produced by the excitation current, and the core has reached full magnetic saturation. At this point, the incremental permeability of the iron has fallen to $\mu_0$, and any further increase in magnetizing force results in comparatively little change of the flux density in the core.

In most magnetic circuits, the operating flux density is below the knee of the curve and, since the operating flux density is fairly constant, it is often safe to assume that $\mu_r$ is constant. Figure 4.6 shows how $\mu_r$, $B$ and $H$ vary with one another in a ferromagnetic material.

## 4.10   Hysteresis loop of a ferromagnetic material

Many items of electrical apparatus operate with an alternating magnetic field, with the flux density varying many times (sometimes many hundred times) per second from near saturation in one direction to near saturation in the opposite direction. The electrical transformer is an example of this kind.

We therefore need to know how ferromagnetic materials behave under these conditions. Before we can study this, we must initially

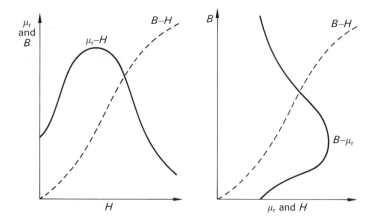

**Fig. 4.6** Graphs of the variation in $\mu_r$, $B$ and $H$ in iron

completely demagnetize the iron. Broadly speaking, this is done by placing the iron in a strong alternating magnetic field, which is gradually reduced to zero. This technique is regularly used in hospitals to demagnetize surgical instruments.

When the core is demagnetized, it is at point T on the curve in Fig. 4.7. Increasing the magnetizing force causes the flux density in the core to follow the conventional magnetization curve (shown by the dotted line) until it reaches point U, at which point the magnetizing force is gradually reduced.

When the magnetizing force reaches zero (point V), we find that there remains a *residual flux density* in the core. This is known as the *remanence* or *remanent flux density*, $B_r$, and is due to the magnetic domains which remain aligned with the original direction of the magnetic field.

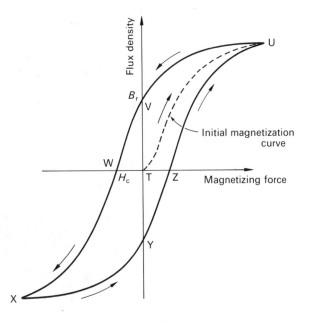

**Fig. 4.7** Hysteresis loop

If the magnetizing force is reversed and increased in value, the magnetic domains in the core begin to realign with the new direction of the magnetic field until, at point W on the curve, the flux density reaches zero. The magnetizing force which causes this to occur is known as the *coercive force*, $H_c$. Further increase in magnetizing force in the reverse direction will, by point X, cause the core to be magnetically saturated once more, but in the reverse direction.

Once again, reducing the magnetizing force to zero and returning it to its original direction causes the curve to follow the path XYZU in Fig. 4.7.

The complete curve UVWXYZU in Fig. 4.7 is known as the **hysteresis loop** or **B–H loop** of the iron [*note*: the word 'hysteresis' means the lagging of the magnetic field behind its cause (the current)].

Materials used for permanent magnets are described as *magnetically hard* materials, and have a high remanence (about 1 T) and a high coercive force (about 50 000 A/m), and their *B–H* curve is 'square' or 'fat'. Electromagnets, which must lose their magnetic field when the excitation current is cut off, are made from *magnetically soft* materials, and have a 'narrow' *B–H* loop. These materials have a high value of saturation flux density and a low coercive force.

## 4.11 Hysteresis loss

Each time the magnetic field of an electromagnet is reversed, the magnetic domains in the core are forced to reverse, with a consequent loss of energy. If the field alternates, as it does in electrical power equipment, energy is expended during each cycle of alternations. The energy consumed in the continued reversal of the magnetic domains appears mainly in the form of heat in the magnetic material, and is known as *hysteresis loss*.

The amount of energy consumed from this cause depends on several factors including the frequency, $f$, of reversal of the magnetic flux, and on the maximum flux density, $B_m$, in the material. The *hysteresis power loss*, $P_h$, is given by

$$P_h = kvfB_m^n \qquad [4.5]$$

where $k$ is the hysteresis coefficient, $v$ the volume of the iron, and $n$ the *Steinmetz index* which has a value in the range 1.6–2.2, and is typically 1.7.

It can be shown by dimensional analysis that the area of the *B–H* loop is proportional to the energy consumed per cycle of the loop, so that the hysteresis power loss is given by

$$P_h = vf \times \text{area of the } B\text{–}H \text{ loop}$$

If the *B–H* loop is drawn on graph paper, the energy loss per cubic metre of iron can be calculated from

$$\text{Area of } B\text{–}H \text{ loop (e.g. cm}^2) \times \text{scale factor for } B \text{ (e.g. T/cm)}$$

$$\times \text{scale factor for } H \text{ (e.g. (A/m)/cm)}$$

***Worked example 4.4*** A specimen of iron is tested at a frequency of 50 Hz, the volume of the specimen being 0.002 m³. The B–H curve for the material is drawn on graph paper, and is found to have an area of 40 cm². If the scale factor for B is 2 (T/cm), and that for H is 4 (A/m)/cm, calculate (a) the energy loss per cycle and (b) the hysteresis power loss. (c) If the hysteresis coefficient is 92 and the core has a maximum flux density of 1.8 T, determine the value of the Steinmetz coefficient of the iron.

*Solution* (a) The energy loss per cycle is

Volume of iron × energy loss per cubic metre

$$= 0.002 \times (40 \times 2 \times 4) = 0.64 \, \text{J}$$

(b) The hysteresis power loss is

$$P_h = vf \times \text{area of } B\text{–}H \text{ loop}$$

$$= f \times \text{energy loss per cycle} = 50 \times 0.64 = 32 \, \text{W}$$

(c) From eqn [4.5]

$$P_h = kvfB_m^n = 92 \times 0.002 \times 50 \times 1.8^n = 9.2 \times 1.8^n$$

or

$$1.8^n = 32/9.2 = 3.478$$

Taking logarithms of both sides gives

$$n \log 1.8 = \log 3.478$$

that is

$$n = \log 3.478 / \log 1.8 = 2.12$$

## 4.12 Reluctance, *S*

For the purposes of analysis, the magnetic circuit is an almost exact analogy of the electric circuit. In the magnetic circuit we can think of m.m.f. producing a magnetic flux in a magnetic circuit of **reluctance**, *S*. Reluctance is the magnetic circuit analogue of resistance in the electrical circuit. That is, 'Ohm's law' for the magnetic circuit can be quoted in the form

$$F = \Phi S \qquad\qquad [4.6]$$

where *F* is the m.m.f. (ampere turns), $\Phi$ the magnetic flux (Wb) and *S* the reluctance of the magnetic circuit (ampere turns per weber or A/Wb). The relationships between the magnetic and electric circuits are as follows:

| Magnetic circuit | Electric circuit |
| --- | --- |
| Magnetomotive force, $F$ | Electromotive force, $E$ |
| Flux, $\Phi$ | Current, $I$ |
| Reluctance, $S$ | Resistance, $R$ |
| $F = \Phi S$ | $E = IR$ |

The reluctance of a magnetic circuit can be determined in terms of its physical size and its permeability, as follows. From eqn [4.2]

$$F = Hl$$

where $l$ is the length of the magnetic circuit. Also from eqns [4.1] and [4.3]

$$\Phi = BA = \mu HA$$

where $A$ is the cross-sectional area of the magnetic circuit. Equation (4.6) tells us that

$$S = \frac{F}{\Phi} = \frac{Hl}{\mu HA} = \frac{l}{\mu A} = \frac{l}{\mu_r \mu_0 A} \qquad [4.7]$$

*Worked example 4.5*    A steel ring has a mean circumference of 0.3 m and a cross-sectional area of 0.002 m². If the relative permeability of the ring at the operating flux density is 650, calculate the current which must flow in a coil of 1000 turns wound uniformly on the core to produce a flux of 2 mWb in the iron.

*Solution*    From eqn [4.7]

$$S = \frac{l}{\mu_r \mu_0 A} = \frac{0.3}{650 \times 4\pi \times 10^{-7} \times 0.002}$$

$$= 0.1836 \times 10^6 \ \text{A/Wb}$$

Also, from eqn [4.6]

$$F = \Phi S = 2 \times 10^{-3} \times 0.1836 \times 10^6 = 367.2 \ \text{ampere turns}$$

$$= NI$$

hence

$$I = 367.2/1000 = 0.3672 \ \text{A}$$

## 4.13 Series-connected magnetic circuit

The analogy between the magnetic circuit and the electric circuit is sufficiently close to allow us to represent the magnetic circuit in much the same way as we represent the electric circuit. We may therefore represent series-connected reluctances as shown in Fig. 4.8.

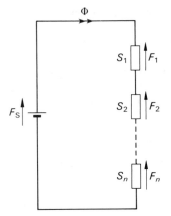

**Fig. 4.8** Reluctances in series

While KVL applies to electric circuits we may, by analogy, apply it to the magnetic circuit in Fig. 4.8 as follows:

$$F_S = F_1 + F_2 + \cdots + F_n$$

and if $S_E$ is the equivalent reluctance of the series-connected magnetic circuit, then

$$\Phi S_E = \Phi S_1 + \Phi S_2 + \cdots + \Phi S_n = \Phi(S_1 + S_2 + \cdots + S_n)$$

where $S_n$ is the reluctance of the $n$th element in the magnetic circuit. That is, the equivalent reluctance of the complete circuit is

$$S_E = S_1 + S_2 + \cdots + S_n \qquad [4.8]$$

Since each element carries the same magnetic flux, then

$$\Phi = \frac{F_S}{S_E} = \frac{F_n}{S_n}$$

That is, the m.m.f. across the $n$th element is

$$F_n = F_S \frac{S_n}{S_E} \qquad [4.9]$$

The reader will observe that eqn [4.8] has a similar form to the equivalent resistance of a series electrical circuit, and that eqn [4.9] has the same general form as the potential drop across the $n$th element in a series circuit.

***Worked example 4.6***  The iron core of a magnetic circuit has an air gap in it, as shown in Fig. 4.9(a). The length, $l_1$, of the iron is 0.3 m, its cross-sectional area is 0.001 m² and its relative permeability is 650; the length, $l_2$, of the air gap is 1 mm.

Calculate the current which must flow in a coil of 1500 turns wound uniformly around the iron to produce a flux in the air gap of 1.2 mWb. Assume that all the flux in the iron passes through the air gap.

*Solution*  The reluctance of the iron path is

$$S_1 = \frac{l_1}{\mu_r \mu_0 A} = \frac{0.3}{650 \times 4\pi \times 10^{-7} \times 0.001}$$

$$= 0.367 \times 10^6 \text{ A/Wb}$$

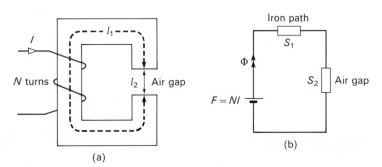

**Fig. 4.9** Diagram for Worked Example 4.6

Since the air gap has the same cross-sectional area, its reluctance is

$$S_2 = \frac{l_2}{\mu_0 A} = \frac{10^{-3}}{4\pi \times 10^{-7} \times 0.001} = 0.796 \times 10^6 \text{ A/Wb}$$

The reader should note that, although the length of the air gap is only 0.3 per cent of the length of the iron part, its reluctance is more than twice that of the iron! The total reluctance of the magnetic path is

$$S_T = S_1 + S_2 = 1.163 \times 10^6 \text{ A/Wb}$$

The m.m.f. needed to produce the magnetic flux is

$$F = \Phi S = 1.2 \times 10^{-3} \times 1.163 \times 10^6 = 1396 \text{ ampere turns}$$

$$= NI$$

Hence

$$I = 1396/1500 = 0.931 \text{ A}$$

***Worked example 4.7***  A magnetic circuit has three series-connected parts, whose dimensions are as follows:

| Part | Length (m) | Area (m²) |
|------|------------|-----------|
| A | 0.4 | $2.73 \times 10^{-4}$ |
| B | 0.15 | $4 \times 10^{-4}$ |
| C | $10^{-3}$ | $3 \times 10^{-4}$ |

If magnetic leakage can be ignored, calculate the current in a coil of 1000 turns wound uniformly on the magnetic circuit to give a flux of

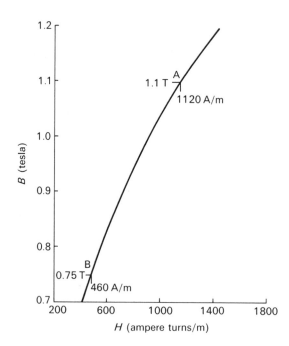

**Fig. 4.10** Diagram for Worked Example 4.7

0.3 mWb in part C, which is an air gap. The $B$–$H$ curves of parts A and B is shown in Fig. 4.10.

*Solution*      In this case we are provided with the $B$–$H$ curve for the reason that $\mu_r$ varies with $B$ and $H$ along the curve (see also Fig. 4.6). The flux density in part A of the iron circuit is

$$B_A = \Phi_A/A_A = 0.3 \times 10^{-3}/2.73 \times 10^{-4} = 1.1 \text{ T}$$

The corresponding point is marked A on Fig. 4.10, and we see from the curve that the corresponding value of $H$ is $H_A = 1120$ ampere turns per metre.

The flux density in part B is

$$B_B = \Phi_B/A_B = 0.3 \times 10^{-3}/4 \times 10^{-4} = 0.75 \text{ T}$$

and $H_B = 460$ ampere turns per metre (see Fig. 4.10). Consequently

$$F_A = H_A l_A = 1120 \times 0.4 = 448 \text{ ampere turns}$$

$$F_B = H_B l_B = 460 \times 0.15 = 69 \text{ ampere turns}$$

The reluctance of the air gap (part C) is

$$S_C = \frac{l_C}{\mu_0 A_C} = \frac{10^{-3}}{4\pi \times 10^{-7} \times 3 \times 10^{-4}}$$

$$= 2.65 \times 10^6 \text{ ampere turns per weber}$$

hence

$$F_C = \Phi S_C = 0.3 \times 10^{-3} \times 2.65 \times 10^6 = 795 \text{ ampere turns}$$

The total m.m.f. requirement for the circuit is

$$F_T = F_A + F_B + F_C = 448 + 69 + 795$$

$$= 1312 \text{ ampere turns}$$

and the excitation current in the coil is

$$I = F_T/N = 1312/1000 = 1.312 \text{ A}$$

**4.14   Parallel magnetic circuit**

Consider the two-branch magnetic circuit in Fig. 4.11(a), and its equivalent circuit in Fig. 4.11(b). Since the branches are in parallel, each has the same m.m.f. across it, that is

$$F_T = \Phi_1 S_1 = \Phi_2 S_2 = \Phi_T S_E$$

where $\Phi_1$ and $\Phi_2$ are the flux in branches 1 and 2, respectively, and $S_1$ and $S_2$ the respective reluctances; $\Phi_T$ is the total flux in the circuit, and $S_E$ the equivalent reluctance of the parallel circuit. Since we can apply Kirchhoff's laws to the magnetic circuit, we can say that

$$\Phi_T = \Phi_1 + \Phi_2 = \frac{F_T}{S_1} + \frac{F_T}{S_2} = \frac{F_T}{S_E}$$

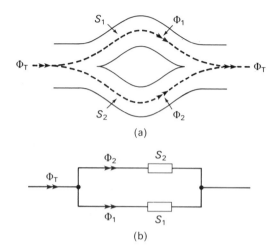

**Fig. 4.11** Parallel magnetic circuit

or

$$\frac{F_T}{S_E} = F_T\left[\frac{1}{S_1} + \frac{1}{S_2}\right]$$

That is

$$\frac{1}{S_E} = \frac{1}{S_1} + \frac{1}{S_2}$$

or

$$S_E = \frac{S_1 S_2}{S_1 + S_2}$$

If we extend the discussion to $n$ parallel-connected magnetic circuits we may say that

$$\Phi_T = \Phi_1 + \Phi_2 + \cdots + \Phi_n$$

and

$$\frac{1}{S_E} = \frac{1}{S_1} + \frac{1}{S_2} + \cdots + \frac{1}{S_n}$$

## 4.15 Magnetic leakage and fringing

In practice, not all the magnetic flux produced by a coil is 'useful', because a certain amount leaks away from the magnetic circuit. A simple example is shown in Fig. 4.12(a), in which the 'useful' flux is that which reaches the air gap. Some of the flux 'leaks' away from the main path; this is shown in the figure as the *leakage flux*, $\Phi_L$. Some more of the flux reaches the magnet poles, but fringes the useful area, and is known as the *fringing flux*, $\Phi_F$.

The equivalent circuit is shown in Fig. 4.12(b). Here $\Phi_T$ is the total flux provided by the coil, $\Phi_L$ the leakage flux, $\Phi_F$ the fringing

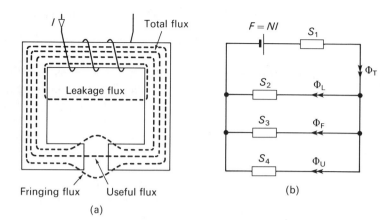

**Fig. 4.12**  Magnetic leakage and fringing

(a)

(b)

flux, and $\Phi_U$ the useful flux in the air gap. The ratio of the total flux to the useful flux is known as the **leakage coefficient**, and is given by

$$\text{Leakage coefficient} = \frac{\text{total flux}}{\text{useful flux}} = \frac{\Phi_T}{\Phi_U}$$

The value of the leakage coefficient is rarely as low as unity and, in a practical magnetic circuit, it may be as high as 1.4.

*Worked example 4.8*  A magnetic circuit has an iron path of reluctance $0.3 \times 10^6$ A/Wb, with an air gap of reluctance $0.7 \times 10^6$ A/Wb. If the magnetic flux in the air gap of 1.6 mWb is produced by a coil of 1000 turns wound uniformly over the magnetic path, calculate the current in the coil if the leakage coefficient is (a) unity (i.e. no leakage) and (b) 1.3.

*Solution*  (a) If the leakage coefficient is unity, the magnetic flux in the iron is the same as it is in the air gap, and the total m.m.f. is

$$F = \Phi(S_{\text{iron}} + S_{\text{air}})$$

$$= 1.6 \times 10^{-3}(0.3 \times 10^6 + 0.7 \times 10^6)$$

$$= 1600 \text{ ampere turns}$$

The current in the coil is

$$I = F/N = 1600/1000 = 1.6 \text{ A}$$

(b) When the leakage coefficient is 1.3, the magnetic flux in the iron is

$$\Phi_{\text{iron}} = 1.3 \times 1.6 \text{ mWb} = 2.08 \text{ mWb}$$

and the total m.m.f. is

$$F = \Phi_{\text{iron}} S_{\text{iron}} + \Phi_{\text{air}} S_{\text{air}}$$

$$= (2.08 \times 10^{-3} \times 0.3 \times 10^6) + (1.6 \times 10^{-3} \times 0.7 \times 10^6)$$

$$= 624 + 1120 = 1744 \text{ ampere turns}$$

The current in the coil is

$$I = F/N = 1744/1000 = 1.744 \text{ A}$$

**4.16 Magnetic screening**

Certain types of electrical and electronic apparatus are sensitive to the effect of strong magnetic fields. In such cases it is necessary to shield the apparatus from the field by enclosing it in a metal screen which has a very low reluctance (or high relative permeability). This effectively places the apparatus in a magnetic 'short circuit', so that the magnetic field within the screen is practically zero.

Typical apparatus requiring this treatment include certain types of cathode-ray tube, and sensitive measuring instruments such as galvanometers.

**Problems**

**4.1**   A coil of 450 turns is wound uniformly over a magnetic circuit which is 30 cm long, and it carries a current of 0.6 A. Calculate the m.m.f. of the coil and the magnetizing force.

[270 ampere turns; 900 ampere turns per metre]

**4.2**   A steel ring of uniform cross-sectional area and mean diameter 20 cm, is wound with a coil of 1000 turns. If the magnetizing force is 5000 A/m, calculate the current in the coil.

[6.98 A]

**4.3**   An iron circuit has a cross-sectional area of 500 mm$^2$ and length 30 cm. If the flux density in the core is 1.25 T, and the relative permeability of the iron is 1800, determine the reluctance of the magnetic circuit, and the m.m.f. needed to produce the flux.

[0.265 × 10$^6$ A/Wb; 165.8 ampere turns]

**4.4**   A coil carrying a current of 1.5 A is uniformly wound on an iron toroid of cross-sectional area 150 mm$^2$ and mean diameter 12 cm. If the coil has 200 turns of wire on it, and the relative permeability of the iron is 1200, calculate the magnetic flux in the iron.

[0.18 nWb]

**4.5**   If an air gap of length 1 mm is introduced into the iron circuit in problem 4.4, calculate the value of the current in the coil which gives a magnetic flux of 0.2 mWb in the air gap if (a) there is no leakage or fringing and (b) if the leakage factor is 1.25.

[(a) 6.97 A; (b) 7.39 A]

**4.6**   A 140-turn coil is uniformly wound on an iron core of length 78 cm and cross-sectional area 1000 mm$^2$. If the magnetic flux in the core is 1 mWb and the current in the coil is 8 A, determine (a) the m.m.f. produced by the coil, (b) the magnetizing force and (c) the relative permeability of the iron.

[(a) 1120 ampere turns; (b) 1436 ampere turns per metre; (c) 554.2]

**4.7**   A coil of 250 turns is wound uniformly over an iron ring of mean diameter 120 mm and cross-sectional area 200 mm². The coil has a resistance of 10 Ω and is connected to a 25 V d.c. supply. Calculate (a) the steady value of current in the coil, (b) the m.m.f. produced by the coil, (c) the magnetizing force at the mean circumference, (d) the reluctance of the iron circuit given that its relative permeability is 2000 and (e) the flux density in the iron.
[(a) 2.5 A; (b) 625 ampere turns; (c) 1658 ampere turns per metre; (d) 0.75 × 10⁶ A/Wb; (e) 4.17 T]

**4.8**   An iron circuit has three parts, all in series with one another, the length of each part being 0.25 m. The cross-sectional area of one part is 250 mm², that of the second is 500 mm², and that of the third is 1000 mm². If the relative permeability of the first part is 400, and that of the other two parts is 600, calculate the m.m.f. required to produce a magnetic flux of 0.5 mWb in the iron circuit. Neglect leakage and fringing.

[1492 ampere turns]

**4.9**   The magnetic circuit in Fig. 4.13 has a 308-turn coil wound on the centre limb. The area of each outer limb is 4 cm², and that of the centre limb is 8 cm². Given the following data for the iron circuit, calculate the current in the coil to give a flux of 0.42 mWb in each outer limb. Neglect leakage and fringing.

| $B$ (T)   | 0.9 | 1.1 | 1.2 | 1.25 |
|-----------|-----|-----|-----|------|
| $H$ (A/m) | 350 | 650 | 975 | 1200 |

[0.5 A]

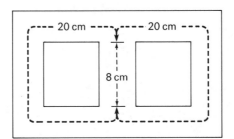

**Fig. 4.13**

**4.10**   A series magnetic circuit has an iron part of length 0.4 m and an air gap of length 1 mm, the cross-sectional area of the iron being 550 mm², and the relevant part of the *B–H* characteristic of the iron is

| $H$ (A/m) | 500 | 1000 | 2000 | 5000 |
|-----------|-----|------|------|------|
| $B$ (T)   | 1.2 | 1.35 | 1.45 | 1.51 |

If the coil, which is wound uniformly on the iron, has 500 turns of wire, estimate the current needed in it to produce a flux of 0.7 mWb in the air gap if the leakage coefficient is (a) unity, (b) 1.15.

[(a) 2.6 A; (b) 3.8 A]

**4.11** An inductor has an iron circuit with two 1 mm air gaps in it, and a coil of 10 000 turns of wire wound uniformly on it. The cross-sectional area of the iron is 10 cm² and its mean length is 50 cm. Calculate the current needed to establish a flux of 1.2 mWb in the magnetic circuit. The data for the relevant part of the $B$–$H$ curve for the iron are given below. Neglect magnetic leakage.

| $B$ (T) | 0.6 | 1.05 | 1.375 | 1.5 | 1.55 | 1.6 |
|---------|-----|------|-------|-----|------|-----|
| $H$ (A/m) | 100 | 200 | 400 | 800 | 1000 | 1400 |

[0.205 A]

# 5 Electromagnetism

## 5.1 Introduction

In this chapter we meet one of the most important topics in the field of electrical and electronic engineering. It is here that we study the principle of electromagnetic induction, and meet with the basic features of self-induction, mutual induction and induction by motion. These principles cover the action of all types of transducer, motor, generator and transformer.

## 5.2 Electromagnetic induction

In Chapter 4 we saw that current flow in a wire established a magnetic field around the wire, and when the current changed in value, the associated magnetic flux also changed. The converse is also true, that is

**if the magnetic flux linking with a wire or coil changes in value, an e.m.f. is induced in the wire or coil.**

If the electric circuit connected to this 'induced' e.m.f. forms a closed loop, then an induced current flows in the loop. There are three principal methods of inducing an e.m.f., which are described below.

### 5.2.1 Self-induction

If the current in a wire or coil increases in value, the magnetic flux associated with the current also increases in value. This increase in flux also links with the wire through which the current itself is passing and, as stated above, has the effect of inducing a voltage in the wire. That is to say, an e.m.f. is **self-induced** in the wire.

### 5.2.2 Induction by motion

If a conductor moves through a magnetic field, i.e. between the poles of a magnet, it 'cuts' the magnetic field, and an e.m.f. is induced in the conductor. This e.m.f. is said to be **induced by motion**, and is the principle of the electrical generator.

### 5.2.3 Mutual induction

As outlined in section 5.2.1 above, when the current in a conductor or coil alters, the associated magnetic field also alters. If the changing value of current is in conductor A, then if another conductor (conductor B) is positioned near to it, an e.m.f. is induced in conductor B due to the change in magnetic flux associated with conductor A. The e.m.f. induced in conductor B is said to be a **mutually induced** e.m.f.; by the same argument, if the current in conductor B changes, then an e.m.f. is induced in coil A by mutual induction. The operation of the transformer is based on the principle of mutual induction.

**5.3 Faraday's laws of electromagnetic induction**

Faraday's laws are:
1. **An e.m.f. is induced in a conductor whenever the magnetic field linking with the conductor changes.**
2. **The magnitude of the induced e.m.f. is proportional to the rate of change of the magnetic flux linking with the conductor.**

Faraday's first law casts a broad net over the whole spectrum of electrical and electronic engineering, and is the basis of all three methods of electromagnetic induction outlined in Section 5.2. The second law is more specific, and relates the magnitude of the induced e.m.f. to the rate of change of flux as follows:

Induced e.m.f., $E \propto$ rate of change of flux linkages

By 'flux linkages' we mean the product

Number of turns × flux

That is

$E \propto$ rate of change of $N\Phi$

In the SI system, the units of $\Phi$ have been chosen so that the constant of proportionality is unity, or

$E =$ rate of change of $N\Phi$

The important thing to note in the equation is that $E$ *only changes when $N\Phi$ changes*. The flux linking with a conductor can, in many cases, be very large indeed, but if it does not alter in value then no e.m.f. is induced in the conductor.

Expressed mathematically, if the flux changes by a minute amount $d\Phi$ in a minute length of time, $dt$, then

$$E = N \frac{d\Phi}{dt} \qquad [5.1]$$

A number of textbooks give the induced e.m.f. in eqn [5.1] a negative sign because, it can be argued, the induced e.m.f. opposes the change in flux which produces it (see also Lenz's law in section 5.4). However, by appropriate selection of data, we can use either a positive or negative sign, and we shall use a positive sign in this book.

***Worked example 5.1***  A magnetic flux of 10 mWb links with a coil of 100 turns of wire. If the flux is (a) doubled, (b) halved in a period of 20 ms, determine the average e.m.f. induced in the coil.

*Solution*  (a) In this case

Original magnetic flux = 10 mWb

Final magnetic flux = 20 mWb

hence

Change in flux, d$\Phi$ = final value − original value

= 20 − 10 = 10 mWb

The time interval, d$t$, over which the change occurs is

d$t$ = 20 ms

From eqn [5.1] we see that

$$\text{Induced e.m.f., } E = N\frac{d\Phi}{dt} = 100 \times \frac{10 \times 10^{-3}}{20 \times 10^{-3}}$$

$$= 50 \text{ V}$$

(b) Here we have

Original magnetic flux = 10 mWb

Final magnetic flux = 5 mWb

hence

Change in flux, d$\Phi$ = final value − original value

= 5 − 10 = −5 mWb

hence

$$\text{Induced e.m.f., } E = N\frac{d\Phi}{dt} = 100 \times \frac{-5 \times 10^{-3}}{20 \times 10^{-3}}$$

$$= -25 \text{ V}$$

## 5.4 Lenz's law

Lenz's law tells us about the direction in which an induced e.m.f. acts. It says that

**the induced e.m.f. acts to circulate a current in a direction which opposes the change in flux which produced the e.m.f.**

A physical understanding of Lenz's law is important for anyone wishing to appreciate the operation of electrical equipment.

Referring to Fig. 5.1(a), if the N-pole of the magnet is *moved towards the left-hand end of the coil*, the amount of magnet flux entering the coil is increased. According to Lenz's law, the e.m.f. induced in the coil causes a current to circulate in the coil which opposes the increase in flux entering the left-hand end of the coil. That is, *the induced current in the coil produces an N-pole at the left-hand end of the coil.*

Using any of the methods described in Chapter 4, we see that the induced current leaves terminal X of the coil and enters terminal Y. That is, terminal X is positive with respect to terminal Y.

If the N-pole is moved away from the left-hand end of the coil [see Fig. 5.1(b)], *the amount of flux entering the left-hand end of the coil is reduced.* If the reader applies Lenz's law in the manner described above, it will be found that the induced current flows from terminal Y to terminal X, that is terminal Y is positive with respect to terminal X.

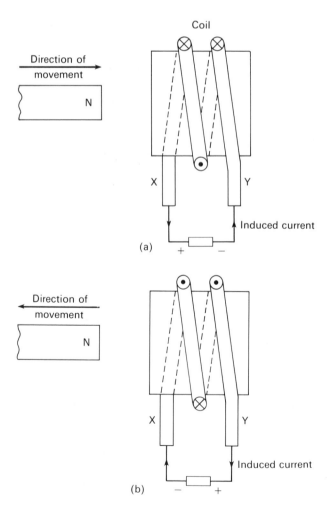

**Fig. 5.1** Lenz's law

## 5.5 Fleming's right-hand rule

Fleming's right-hand rule is a simple and convenient method of predicting *the direction of the e.m.f. induced in a conductor*, and can be thought of as a variation to Lenz's law as follows.

Consider the loop of wire in Fig. 5.2(a), which is moved to the left. The net result is that *the total flux entering the loop of wire is reduced in value*. Lenz's law says that the current induced in the loop circulates in a direction to try to prevent the reduction in flux. That is, *the current induced in the conductor under the pole flows away from the reader and into the page of the book*, i.e. it tries to increase the flux entering the loop.

If the loop of wire is moved to the right [see Fig. 5.2(b)], the amount of flux entering the loop increases. In this case, Lenz's law says that the direction of the induced current in the loop is such that it tries to reduce the total flux entering the loop. Consequently, the current in the conductor under the pole in Fig. 5.2(b) *flows towards the reader*, i.e. it flows out of the page of the book.

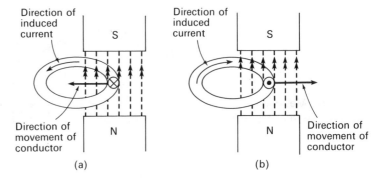

**Fig. 5.2** Direction of e.m.f. induced in a loop of wire

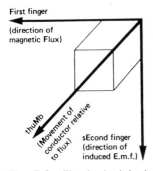

**Fig. 5.3** Fleming's right-hand rule

These effects are summarized by Fleming's right-hand rule as shown in Fig. 5.3, in which the thumb, first finger and second finger of the *right hand* are extended mutually at right angles to one another, and the direction they represent are

First finger – direction of **Flux**
sEcond finger – direction of the induced **E.m.f.**
thuMb – **Movement** of the conductor relative to the flux

Since Fleming's right-hand rule refers to the e.m.f. induced in a conductor when it moves relative to a magnetic field, we think of this rule as *referring to electrical generator action*.

## 5.6 Induced e.m.f. due to motion

The conductor in Fig. 5.4 can be considered to be part of a rudimentary electrical generator. When the conductor is moved from X to Y, the amount of flux cut by the conductor is

Change in flux = flux density × area = $Blx$

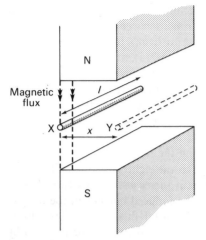

**Fig. 5.4** E.m.f. induced in a conductor due to motion in a magnetic field

where $l$ is the *active length* of the conductor in the magnetic field. If the distance between X and Y is covered in $t$ seconds, the linear velocity perpendicular to the flux is $v = x/t$. From eqn [5.1], we know that the induced e.m.f. in $N$ turns of wire is $E = N \, d\Phi/dt$ and, since we are dealing with one turn of wire, the induced e.m.f. is

$$E = \frac{d\Phi}{dt} = \frac{Blx}{t} = Blv \qquad\qquad [5.2]$$

The case where the conductor moves at an angle $\theta$ to the magnetic field is shown in Fig. 5.5. If the linear velocity of the conductor is $v$, then the velocity of the conductor perpendicular to the field is $v \sin \theta$, so that the induced e.m.f. in the conductor is

$$E = Blv \sin \theta \qquad\qquad [5.3]$$

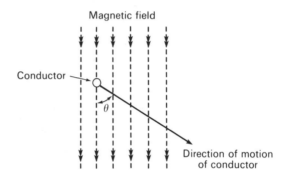

**Fig. 5.5** Movement of a conductor at an angle to a magnetic field

*Worked example 5.2* A conductor of active length 0.2 m moves at a linear velocity of 400 m/s. If it moves through a flux density of 0.05 T, calculate the average e.m.f. induced in the conductor when it moves (a) perpendicular to the magnetic field, (b) at 60° to the field and (c) at 30° to the field.

*Solution* From the data given $B = 0.05$ T, $l = 0.2$ m and $v = 400$ m/s.
(a) When the conductor moves perpendicular to the field then, from eqn [5.2]

$$E = Blv = 0.05 \times 0.2 \times 400 = 4 \text{ V}$$

or, from eqn [5.3]

$$E = Blv \sin \theta = 0.05 \times 0.2 \times 400 \times \sin 90° = 4 \text{ V}$$

(b) In this case, $\theta = 60°$, hence

$$E = Blv \sin \theta = 0.05 \times 0.2 \times 400 \times \sin 60° = 3.46 \text{ V}$$

(c) Here $\theta = 30°$ and

$$E = Blv \sin \theta = 0.05 \times 0.2 \times 400 \times \sin 30° = 2 \text{ V}$$

## 5.7 Eddy current and eddy current loss

As outlined above, when conducting material moves through or 'cuts' a magnetic field, an e.m.f. is induced in it and, if the circuit is complete, a current flows in the material.

In an electrical machine, the 'useful conductors' (copper or aluminium) are surrounded by the iron circuit, which is itself a conductor. Consequently, a current is also induced in the iron, and this current is known as an **eddy current**. In the case of a transformer (see Chapter 10 for details), the iron is stationary and the magnetic flux is alternating; the net result is the same, and eddy currents are induced in the iron.

Since iron has electrical resistance, the eddy current produces a power loss (an $I^2R$ loss) known as the **eddy current loss**. This has the effect of reducing the electrical efficiency of the machine, and the power loss is expended in heating up the machine. It can be shown that the eddy current loss per unit volume of iron is given by the equation

$$P_e = kf^2 B_{max}^2 \text{ watts per cubic metre}$$

where $k$ is a constant, $f$ the frequency of alternations of the magnetic flux, and $B_{max}$ the maximum value of the flux density.

## 5.8 The principle of electric motor action

The torque produced by an electric motor is due to the force acting on current-carrying conductors in the magnetic field of the motor.

We can use one of the rules described in Chapter 4 to predict the direction of the magnetic flux produced by a current-carrying conductor. Since the current in the conductor in Fig. 5.6(a) is out of the page, the magnetic field acts in a counter-clockwise direction around the conductor. When it is placed between the poles of a permanent magnet, as shown in Fig. 5.6(b), the magnetic field of the current-carrying conductor distorts the uniform magnetic field between the poles of the magnet in the manner shown.

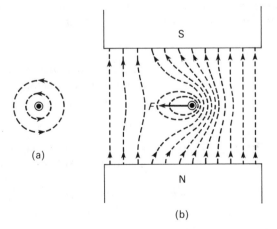

**Fig. 5.6** (a) Field produced by a current-carrying conductor and (b) the torque acting on a current-carrying conductor in a magnetic field

It will be seen that the magnetic field on the left-hand side of the conductor opposes the field of the permanent magnet, while it assists it on the right-hand side. That is, the magnetic field on the left-hand side of the conductor reduces, and that on the right-hand side increases. The current-carrying conductor experiences a force causing it to move from the stronger magnetic field to the weaker field; that is, the force acts to move the conductor in Fig. 5.6(b) to the left.

It is left as an exercise for the reader to verify that the direction of the mechanical force on the conductor is reversed if *either* the direction of the current is reversed *or* the direction of the magnetic field is reversed. If both the current and magnetic field are simultaneously reversed, the direction of the force is unchanged.

## 5.9 Fleming's left-hand rule

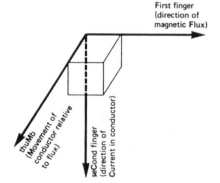

**Fig. 5.7** Fleming's left-hand rule

The direction of the force acting on a current-carrying conductor in a magnetic field is summarized by **Fleming's left-hand rule** as follows (see also Fig. 5.7):

> First finger – direction of magnetic **Flux**
> seCond finger – direction of the **Current**
> thuMb – direction of **Motion** of the conductor

Electric motors and analogue measuring instruments depend for their operation on the force of a current-carrying conductor; we therefore think of the left-hand rule as applying *to motor action*.

It may be useful to recall that, in the UK, *motors* drive on the *left-hand* side of the road.

## 5.10 Magnitude of the force on a current-carrying conductor

It can be verified experimentally that the force acting on a current-carrying conductor in a magnetic field is

$$F = BIl \qquad\qquad [5.4]$$

where $F$ is the force on the conductor (N), $B$ the flux density (T) of the magnetic field, $I$ the current (A) in the conductor, and $l$ the active length (m) of the conductor in the magnetic field.

*Worked example 5.3*  Calculate the force acting on a conductor of length 0.25 m, which carries a current of 100 A in a magnetic field of flux density 0.2 T.

*Solution*  From eqn [5.4]

$$F = BIl = 0.2 \times 100 \times 0.25 = 5\,\text{N}$$

**5.11 Force between parallel current-carrying conductors**

It was shown in section 5.10 that a current-carrying conductor in a magnetic field experiences a force and, furthermore, the direction of the force can be predicted by Fleming's left-hand rule (see section 5.9).

If we consider the case where two conductors carry current in opposite directions [see Fig. 5.8(a)], the current carried by conductor A produces a magnetic field in an anticlockwise direction around the conductor. Since the current in conductor B lies in this field, it can be argued (using Fleming's left-hand rule) that conductor B experiences a force away from conductor A. By a similar reasoning, it can also be shown that conductor A experiences a force away from conductor B. That is

> **parallel conductors which carry current in opposite direction repel each other.**

Following this general line of argument, it can be shown [see Fig. 5.8(b)] that

> **parallel conductors which carry current in the same direction attract each other.**

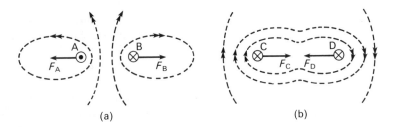

**Fig. 5.8** Force between parallel current-carrying conductors

(a)                     (b)

**5.12 Torque developed by a simple current-carrying coil of wire**

The two conductors in Fig. 5.9 form a loop of wire which carries a current in a magnetic field. The force on each conductor is $F = BIl$, where the terms have been defined earlier. The torque developed by each conductor is $BIlr$, where $r$ is the radius of the

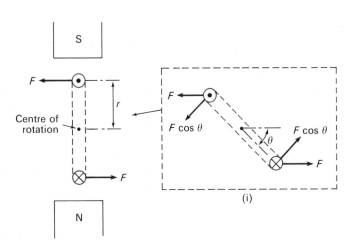

**Fig. 5.9** Torque developed by a current-carrying loop of wire

loop of wire. Since the torques produced by the two conductors assist one another, the total torque for the loop of wire is

$$T = 2BIlr \qquad\qquad [5.5]$$

This torque causes the loop to rotate, as shown in inset (i). However, the force $F$ on each conductor remains perpendicular to the magnetic field, so that the component of the force which produces a torque about the centre of rotation is now $F \cos \theta$, hence the torque produced by the loop of wire in the inset is

$$T = 2BIlr \cos \theta$$

If the loop of wire is replaced by a coil of $N$ turns, the torque produced is

$$T = 2NBIlr \cos \theta \qquad\qquad [5.6]$$

The interesting point about this equation is that when $\theta = 90°$, i.e. when the coil is horizontal, the torque produced is zero, and it cannot rotate further!

In an electrical machine (see Chapter 14), arrangements are made to ensure that the current in the conductors reverses when the coil reaches horizontal position, so that it can continue to rotate.

*Worked example 5.4*     The coil of a moving-coil instrument has 50 turns of wire on it, and carries a current of 1.5 A. If the coil has a radius of 1 cm and an active length of 2 cm, and the magnetic flux density is 0.5 T, calculate (a) the maximum torque produced and (b) the torque when $\theta = 70°$.

*Solution*     (a) The maximum torque is

$$T_{max} = 2NBIlr = 2 \times 50 \times 0.5 \times 1.5 \times 0.02 \times 0.01$$

$$= 0.015 \text{ N m}$$

(b) When $\theta = 70°$

$$T = T_{max} \cos \theta = 0.015 \cos 70° = 0.005 \text{ N m}$$

## 5.13 Self-inductance, *L*

When a conductor carries a current, a magnetic flux is established around it, and when the current changes in value the flux also changes. Also, as outlined earlier, when a conductor is situated in a changing magnetic field, an e.m.f. is induced in it.

The net effect is that when the current in a conductor changes in value, an e.m.f. is induced in the conductor which carries the current. This property is known as **self-inductance**, or simply as **inductance**. The symbol for inductance is $L$, implying linkages of magnetic flux, and the unit is the *henry* (symbol H). The henry is defined as follows:

**A circuit has an inductance of one henry if an e.m.f. of one**

**volt is induced in the circuit when the current in the circuit changes at the rate of one ampere per second.**

The relationship between the self-induced e.m.f. and the inductance is

$$E = L \times \text{rate of change of current}$$

$$= L\frac{di}{dt} = L\frac{I_2 - I_1}{t_2 - t_1} \tag{5.7}$$

where the current change $di = (I_2 - I_1)$ occurs in a time $dt = (t_2 - t_1)$.

*Worked example 5.5*  (a) Calculate the inductance of a circuit which has a voltage of 10 V induced in it when the current in the circuit changes at the rate of 5 A/s. (b) What is the value of the induced e.m.f. when the current changes at the rate of 20 A/s?

*Solution*  (a) From eqn [5.7]

$$E = L \, di/dt$$

or

$$L = E/(di/dt) = 10/5 = 2 \, \text{H}$$

(b) The induced e.m.f. is

$$E = L \, di/dt = 2 \times 20 = 40 \, \text{V}$$

Clearly, the greater the rate of change of current, the larger the value of the induced e.m.f.

**5.14  Relationship between inductance and number of turns**

If the reluctance of the magnetic circuit on which a coil is wound is constant, the effect of increasing the current in the circuit is to produce a proportional increase in magnetic flux. In such a magnetic circuit, the self-induced e.m.f. can be determined from either of the following equations:

$$E = L\frac{dI}{dt} \quad \text{(see eqn [5.7])}$$

and

$$E = N\frac{d\Phi}{dt} \quad \text{(see eqn [5.1])}$$

Since the two values of e.m.f. are the same, we may say that

$$L\frac{dI}{dt} = N\frac{d\Phi}{dt}$$

or

$$L = N \frac{d\Phi}{dI} \qquad\qquad [5.8]$$

It was shown in Chapter 4 that

$$\Phi = \frac{\text{m.m.f.}}{\text{magnetic circuit reluctance}}$$

or

$$d\Phi = \frac{\text{change in m.m.f.}}{\text{reluctance}} = \frac{N \times \text{change in current}}{\text{reluctance}}$$

$$= \frac{N\,dI}{S}$$

where $N$ is the number of turns on the coil. Substituting this in eqn [5.8] gives

$$L = N \frac{N\,dI/S}{dI} = \frac{N^2}{S} \qquad\qquad [5.9]$$

That is, if the reluctance does not vary with magnetic flux (as in the case of an air-cored coil), then

$$L \propto N^2$$

so that doubling the number of turns on a coil increases its inductance by $2^2 = 4$ times.

*Worked example 5.6*    An air-cored coil of 1000 turns has an inductance of 0.1 mH. Calculate (a) the reluctance of the magnetic circuit, (b) the inductance of the coil if the number of turns is increased to 1200 turns. Assume that the reluctance of the core is constant.

*Solution*    (a) From eqn [5.9]

$$S = N^2/L = 1000^2/0.1 \times 10^{-3} = 1 \times 10^{10} \text{ A/Wb}$$

(b) If the new inductance is $L_2$, then $L_2 \propto N_2^2$, or

$$\frac{L_2}{L_1} = \frac{N_2^2}{N_1^2}$$

or

$$L_2 = L_1(N_2^2/N_1^2) = 0.1 \times 10^{-3}(1200^2/1000^2)$$

$$= 0.144 \times 10^{-3} \text{ H} \quad \text{or} \quad 0.144 \text{ mH}$$

Alternatively we may write (see eqn [5.9])

$$L_2 = N_2^2/S = 1200^2/1 \times 10^{10} = 0.144 \times 10^{-3} \text{ H}$$

**5.15  Energy stored in a magnetic field**

While a magnetic field is being built up, energy is stored in the field. When the field is reduced, energy is returned to the circuit. We can deduce an equation for the amount of energy stored in the field by either of two methods as follows.

**Intuitive derivation of the equation for stored energy**

If the current in a circuit of inductance $L$ henrys increases at a uniform rate from zero to $I$ amperes in $t$ seconds, then the *average current* in the circuit is $I/2$ amperes, and the *average value of induced e.m.f.* is

$$L \times (\text{rate of change of current}) = LI/t$$

The *average energy* consumed by the inductance is therefore

$$W = EIt = \frac{LI}{t} \times \frac{I}{2} \times t = \tfrac{1}{2}LI^2 \qquad [5.10]$$

**Derivation by calculus**

In the general case, the current does not increase at a uniform rate, and the e.m.f. induced in the inductor in time $dt$ is $e = L\,di/dt$, and the energy change during time $dt$ is

$$w = ei\,dt = L\frac{di}{dt} \times i \times dt = Li\,di$$

The total energy consumed during the time that the current changes from zero to $I$ is

$$W = \int_0^I Li\,di = L[\tfrac{1}{2}i^2]_0^I = \tfrac{1}{2}LI^2$$

*Worked example 5.7*   A coil of resistance $10\,\Omega$ is connected to a 20 V d.c. supply, the inductance of the coil being 5 H. Calculate the steady value of the current in the coil, and the energy stored in the magnetic circuit when the current has reached its steady value.

*Solution*   It takes a little time for the current to reach its steady-state value (in this case it takes about 1.25 s – for details see Chapter 13, 'Tranients in electrical circuits'), but when it does the final current is

$$I = E/R = 20/10 = 2\,\text{A}$$

From eqn [5.10] we see that the energy stored in the field is

$$W = \tfrac{1}{2}LI^2 = \tfrac{1}{2} \times 5 \times 2^2 = 10\,\text{J}$$

**5.16   Mutual inductance, *M***

When the current in a coil of $N_1$ turns (see Fig. 5.10) changes in value, the change in magnetic flux induces a voltage in any conductor or coil in its magnetic field. This induced e.m.f. is known as a *mutually induced e.m.f.*, and the two coils are said to be *mutually coupled*. The reader should note that a mutually induced voltage exists only when the mutual flux *changes in value*; that is, if the mutual flux has a steady value, the mutually induced e.m.f. is zero. This situation is reflected in Fig. 5.10, where a mutually induced e.m.f. is produced in the coil of $N_2$ turns when the current in $N_1$ changes at the rate of $dI_1$ in $dt$ seconds.

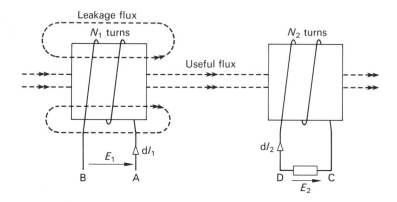

**Fig. 5.10**   Mutual inductance

The unit of mutual inductance, *M*, is the *henry*, and

**if two coils have a mutual inductance of one henry between them, then an e.m.f. of one volt is induced in one coil when the current in the other coil changes at the rate of one ampere per second.**

The coil which is connected to the current source ($N_1$ in Fig. 5.10) is known as the *primary coil* or *primary winding*, and the coil which has the mutual e.m.f. induced in it ($N_2$ in Fig. 5.10) is the *secondary coil* or *secondary winding*. The coupled circuit in Fig. 5.10 is a practical circuit, in which a proportion of the magnetic flux produced by the primary winding does not link with the secondary winding; this proportion of the flux is known as the *leakage flux* (see also Section 5.17).

When coils are wound on a common iron circuit, as in the case of a *power transformer*, practically all the flux produced by the primary winding links with the secondary winding; in this case, the coils are said to be *closely coupled*. When the coils have an air core, only a proportion of the flux from the primary winding links with the secondary winding; in this case, the coils are said to be *loosely coupled*.

We can determine the polarity of the e.m.f. induced in the secondary coil in Fig. 5.10 using Lenz's law as follows. When current enters terminal A of the primary winding (let us assume that the current is rising in value), the flux produced by the primary winding (which is also rising in value) leaves the right-hand end of the primary winding and enters the left-hand end of the secondary winding.

Lenz's law tells us that the e.m.f. induced in the secondary winding produces a current which opposes the change which produces it. That is, the current induced in the secondary winding produces a flux which *leaves* the right-hand end of the secondary winding. Applying the rules described in Chapter 4, we see that the current in the secondary winding enters terminal D and leaves terminal C. So far as the load connected to the secondary winding is concerned, terminal C is positive with respect to terminal D.

Mathematically, the induced e.m.f. is related to the change in primary current as follows:

$$E_2 = M \times \text{rate of change of primary current}$$

$$= M \frac{dI_1}{dt} \tag{5.11}$$

If the useful flux which links the two windings is $\Phi_2$, then the e.m.f. induced in $N_2$ is

$$E_2 = N_2 \frac{d\Phi_2}{dt} \tag{5.12}$$

Since the value of $E_2$ is the same in both equations, then

$$M \frac{dI_1}{dt} = N_2 \frac{d\Phi_2}{dt}$$

or

$$\text{Mutual inductance, } M = N_2 \frac{d\Phi_2}{dI_1} \tag{5.13}$$

*Worked example 5.8*    The mutual inductance between two coils is 0.2 H. If the current in the primary winding rises from 0.1 to 0.6 A in 5 ms, (a) calculate the average value of the induced e.m.f. in the secondary winding in this period o time, and (b) if the secondary winding has 450 turns, determine th change in magnetic flux which links with the winding.

*Solution*    (a) From eqn [5.11]

$$E_2 = M \, dI_1/dt = 0.2 \times (0.6 - 0.1)/5 \times 10^{-3}$$

$$= 20 \text{ V}$$

(b) Since $N_2 = 450$ turns then, from eqn [5.13]

$$d\Phi_2 = M \, dI_1/N_2 = 0.2 \times (0.6 - 0.1)/450$$

$$= 0.222 \times 10^{-3} \text{ Wb} \quad \text{or} \quad 0.222 \text{ mWb}$$

**5.17 Coupling coefficient**

Suppose that the primary winding in Fig. 5.10 produces flux $\Phi_1$, and that the flux entering the secondary winding is $\Phi_2 = k\Phi_1$, where $k$ is a factor known as the *magnetic coupling coefficient* or simply as the *coupling coefficient*, whose value lies in the range $0 \leqslant k \leqslant 1$. As with self-inductance, the mutual inductance between the windings is equal to the flux linkages per ampere, that is

$$M = \frac{N_2 \Phi_2}{I_1} = \frac{N_2 k \Phi_1}{I_1}$$

If the function of the two coils is reversed, that is the coil with $N_2$ turns becomes the primary winding, and the coil with $N_1$ turns is the secondary winding, the flux entering $N_1$ is $\Phi_1 = k\Phi_2$. Using a similar argument to the above we get

$$M = \frac{N_1 \Phi_1}{I_2} = \frac{N_1 k \Phi_2}{I_2}$$

Multiplying the above equations together gives

$$M^2 = \frac{N_2 k \Phi_1}{I_1} \times \frac{N_1 k \Phi_2}{I_2} = k^2 \frac{N_1 \Phi_1}{I_1} \times \frac{N_2 \Phi_2}{I_2}$$

$$= k^2 L_1 L_2$$

where $L_1$ is the self-inductance of the primary winding and $L_2$ the self-inductance of the secondary winding. The magnetic coupling coefficient between the two windings is therefore

$$k = \frac{M}{\sqrt{(L_1 L_2)}} \qquad\qquad [5.14]$$

**5.18 Series-connected magnetically coupled coils**

When two magnetically coupled coils are connected in series, the mutual inductance between the coils has the effect of modifying the overall inductance of the circuit.

If the current in the circuit is $i$ (see Fig. 5.11), the self-induced e.m.f. in coils $L_1$ and $L_2$ are, respectively, $L_1 \, di/dt$ and $L_2 \, di/dt$. The magnitude of the mutually induced e.m.f. in *each coil* is $M \, di/dt$, but the direction of this e.m.f. (and therefore its magnetic sign) depends on whether the magnetic fluxes associated with the coils assist or oppose one another. The equation for the total induced e.m.f. is therefore

$$e = (L_1 + L_2 \pm 2M) \frac{di}{dt} = L_E \frac{di}{dt}$$

**Fig. 5.11** Series-connected magnetically coupled coils

where $L_E$ is the effective inductance of the series-connected magnetically coupled coils. That is

$$L_E = L_1 + L_2 + 2M$$

If the magnetic fluxes produced by the coils assist one another (*series-aiding* connection), then the '+' sign is used, and if the fluxes produced by the coils oppose one another (*series-opposing* connection), then the '−' sign is used.

**Problems**

5.1    A conductor of length 50 cm moves perpendicularly to a magnetic field of flux density 0.5 T at a speed of 500 m/s. Calculate the induced e.m.f.

[125 V]

5.2    A conductor has an e.m.f. of 20 V induced in it when moving at a speed of 10 m/s perpendicular to a magnetic field of flux density 0.8 T. Determine the length of the conductor.

[2.5 m]

5.3    A conductor of length 25 cm moves at a constant velocity of 5 m/s at an angle of 55° to the line of action of a magnetic field of flux density 0.4 T. Calculate the e.m.f. induced in the conductor.

[0.41 V]

5.4    An aircraft has a wingspan of 50 m and flies at a speed of 1000 km/h. Given that the vertical component of the flux density of the Earth's magnetic field is $4 \times 10^{-5}$ T, calculate the induced voltage between the wing tips.

[0.555 V. *Note*: it is proposed that this principle will be used to power Earth-orbit satellites!]

5.5    An air-cored coil carries a current of 5 A and produces magnetic flux of 10 $\mu$Wb. If the coil is wound with 1500 turns of wire, what is the inductance of the coil?

[3 mH]

5.6    When the current in an air-cored coil changes in value, the magnetic flux in its core changes by 5 mWb in 0.1 s. Calculate the number of turns of wire on the coil if the average value of the induced e.m.f. is 50 V.

[1000]

5.7    The coil of an electric bell has an effective inductance of 2 H. Calculate the e.m.f. induced in the coil when a current of 0.25 A is reduced to zero in 2 ms.

[250 V]

5.8    When the current in an air-cored coil increases from 5 to 20 A in 0.2 s, the average value of induced e.m.f. is 100 V. Determine the inductance of the coil.

[1.33 H]

5.9    An air-cored coil has an inductance of 2 mH. If its length is 0.25 m, and it has 100 turns of wire on it, determine the area of the core of the coil.

[3.98 cm$^2$]

**5.10** If the number of turns on the coil in problem 5.9 is increased by 40 per cent, determine the new inductance of the coil.

[3.92 mH]

**5.11** When a current of 10 A flowing through coil A is reversed, the change in magnetic flux linkage with an adjacent coil B amounts to 20 mWb-turns. What is the mutual inductance between the coils?

[2 mH]

**5.12** If the self-inductance of coil X is 0.2 H and that of coil Y is 0.6 H, determine the coefficient of magnetic coupling between them given that the mutual inductance between the coils is 100 mH.

[0.289]

**5.13** A coil of 1000 turns is wound uniformly on a non-magnetic ring of mean diameter 15 cm and cross-sectional area 5 cm$^2$. A second coil of 500 turns is wound over the top of the first coil. Given that the magnetic coupling coefficient between the coils is 0.7, calculate the mutual inductance between the coils.

[0.466 mH]

**5.14** A straight conductor of length 50 cm carries a current of 30 A perpendicular to a magnetic field of flux density 0.5 T. Determine the force acting on the conductor.

[7.5 N]

**5.15** The coil of a galvanometer has 25 turns of wire and carries a current of 20 mA. The active length in the magnetic field of each side of the coil is 2 cm, the flux density being 0.15 T. Calculate the total force acting on the coil.

[3 mN]

**5.16** A coil of inductance 2 H and resistance 1 Ω is connected to a 10 V battery having negligible internal resistance. When the current has reached its steady value, calculate the energy stored in the magnetic field.

If, due to ageing, the internal resistance of the battery rises to 0.5 Ω, determine the new value of energy stored in the magnetic field.

[100 J; 44.44 J]

# 6 Alternating current and voltage

## 6.1 Introduction to alternating waveforms

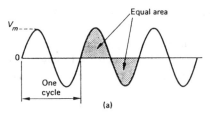

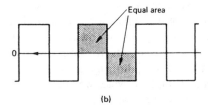

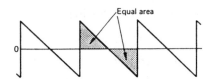

**Fig. 6.1** Examples of alternating waveforms: (a) a singusoidal wave; (b) a square or rectangular wave; (c) a sawtooth wave

An **alternating wave** is one which alternates periodically about zero value, several examples being shown in Fig. 6.1. The waveforms shown may represent voltage, or current, or power, etc. Two features which typify an alternating wave are as follows:

1. It has a periodic time, $T$.
2. Its mathematical average value is zero.

The wave completes $f$ cycles in one second, and the **periodic time**, $T$, of the wave is

$$T = 1/f \text{ hertz (Hz)} \qquad [6.1]$$

As outlined in (2) above, the mathematical average value of the alternating wave is zero, and this means that *the area above the zero line in each complete cycle in each graph in Fig. 6.1 is equal to the area below the zero line in each cycle.* Engineers sometimes have a slightly different understanding of average value in this context, and this is fully explained in section 6.4.

Strictly speaking, the abbreviation a.c. means alternating current; however, engineers interpret the abbreviation to simply mean 'alternating'. That is, we refer to an alternating voltage as an 'a.c. voltage', and an alternating current as an 'a.c. current'. A circuit which is excited by an alternating source is described as an 'a.c. circuit', and a machine which is driven by an alternating source is called an 'a.c. machine'.

## 6.2 A simple alternator

Consider for the moment the single-loop alternator in Fig. 6.2. At the instant shown in Fig. 6.2(a), the conductors AB and CD pass under the centre of the poles and cut the flux at the maximum rate, consequently the maximum voltage is induced in each conductor.

At the same time in Fig. 6.2(a), one half of the conductor linking B to C moves to the left in the magnetic field, and the other half moves to the right; that is, wire BC has two equal e.m.f.s induced in it which act in opposite directions, i.e. the net e.m.f. induced in BC is zero! Similarly, we can argue that no net e.m.f. is induced

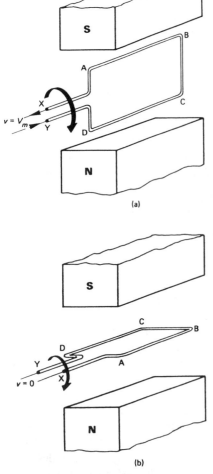

**Fig. 6.2** A simple single-loop alternator: (a) the conductors cutting the flux at the maximum rate to give the maximum output voltage; (b) the conductors moving parallel to the flux, when the output voltage is zero

in the two radial conductors at the other end of the loop. Clearly, at this time the *maximum voltage* or *peak voltage*, $V_m$, is induced in the loop of wire.

Assuming that a resistive load is connected to the loop of wire then, by Fleming's right-hand rule, the direction of the induced current in Fig. 6.2(a) is from B to A and from D to C. That is, current enters the loop at D and leaves at A; so far as the external circuit is concerned, *point A is instantaneously positive with respect to point D*.

When the loop of wire has rotated through 90° in a clockwise direction, as shown in Fig. 6.2(b), the conductors AB and CD move parallel to the magnetic field, and no voltage is induced in either conductor at this time. That is, the instantaneous voltage induced in the coil is zero.

As the loop of wire rotates further, the conductor AB comes under the influence of the N-pole, and conductor CD begins to move across the S-pole. That is, the direction of the e.m.f. induced in the coil is reversed.

If we plot the induced e.m.f. in the loop to a base of time, the net result is the voltage sine wave shown in Fig. 6.3 (see also section 6.3). When the loop of wire is in the position corresponding to Fig. 6.2(a), the induced e.m.f. is $+V_m$ on the waveform in Fig. 6.3(b). By the time that the loop has rotated to the position in Fig. 6.2(b), the induced e.m.f. is zero [see Fig. 6.3(c)]. If the argument is continued for the complete cycle, we see that the induced voltage waveform is (ideally) a sine wave.

The *peak-to-peak voltage* (see Fig. 6.3) is the difference in potential between the two peak voltages and, for Fig. 6.3, is

$$\text{Peak-to-peak voltage} = V_m - (-V_m) = 2V_m$$

$$= 2 \times \text{peak voltage} \qquad [6.2]$$

One cycle of the alternating waveform corresponds to 360° (electrical). In many cases we use the angular measure of the *radian* (abbreviated to *rad*); since there are $2\pi$ radians in a complete circle or cycle, then

$$\frac{\text{Angle in radians}}{2\pi} = \frac{\text{angle in degrees}}{360}$$

or

$$\text{Angle in degrees} = \text{angle in radians} \times 180/\pi$$

$$= \text{angle in radians} \times 57.3 \qquad [6.3]$$

hence

$$1 \text{ rad} = 57.3°$$

While the construction of the machine in Fig. 6.2 is simple, we must solve the problem of getting the current away from the wire

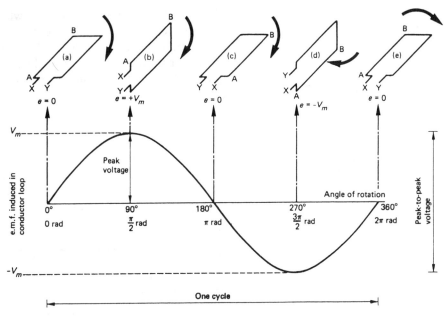

**Fig. 6.3** One complete cycle of a
sine wave of voltage

loop while it is rotating. Engineers have devised a simple solution, namely the slip-rings and brushes in Fig. 6.4. The slip-rings are made from metal (copper, brass or steel), and the brushes are made from graphite to give them a long life and low contact resistance. In Fig. 6.4, brush A′ is in constant contact with end A of the loop (see also Fig. 6.2), and brush D′ is connected to end D of the loop.

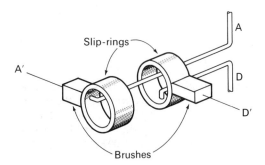

**Fig. 6.4** Slip-rings and brushes

In the case of an alternator, we refer to the rotating member (the loop of wire in Fig. 6.2) as the *rotor*, and to the stationary member (the magnetic field in Fig. 6.2) as the *stator*.

In many cases, the situation in Fig. 6.2 is reversed, and the winding in which the e.m.f. is induced is on the stator of the machine, and the magnetic field system is on the rotor. None-the-less, such a machine will have a set of slip-rings because the magnetic field is produced by a coil which is energized or *excited* by direct current. The direct current used for this purpose may either be obtained from a rectifier system energized from the mains, or from a d.c. generator (known as the *exciter*) which is physically mounted on the same shaft as the alternator rotor.

**6.3 The sine wave**

Since the *sinusoidal wave* or *sine wave* is the waveform most frequently encountered in electrical and electronic engineering, we will look at it in some detail. If the loop of wire in Fig. 6.2 rotates at a speed of $\omega$ rad/s, then in a time of $t$ seconds it has rotated through an angle of

$$\theta = \text{rotational speed (rad/s)} \times \text{time (s)}$$

$$= \omega t \text{ rad}$$

During the periodic time, $T$ seconds, the loop will have rotated through one complete cycle, that is

$$2\pi = \omega T = \omega/f$$

where $f$ is the frequency of the wave in Hz. Hence

$$\omega = 2\pi f \text{ rad/s} \tag{6.4}$$

That is

$$\theta = \omega t = 2\pi f t \text{ rad} \tag{6.5}$$

The instantaneous value, $v$, of the sine wave at angle $\theta$ is

$$v = V_m \sin \theta = V_m \sin \omega t = V_m \sin 2\pi f t$$

$$= V_m \sin \frac{2\pi t}{T} \tag{6.6}$$

*Worked example 6.1*   A 120 Hz sine wave has an instantaneous value of 8.66 V when $\theta = 60°$. Determine (a) the maximum value of the wave, (b) the peak-to-peak voltage, (c) the periodic time of the wave, (d) the value of the voltage when $\theta$ is (i) 15°, (ii) 1 rad, (iii) 200°, (iv) 2 rad and when $t$ is (v) 1 ms, (vi) 0.007 s.

*Solution*   (a) From eqn [6.6]

$$v = V_m \sin \theta$$

or

$$V_m = v/\sin \theta = 8.66/\sin 60° = 10 \text{ V}$$

(b) From eqn [6.2]

$$\text{Peak-to-peak voltage} = 2V_m = 20 \text{ V}$$

(c) The periodic time of the wave is calculated from eqn [6.1] as follows:

$$T = 1/f = 1/120 = 0.00833 \text{ s} \quad \text{or} \quad 8.33 \text{ ms}$$

(d) For convenience, we will convert each value into degrees.
(i) When $\theta = 15°$

$$v = V_m \sin \theta = 10 \sin 15° = 2.588 \text{ V}$$

(ii) When $\theta = 1$ rad, then $\theta° = \theta(\text{rad}) \times 57.3 = 57.3°$, hence

$$v = 10 \sin 57.3° = 8.415 \text{ V}$$

(iii) The voltage when $\theta = 200°$ is

$$v = 10 \sin 200° = -3.42 \text{ V}$$

(iv) When $\theta = 2$ rad, then $\theta° = 2 \times 57.3 = 114.6°$, and

$$v = 10 \sin 114.6° = 9.09 \text{ V}$$

(v) Since the periodic time is $T$, then

$$\frac{\theta°}{360} = \frac{t}{T}$$

or

$$\theta° = t \times 360/T$$

When $t = 1$ ms $= 0.001$ s, then

$$\theta° = 0.001 \times 360/0.0083 = 43.22°$$

and

$$v = 10 \sin 43.22° = 6.85 \text{ V}$$

(vi) When $t = 0.007$ s, then

$$\theta° = 0.007 \times 360/0.00833 = 302.52°$$

hence

$$v = 10 \sin 302.52° = -8.43 \text{ V}$$

**6.4 Mean value or average value of an alternating wave**

A mathematician understands the *mean value* or *average value* of an alternating wave to be the total area under the wave (taken over the whole cycle) divided by the 'length' of the base. However, as explained in section 6.1, the mathematical average value of a true alternating waveform is zero!

Engineers use a somewhat modified version of the meaning of 'mean value' or 'average value', and is sometimes known as the 'rectified average value'. It is, in fact, the mean value or average value *taken over one half-cycle of the wave* (usually the positive half-cycle). We can determine its value either:

1. by a graphical method such as the mid-ordinate rule; or
2. by an analytical method such as calculus.

The descriptions used in the following do not refer either to a voltage or a current; this can be taken to imply that a similar technique can be used for either.

**Mid-ordinate method**

Consider the triangular wave in Fig. 6.5. Initially we divide the positive half-cycle into $n$ equal parts (six in the case of Fig. 6.5), having the same space between the *ordinates* (these are shown by the dotted lines in the figure). The *mid-ordinates* ($M_1 - M_6$) are measured, and the 'mean' or 'average' value is taken over the half-cycle as follows:

$$\text{Mean value} = \frac{\text{sum of the values of the mid-ordinates}}{\text{number of mid-ordinates}}$$

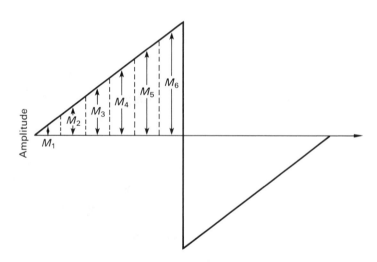

**Fig. 6.5** The mid-ordinate rule for calculating average value

In the case of Fig. 6.5, the mean value is

$$\text{Mean value} = \frac{M_1 + M_2 + M_3 + M_4 + M_5 + M_6}{6} \qquad [6.7]$$

Since we are taking only a relatively small number of samples during the half-cycle, the reader will realize that, at best, we will only obtain an approximate value. However, in some cases, the result is sufficiently accurate.

**Solution by calculus**

In this case we determine the area under the positive half-cycle curve by *integration*, and the average value is obtained by dividing the area by the length of the base. The process of integration can be thought of as being similar to adding the mid-ordinates of the wave, with the difference that the distance between them is so small that we have taken many millions of mid-ordinates over the half-cycle.

If the curve is expressed as a function of time, then the average value is given by

Average value

$$= \frac{\text{area under the graph, i.e. over the period } T/2}{\text{period of the half-cycle, i.e. } T/2}$$

$$= \frac{1}{T/2} \int_0^{T/2} f(t) \, \mathrm{d}t \qquad\qquad [6.8]$$

where $f(t)$ means that the equation is a *function of time*.

Alternatively, if the curve is expressed as a function of angle, the average value is given by

Average value

$$= \frac{\text{area under the graph, i.e. over the period } \pi \text{ rad}}{\text{period of the half-cycle, i.e. } \pi \text{ rad}}$$

$$= \frac{1}{\pi} \int_0^{\pi} f(\theta) \, \mathrm{d}\theta \qquad\qquad [6.9]$$

where $f(\theta)$ means that the equation is expressed as *a function of* $\theta$.

***Worked example 6.2***  Calculate the mean value of the square wave of voltage in Fig. 6.6.

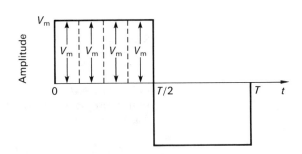

**Fig. 6.6**  Figure for Worked Example 6.2

*Solution*  *Mid-ordinate method*
In this case we divide the positive half-cycle into four parts, as shown in Fig. 6.6, each mid-ordinate having a length $V_m$. From eqn [6.7], the mean value is

$$V_{av} = \frac{V_m + V_m + V_m + V_m}{4} = V_m$$

*Calculus method*

The equation for the positive half-cycle is $f(t) = V_m$ hence, from eqn [6.7]

$$V_{av} = \frac{1}{T/2} \int_0^{T/2} V_m \, dt = \frac{1}{T/2} [V_m t]_0^{T/2}$$

$$= \frac{1}{T/2} \left[ V_m \times \frac{T}{2} - (V_m \times 0) \right] = V_m$$

In this example, both methods give the same result.

***Worked example 6.3***  Determine the average value of a sine wave of maximum value $V_m$.

*Solution*  *Mid-ordinate method*

Let us divide the first half-cycle into six sections, i.e. ordinates occur at intervals of 30°. The first mid-ordinate is therefore at 15°, the next at 45°, the next at 75°, etc. The table of mid-ordinates and their sum is shown in Table 6.1.

The mean value of the sine wave is therefore

$$V_{av} = \frac{\text{sum of mid-ordinates}}{\text{number of mid-ordinates}} = \frac{3.8636 V_m}{6}$$

$$= 0.6439 V_m$$

**Table 6.1**

| Mid-ordinate number | Value | |
|---|---|---|
| 1 | $V_m \sin 15°$ | $= 0.2588 V_m$ |
| 2 | $V_m \sin 45°$ | $= 0.7071 V_m$ |
| 3 | $V_m \sin 75°$ | $= 0.9659 V_m$ |
| 4 | $V_m \sin 105°$ | $= 0.9659 V_m$ |
| 5 | $V_m \sin 135°$ | $= 0.7071 V_m$ |
| 6 | $V_m \sin 165°$ | $= 0.2588 V_m$ |
| Sum | | $3.8636 V_m$ |

*Calculus method*

In this case we will express the equation for the wave as a function of $\theta$ as follows: $f(\theta) = V_m \sin \theta$. Hence, from eqn [6.8]

$$V_{av} = \frac{1}{\pi} \int_0^{\pi} V_m \sin \theta \, d\theta = \frac{V_m}{\pi} \int_0^{\pi} \sin \theta \, d\theta$$

$$= \frac{V_m}{\pi} [-\cos \theta]_0^{\pi}$$

$$= \frac{V_m}{\pi} [1 - (-1)] = \frac{2 V_m}{\pi} = 0.637 V_m$$

This is the correct value, since the calculus method is equivalent to taking many millions of mid-ordinates separated by only the most minute distance. We clearly see, in this case, that the mid-ordinate method leads to an approximate solution.

Also in a resistive circuit, since the current is directly proportional to the voltage across it, we may say that the mean value of the current in a resistor is

$$I_{av} = \frac{2I_m}{\pi} = 0.637I_m$$

## 6.5 The effective value or root-mean-square (r.m.s.) value of an alternating waveform

The *effective value* of an alternating current is the value to be used for electrical heating purposes, i.e. how much power does a particular value of current produce in a pure resistor? What we really need to know is, for example, if a direct current produces 100 W of power in a resistor, then what value of alternating current produces the same power?

The instantaneous heating effect of a current, $i$, in a resistor $R$ is $i^2R$. As the current varies in value over the alternating cycle, so the heating effect varies. Let us consider the simple case of the current waveform in Fig. 6.7, which has $n$ mid-ordinate values of current *taken over the whole cycle*. If this current flows in a resistor $R$, and the effective value of current is $I$, then the overall heating effect is $I^2R$. Also, the average heating effect of all the mid-ordinate values of current is

$$\frac{i_1^2R + i_2^2R + \cdots + i_n^2R}{n}$$

That is

$$I^2R = \frac{i_1^2R + i_2^2R + \cdots + i_n^2R}{n}$$

or

$$I = \sqrt{\left[\frac{i_1^2 + i_2^2 + \cdots + i_n^2}{n}\right]} \qquad [6.10]$$

= square *root* of the *mean* of the sum of the *square* values of the current

= *root-mean-square* (r.m.s.) value of the current

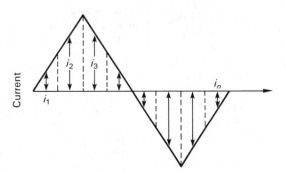

**Fig. 6.7** Root-mean-square (r.m.s.) value of a wave

The reader will note that, in Fig. 6.7, we have taken instantaneous values of current in both positive and negative half-cycles, and have taken the average value over the whole cycle. We can do this because the instantaneous values of current are squared, resulting in a positive value in either half-cycle.

Moreover, it is generally the case that the same value of effective current is obtained if we perform the calculation over one half-cycle or over the complete cycle.

We have argued the case using the mid-ordinate rule, and we can similarly evaluate the r.m.s. value using calculus, as will be illustrated in the following examples.

*Worked example 6.4*  Determine the r.m.s. value of a square wave of current (see also Fig. 6.6 for a square wave of voltage).

*Solution*  *Mid-ordinate method*
Let us assume that we have divided both the positive and negative half-cycles into four equal parts, giving eight mid-ordinates; the first four have a value $I_m$, the second four having the value $-I_m$. From eqn [6.10], the r.m.s. value of the current wave is

$$I = \sqrt{\left[\frac{I_m^2 + I_m^2 + I_m^2 + I_m^2 + (-I_m^2) + (-I_m^2) + (-I_m^2) + (-I_m^2)}{8}\right]}$$

$$= \sqrt{\left[\frac{8I_m^2}{8}\right]} = I_m$$

*Worked example 6.5*  Calculate the r.m.s. value of a sine wave of maximum value $I_m$.

*Solution*  *Mid-ordinate method*
Let us divide each half into 6 parts, giving 12 mid-ordinates for the complete cycle, each being separated by 30° (the first being at 15°). The value of the square of the mid-ordinates is listed in Table 6.2. The r.m.s. value of the wave is therefore

$$I = \sqrt{\left[\frac{6I_m^2}{12}\right]} = 0.7071I_m$$

*Calculus method*
Here we will evaluate the heat generated during the cycle as the current flows through a resistor $R$ as follows:

$$\text{Heat} = \int_0^{2\pi} (I_m \sin \theta)^2 R \ d\theta = I_m^2 R \int_0^{2\pi} \sin^2 \theta \ d\theta$$

$$= I_m^2 R \int_0^{2\pi} \tfrac{1}{2}(1 - \cos 2\theta) \ d\theta$$

$$= \frac{I_m^2 R}{2} [\theta - \tfrac{1}{2} \sin 2\theta]_0^{2\pi} = \pi I_m^2 R$$

**Table 6.2**

| Mid-ordinate | (Current)² value | |
|---|---|---|
| 1 | $(I_m \sin 15°)^2$ | $= 0.067I_m^2$ |
| 2 | $(I_m \sin 45°)^2$ | $= 0.5I_m^2$ |
| 3 | $(I_m \sin 75°)^2$ | $= 0.933I_m^2$ |
| 4 | $(I_m \sin 105°)^2$ | $= 0.933I_m^2$ |
| 5 | $(I_m \sin 135°)^2$ | $= 0.5I_m^2$ |
| 6 | $(I_m \sin 165°)^2$ | $= 0.067I_m^2$ |
| 7 | $(I_m \sin 195°)^2$ | $= 0.067I_m^2$ |
| 8 | $(I_m \sin 225°)^2$ | $= 0.5I_m^2$ |
| 9 | $(I_m \sin 255°)^2$ | $= 0.933I_m^2$ |
| 10 | $(I_m \sin 285°)^2$ | $= 0.933I_m^2$ |
| 11 | $(I_m \sin 315°)^2$ | $= 0.5I_m^2$ |
| 12 | $(I_m \sin 345°)^2$ | $= 0.067I_m^2$ |
| Total | | $6I_m^2$ |

From eqn [6.10], the average heating effect over the complete cycle is

$$\frac{\text{Total heat}}{\text{Length of base}} = \frac{\pi I_m^2 R}{2\pi} = \tfrac{1}{2} I_m^2 R$$

If $I$ is the effective value of the current, then

$$I^2 R = \tfrac{1}{2} I_m^2 R$$

That is, the r.m.s. value of a *sine wave* of current is

$$I = \frac{I_m}{\sqrt{2}} = 0.7071 I_m$$

*Note*: By Ohm's law, the voltage across the resistor is directly proportional to the current, so that the r.m.s. value of a voltage sine wave is

$$V = \frac{V_m}{\sqrt{2}} = 0.7071 V_m$$

## 6.6 Form factor and peak factor

The **form factor** and the **peak factor** (or **crest factor**) are figures of merit which give an indication of the shape of a wave.

For two waves to have an identical waveshape, they must both have the same form factor and the same peak factor.

The factors are defined as follows:

$$\text{Form factor} = \frac{\text{r.m.s. value of wave}}{\text{mean value of wave}} = \frac{I}{I_{av}} \qquad [6.11]$$

$$\text{Peak factor} = \frac{\text{peak value of wave}}{\text{r.m.s. value of wave}} = \frac{I_m}{I} \qquad [6.12]$$

***Worked example 6.6***  Determine the form factor and the peak factor for (a) a square wave and (b) a sine wave.

*Solution*  In the following we will assume that we are dealing with a current wave.

(a) For a square wave (see also previous worked examples)

$$\text{Peak value} = I_m$$
$$\text{r.m.s. value} = I_m$$
$$\text{Mean value} = I_m$$

hence

$$\text{Form factor} = \frac{\text{r.m.s. value}}{\text{mean value}} = \frac{I_m}{I_m} = 1$$

$$\text{Peak factor} = \frac{\text{peak value}}{\text{r.m.s. value}} = \frac{I_m}{I_m} = 1$$

(b) For a sine wave

$$\text{Peak value} = I_m$$
$$\text{r.m.s. value} = 0.7071 I_m$$
$$\text{Mean value} = 0.637 I_m$$

That is

$$\text{Form factor} = \frac{\text{r.m.s. value}}{\text{mean value}} = \frac{0.7071I_\text{m}}{0.637I_\text{m}} = 1.11$$

$$\text{Peak factor} = \frac{\text{peak value}}{\text{r.m.s. value}} = \frac{I_\text{m}}{0.7071I_\text{m}} = 1.414$$

The values in (b) above are unique to a sinusoidal wave, i.e. they apply equally to a cosine wave as they do to a sine wave.

**6.7 Phasors and phase angle**

The magnitude of a sinusoidally varying quantity is continually changing, and we need a method by which we can refer to 'fixed' values rather than varying values when describing alternating current circuits. Phasors provide a basis for doing this.

Consider the radial line in Fig. 6.8(a), which rotates in a counter-clockwise direction at an angular velocity of $\omega$ rad/s. If we draw a graph of the vertical displacement of the line, the result is the sine wave in the figure. Moreover, if the length of the line is $V_\text{m}$, then the sine wave has a maximum value $V_\text{m}$. At angle $\theta_1$, the instantaneous value of the sine wave is $V_\text{m} \sin \theta_1$, and at angle $\theta_2$ it is $V_\text{m} \sin \theta_2$, etc.

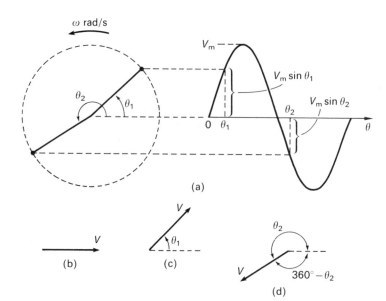

**Fig. 6.8** (a) A sine wave is the graph traced out by the vertical displacement of a rotating line. Phasor representation at (b) $\theta = 0$, (c) $\theta_1$ and (d) $\theta_2$

Engineers represent a sinusoidally varying quantity by a line described as a *phasor*, the length of the line being scaled to represent the *r.m.s. value* of the sine wave; that is, its length is $1/\sqrt{2}$ or 0.7071 of the maximum value of the sine wave. Moreover, the phasor is drawn at some particular angle, and represents the angle that the rotating line in Fig. 6.8(a) has reached at some point in time.

For example, the phasor $V$ in Fig. 6.8(b) represents the sine wave at $\theta = 0°$ (where $V = V_m/\sqrt{2} = 0.7071 V_m$). The phasor $V$ in Fig. 6.8(c) represents the same sine wave at angle $\theta_1$, and the phasor $V$ in Fig. 6.8(d) represents the sine wave at $\theta_2$; in both of these cases $V = V_m/\sqrt{2}$.

We normally give the phase relationship between two phasors in terms of the *phase angle* between the two. If no other voltage or current exists, then the phase angle of the phasor is generally given with respect to the positive horizontal axis, which is known as the *reference direction*. If the phasor lies in the reference direction, as it does in Fig. 6.8(b), the phasor is described as the *reference phasor*.

The two sine waves in Fig. 6.9(b) are drawn out by the rotating lines in Fig. 6.9(a), the lines having a phase angle of $\phi$ between them. The way in which we describe the phase relationship between the rotating lines is the way in which an observer sees them from a fixed position. Imagine for the moment that we are looking inwards from the right towards the centre of the rotating lines in Fig. 6.9(a). The line representing the voltage would pass us before the line representing the current. We therefore say that, for Fig. 6.9

**the voltage leads the current by $\phi$.**

Alternatively we may say

**the current lags behind the voltage by $\phi$.**

The diagrams in Figs 6.9(c)–(e) are known as *phasor diagrams*, which show the relationship between the phasors, and any of the three phasor diagrams can be used to describe the operation of

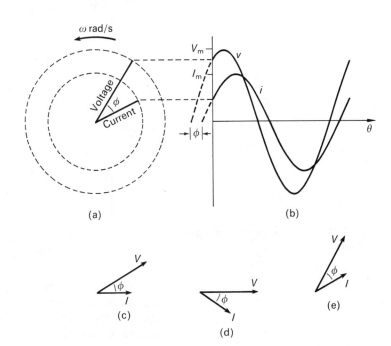

**Fig. 6.9**  Phase angle difference, $\phi$, between two phasors

the circuit. If we choose to draw the current in the reference direction, then the phasor diagram of the circuit is as shown in Fig. 6.9(c); if the voltage is drawn in the reference direction, then the phasor diagram in Fig. 6.9(d) is used. If neither the voltage nor the current is used as the reference phasor, then one possible phasor diagram is shown in Fig. 6.9(e).

**6.8  Addition of phasors**

In alternating current circuits, quantities such as e.m.f.s and p.d.s which are connected in series may have differing magnitudes and phase angles between them; to obtain the total voltage in the circuit, we need to *add the voltage phasors together*. Similarly, the total alternating current drawn by a parallel circuit is the *phasor sum* of the currents in each of the branches.

*In order to be able to add phasors together, it is vital that they are all of the same frequency.* Some circuits have voltages (or currents) in them which are at different frequencies (see Section 6.10) and, in this case, we cannot add the phasors together to give a resultant value.

Phasors can be added either graphically or mathematically, but the reader is advised that, at best, a graphical solution is only an approximate one. If a graphical solution is adopted, use the largest possible diagram to give the best possible result.

Suppose that the two voltages represented by the phasors $V_1$ and $V_2$ in Fig. 6.10, are connected together in a series circuit (in this case $V_1$ leads $V_2$, or $V_2$ lags behind $V_1$). The resultant voltage, $V_S$, is given by

$$V_S = phasor \; sum \; of \; V_1 \; and \; V_2$$

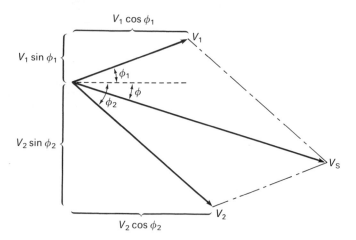

**Fig. 6.10**  Addition of two voltage phasors

Graphically, the value of $V_S$ is the diagonal of the parallelogram formed by the phasors $V_1$ and $V_2$ (see Fig. 6.10). We can measure the magnitude of $V_S$, and its phase angle $\phi$ relative to the reference direction, directly from the diagram.

Mathematically, we can accurately calculate the magnitude and phase angle of $V_S$ by resolving the phasors into their horizontal and vertical components. The horizontal component of a phasor is often said to be the *real component*, and the vertical component is said to be the *quadrature component* or *imaginary component*; the expressions 'real' and 'imaginary' have a mathematical background but, strictly speaking, both components are as 'real' as one another, the difference between them being simply their relative direction.

The horizontal component of $V_1$ is $V_1 \cos \phi_1$, and that of $V_2$ is $V_2 \cos \phi_2$. The vertical component of $V_1$ is $V_1 \sin \phi_1$, and that of $V_2$ is $V_2 \sin \phi_2$. The total horizontal component, $V_h$, of the total voltage $V_S$ is

$$V_h = V_1 \cos \phi_1 + V_2 \cos \phi_2$$

and the vertical component, $V_v$, of the total voltage $V_S$ is

$$V_v = V_1 \sin \phi_1 + V_2 \sin \phi_2$$

However, we must carefully note in this case that $V_2 \sin \phi_2$ has a negative value since $\phi_2$ is negative (see also Worked Example 6.7).

The *magnitude* or *modulus* of $V_S$ is calculated by Pythagoras' theorem as follows:

$$V_S = \sqrt{(V_h^2 + V_v^2)} \qquad\qquad [6.13]$$

and the *phase angle* or *argument*, $\phi$, of $V_S$ with respect to the reference direction can be calculated from any of the following expressions:

$$\phi = \tan^{-1}(V_v/V_h) = \sin^{-1}(V_v/V_S) = \cos^{-1}(V_h/V_S) \qquad [6.14]$$

The reader should take care when evaluating these expressions using a calculator, because the answer provided may be in error. For example, if $V_v = -1$ and $V_h = -1$, a calculator will give an answer of $45°$ when using the expression $\phi = \tan^{-1}(V_v/V_h)$; the correct answer is $45° \pm 180° = 225°$ or $-135°$.

*Worked example 6.7*   A voltage of 10 V r.m.s. at a phase angle of $20°$ is connected in series with a 15 V r.m.s. at a phase angle of $-40°$. What is the resultant voltage?

*Solution*   In this case $V_1 = 10$ V, $\phi_1 = 20°$, $V_2 = 15$ V and $\phi_2 = -40°$ (see also Fig. 6.10). The resolved components of the two voltages are

Horizontal component of $V_1 = V_1 \cos \phi_1$

$$= 10 \cos 20° = 9.397 \text{ V}$$

Horizontal component of $V_2 = V_2 \cos \phi_2$

$$= 15 \cos(-40°) = 11.491 \text{ V}$$

Vertical component of $V_1 = V_1 \sin \phi_1$

$$= 10 \sin 20° = 3.42 \text{ V}$$

Vertical component of $V_2 = V_2 \sin \phi_2$

$$= 15 \sin(-40°) = -9.64 \text{ V}$$

The total horizontal component, $V_h$, of the sum of the two voltages is

$$V_h = V_1 \cos \phi_1 + V_2 \cos \phi_2 = 9.397 + 11.491$$

$$= 20.888 \text{ V}$$

and the total vertical component, $V_v$, of the sum of the two voltages is

$$V_v = V_1 \sin \phi_1 + V_2 \sin \phi_2 = 3.42 + (-9.64)$$

$$= -6.22 \text{ V}$$

The reader should note that, since we are *adding the phasors*, we *add* $V_2 \sin \phi_2$ to $V_1 \sin \phi_1$ to give $V_v$; that is, $V_2 \sin \phi_2 = -9.64$.
The magnitude of $V_S$ is calculated by Pythagoras' theorem as follows:

$$V_S = \sqrt{(V_h^2 + V_v^2)} = \sqrt{(20.888^2 + (-6.22)^2)} = 21.79 \text{ V}$$

and the phase angle is calculated using any of the expressions in eqn [6.14]. Since $V_v$ has a negative sign, care should be exercised when using the expression $\phi = \cos^{-1}(V_h/V_S)$ because the cosine of the angle is positive. In this case we choose

$$\phi = \tan^{-1}(V_v/V_h) = \tan^{-1}(-6.22/20.888) = -16.58°$$

That is, $V_S$ *lags behind the reference direction* by 16.58°. Alternatively, we may say that it *lags behind* $V_1$ by $(20 + 16.58)° = 36.58°$, or it *leads* $V_2$ by $(40 - 16.58)° = 23.42°$.

***Worked example 6.8*** A three-branch parallel a.c. circuit has the following current flowing in each branch:

Branch 1: 10 A at angle 0°

Branch 2: 30 A at angle 40°

Branch 3: 20 A at angle −20°

Calculate the total current drawn from the supply.

*Solution* In this case $I_1 = 10 \text{ A}$, $\phi_1 = 0°$, $I_2 = 30 \text{ A}$, $\phi_1 = 40°$, $I_3 = 20 \text{ A}$ and $\phi_3 = -20°$. The resolved components of the branch currents are

Horizontal component of $I_1 = 10 \cos 0° = 10 \text{ A}$

Horizontal component of $I_2 = 30 \cos 40° = 22.98 \text{ A}$

Horizontal component of $I_3 = 20 \cos(-20°)$

$$= 18.79 \text{ A}$$

Vertical component of $I_1 = 10 \sin 0° = 0 \text{ A}$

Vertical component of $I_2 = 30 \sin 40° = 19.28 \text{ A}$

Vertical component of $I_3 = 20 \sin(-20°) = -6.84$ A

The total horizontal component, $I_h$, is

$$I_h = 10 + 22.98 + 18.79 = 51.77 \text{ A}$$

and the total vertical component, $I_v$, is

$$I_v = 0 + 19.28 + (-6.84) = 12.44 \text{ A}$$

The magnitude of the current drawn from the supply is

$$I_S = \sqrt{(I_h^2 + I_v^2)} = \sqrt{(51.77^2 + 12.44^2)} = 53.24 \text{ A}$$

and its phase angle with respect to the reference direction is

$$\phi = \tan^{-1}(I_v/I_h) = \tan^{-1}(12.44/51.77) = 13.51°$$

## 6.9 Subtraction of phasors

In order to subtract one phasor from another, we need to *add the 'negative' value of the phasor to be subtracted.* That is

$$V_1 - V_2 = V_1 + (-V_2)$$

We must therefore *reverse the phasor $V_2$*, as shown in Fig. 6.11, and add it to the $V_1$ phasor. For example, if $V_1$ is a voltage of 20 V whose phase angle relative to the reference direction is 10°, and the magnitude of $V_2$ is 30 V at 45° relative to the reference direction, then the *phasor difference* $(V_1 - V_2)$ is calculated as follows (see also Fig. 6.12):

Horizontal component of $V_1 = V_1 \cos \phi_1$
$$= 20 \cos 10° = 19.7 \text{ V}$$

Horizontal component of $V_2 = V_2 \cos \phi_2$
$$= 30 \cos 45° = 21.21 \text{ V}$$

Horizontal component of $-V_2 = -21.21$ V

Alternatively, the horizontal component of $-V_2$ is
$$30 \cos(45 + 180)° = 30 \cos 225° = -21.21 \text{ V}$$

The vertical component is calculated as follows:

Vertical component of $V_1 = V_1 \sin \phi_1$
$$= 20 \sin 10° = 3.47 \text{ V}$$

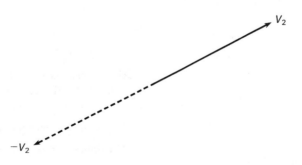

**Fig. 6.11** The 'negative' value of a phasor

$$\text{Vertical component of } V_2 = V_2 \sin \phi_2$$

$$= 30 \sin 45° = 21.21 \text{ V}$$

$$\text{Vertical component of } -V_2 = -21.21 \text{ V}$$

Alternatively, the vertical component of $-V_2$ is

$$30 \sin(45 + 180)° = 30 \sin 225° = -21.21 \text{ V}$$

The horizontal component, $V_h$, of the voltage difference $(V_1 - V_2)$ is

$$19.7 + (-21.21) = -1.51 \text{ V}$$

and the vertical component, $V_v$, is

$$3.47 + (-21.21) = -17.74 \text{ V}$$

Consequently, the magnitude of the difference in voltage is

$$(V_1 - V_2) = \sqrt{(V_h^2 + V_v^2)}$$

$$= \sqrt{((-1.51)^2 + (-17.74)^2)} = 17.8 \text{ V}$$

and its phase angle with respect to the reference direction is

$$\phi = \tan^{-1}(V_v/V_h) = \tan^{-1}(-17.74/(-1.51))$$

$$= -94.87°$$

*Warning*: the reader must exercise engineering and mathematical judgement here, because the angle is in the third quadrant. A calculator may give an answer which suggests that the angle is 85.13°!

The corresponding phasor diagram is shown in Fig. 6.12.

*Worked example 6.9*  A three-branch a.c. parallel circuit draws a current of 55 A at a phase angle of 16.7° with respect to the reference direction. If the current in one branch is 10 A in the reference direction, and the current in the second branch is 30 A at a phase angle of 40° to the reference direction, calculate the magnitude and phase angle of the current in the third branch.

*Solution*  Applying KCL to the circuit, the supply current is given by

$$I_S = \text{phasor sum } (I_1 + I_2 + I_3)$$

Hence

$$I_3 = \text{phasor difference } (I_S - (I_1 + I_2))$$

Clearly, it is necessary to calculate the *phasor sum* $(I_1 + I_2)$, and subtract this form $I_S$ by the phasor difference method. The phasor addition proceeds as follows:

$$\text{Horizontal component of } I_1 = 10 \cos 0° = 10 \text{ A}$$

$$\text{Horizontal component of } I_2 = 30 \cos 40° = 22.98 \text{ A}$$

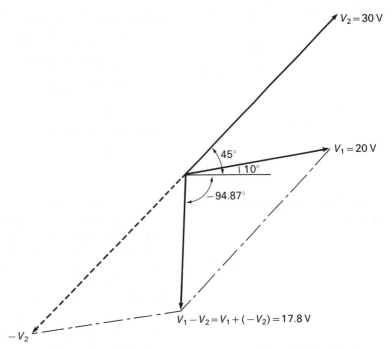

**Fig. 6.12** Subtraction of phasors

Vertical component of $I_1 = 10 \sin 0° = 0$ A

Vertical component of $I_2 = 30 \sin 40° = 19.28$ A

Hence

Horizontal component of $(I_1 + I_2) = 10 + 22.98$

$$= 32.98 \text{ A}$$

Vertical component of $(I_1 + I_2) = 0 + 19.28$

$$= 19.28 \text{ A}$$

therefore

Horizontal component of $-(I_1 + I_2) = -32.98$ A

Vertical component of $-(I_1 + I_2) = -19.28$ A

The horizontal component of $I_S$ is

$$55 \cos 16.7° = 52.68 \text{ A}$$

and the vertical component of $I_S$ is

$$55 \sin 16.7° = 15.8 \text{ A}$$

Therefore the horizontal component of $I_3$ is given by

Horizontal component of $(I_S + (-(I_1 + I_2)))$

$$= 52.68 + (-32.98) = 19.7 \text{ A}$$

and the vertical component of $I_3$ is given by

$$\text{Vertical component of } (I_S + (-(I_1 + I_2)))$$

$$= 15.8 + (-19.28) = -3.48 \, \text{A}$$

The magnitude of $I_3$ is

$$I_3 = \sqrt{(19.7^2 + (-3.48)^2)} = 20 \, \text{A}$$

and its phase angle is calculated as follows:

$$\phi = \tan^{-1}(I_v/I_h) = \tan^{-1}(-3.48/19.7) = -10°$$

**6.10  Harmonics and complex waves**

A circuit element such as a resistor, a capacitor, or an air-cored coil has a *linear voltage–current characteristic*, and the current flowing through it is proportional to the voltage across it.

However, some elements such as an iron-cored coil, a fluorescent lamp, and many semiconductor devices have a *non-linear voltage–current characteristic*, and the current flowing through them is not proportional to the voltage across them. The net result in this case is that, when a sinusoidal voltage is applied to the device, the current waveform is non-sinusoidal. These waveforms are said to be *complex waves*, and can be constructed or *synthesized* by adding together a number of sine waves of differing magnitude, frequency and phase angle. Electronic music is produced in an electronic synthesizer.

A complex waveform has a *fundamental frequency*, which is the frequency of the lowest repetitive sine wave associated with the complex wave. For example, the fundamental frequency of the complex wave in Fig. 6.13(c) is the frequency of the sine wave in Fig. 6.13(a). Associated with the fundamental frequency are one or more *harmonic frequencies*, whose frequency is an integral (whole number) multiple of the fundamental frequency. If the fundamental frequency is $f_1$ hertz, then

Second harmonic frequency, $f_2 = 2f_1$

Third harmonic frequency, $f_3 = 3f_1$

Fourth harmonic frequency, $f_4 = 4f_1$

Tenth harmonic frequency, $f_{10} = 10f_1$, etc.

In most apparatus used in electrical engineering, the harmonics produced generally diminish in magnitude as the order of the harmonic increases. That is, the magnitude of the second harmonic tends to be smaller than that of the fundamental, the magnitude of the third harmonic is even smaller, etc.

In some cases, the harmonic frequencies are in phase with the fundamental frequency component of the complex wave, i.e. they begin their cycle simultaneously. However, this is not generally the case. In Fig. 6.13(b), the third harmonic is in phase with the

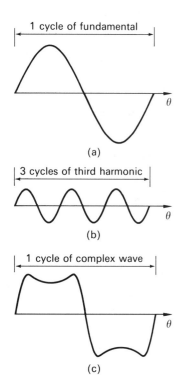

1 cycle of fundamental

$\theta$

(a)

3 cycles of third harmonic

$\theta$

(b)

1 cycle of complex wave

$\theta$

(c)

**Fig. 6.13**  (a) Fundamental frequency and (b) a third harmonic frequency; (c) the resulting complex wave produced by adding waveform (a) to waveform (b)

fundamental frequency, but has a smaller magnitude. The waveform in Fig. 6.13(c) is the sum of the waveforms in diagrams (a) and (b).

In Chapter 8 we will look at a phenomenon known as resonance, and one of the troublesome aspects of resonance is *selective resonance*, which occurs when resonance coincides with one of the harmonics in a complex wave. We will return to this in Chapter 8.

**Problems**

**6.1**   An alternating wave has a periodic time of (a) 50 ms, (b) 50 $\mu$s. Determine in each case (i) the frequency in Hz and (ii) the angular frequency in rad/s.

[(a) (i) 20 Hz, (ii) 125.7 rad/s; (b) (i) 20 kHz, (ii) 125 664 rad/s]

**6.2**   The angular frequency of an alternating wave is 1571 rad/s. Calculate (a) its frequency in Hz and (b) its periodic time.

[(a) 250 Hz; (b) 4 ms]

**6.3**   The instantaneous value of the current in a circuit is 1.5 A at time 1 ms after the commencement of the wave. If the periodic time of the wave is 10 ms, calculate the ~~peak-to-peak~~ peak value of the wave.

[2.55 A]

**6.4**   An alternating voltage is represented by the expression $v = 20 \sin 1257t$ volts. Calculate (a) the frequency of the wave, (b) its periodic time and (c) the time taken from $t = 0$ to the voltage becoming $+15$ V for (i) the first and (ii) the second time.

[(a) 200 Hz; (b) 5 ms; (c) (i) 0.675 ms, (ii) 1.825 ms]

**6.5**   The following values refer to one-half of an alternating current wave, both halves being symmetrical about the zero axis. Plot the waveform and determine (a) its mean value, (b) its r.m.s. value, (c) its form factor and (d) its peak factor.

| Time (ms) | 0 | 1 | 2 | 3 | 4 | 5 | 6 | 7 | 8 |
|---|---|---|---|---|---|---|---|---|---|
| Current (mA) | 0 | 23 | 40 | 50 | 42 | 33 | 25 | 16 | 0 |

[(a) 28.63 mA; (b) 31.62 mA; (c) 1.104; (d) 1.58]

**6.6**   A sinusoidal alternating voltage waveform has a peak-to-peak value of 6 V. Draw the waveform to scale, the periodic time being 100 ms. Using the mid-ordinate rule, estimate (a) its r.m.s. value, (b) its mean value.

[(a) 2.12 V; (b) 1.91 V]

**6.7**   Data for the first half-cycle of two alternating waves A and B are given below. Plot the waves, and estimate the form factor for each wave.

| $\theta$ (deg) | Wave A | Wave B | $\theta$ (deg) | Wave A | Wave B |
|---|---|---|---|---|---|
| 0 | 4 | 50 | 99 | 2 | 27.5 |
| 18 | 10 | 50 | 108 | 1 | 43 |
| 36 | 37 | 50 | 126 | 0 | 50 |
| 54 | 50 | 50 | 144 | 1 | 50 |
| 72 | 37 | 43 | 162 | 2 | 50 |
| 81 | 10 | 27.5 | 180 | 4 | 50 |
| 90 | 4 | 0 | | | |

[A, 1.1751 B, 1.042]

**6.8** Two alternating currents are represented by

$$i_1 = 50 \sin \theta \quad \text{and} \quad i_2 = 75 \cos \theta$$

Add the currents and determine (a) the r.m.s. value of the sum and (b) the phase angle of the sum with respect to $i_1$.

[(a) 63.74 A; (b) 56.3° (leading)]

**6.9** The following alternating currents are added together in a circuit:

$$i_1 = 130 \sin \omega t \text{ amperes}$$

$$i_2 = 120 \sin(\omega t - \pi/4) \text{ amperes}$$

$$i_4 = 150 \sin(\omega t + \pi/3) \text{ amperes}$$

Determine the resultant current in the same form, and calculate its r.m.s. value.

[$293.5 \sin(\omega t + 0.154 \text{ rad}) A$; 207.5 A]

**6.10** If $v_1 = 100 \sin(\theta - 45°) V$ and $v_2 = 150 \sin(\theta + 60°) V$, determine expressions for (a) $v_1 + v_2$, (b) $v_1 - v_2$ and (c) $v_2 - v_1$.
[(a) 157.3 $\sin(\theta + 22.15°) V$; (b) 200.7 $\sin(\theta + 268.8°) V$; (c) 200.7 $\sin(\theta + 88.77°) V$]

**6.11** A sinusoidal alternating voltage has a peak value of 100 V. Calculate the instantaneous value at the following angles after the start of the wave: 0°, 45°, 90°, 160°, 220°, 330° and 360°.

[0 V; 70.7 V; 100 V; 34.2 V; −64.3 V; −50 V; 0 V]

**6.12** Calculate the periodic time corresponding to the following frequency: 50 Hz, 52 kHz, 6 MHz, 8 GHZ.

[0.02 s; 19.23 $\mu$s; 0.1667 $\mu$s; 0.125 ns]

**6.13** Calculate the frequency corresponding to the following periodic time: 2 ms, 5 $\mu$s, 10 ns, 5 ps.

[500 Hz; 200 kHz; 100 MHz; 200 GHz]

**6.14** Determine the mean value, the r.m.s. value and the form factor of the following current waveform which varies linearly between the points given. The second half-cycle is a mirror image of the first half-cycle.

Current (A) 0 3.6 8.4 14 19.4 22.5 25 25.2  23 15.6 9.4 4.2  0
Angle (deg) 0 15  30  45 60   75   90 105 120 135 150 165 180

[14.2; 16.4; 1.155]

**6.15** The periodic time of an alternating current waveform is $T$. The instantaneous value of current between $t = 0$ and $t = T/6$ is 50 A, its value between $t = T/6$ and $t = T/4$ is 20 A, and between $t = T/4$ and $t = T/2$ is zero. The waveshape is inverted in the second half-cycle. Calculate the r.m.s. value and the average value of the wave.

[30 A; 20 A]

**6.16** The following voltages are connected in series:

$$v_1 = 100 \sin \theta \qquad\qquad v_2 = 50 \sin(\theta + 60°)$$

$$v_3 = 60 \sin(\theta - \pi/4) \qquad v_4 = 80 \sin(\theta + \pi/2)$$

Express the sum of the voltages in a similar form.

$$[188 \sin(\theta + 25°)]$$

**6.17** An alternating waveform has a peak value of 500 V. If the form factor of the wave is 1.16 and the peak factor is 1.6, calculate the r.m.s. value and the mean value of the wave.

$$[312.5\,\text{V}; 269.4\,\text{V}]$$

**6.18** A two-branch parallel circuit draws an r.m.s. current of 50 A at a phase angle of 0°. If the current in one branch is

$$i_1 = 70.7 \sin(314.2t - \pi/3)\,\text{A}$$

determine an expression for the current in the other branch. What is the supply frequency?

$$[70.7 \sin(314.2t + \pi/3)\,\text{A}; 50\,\text{Hz}]$$

**6.19** Two branches A and B are connected in parallel. The current drawn by branch A is 12 A r.m.s. in the reference direction, and the current drawn by branch B is 20 A lagging by 30°. Determine the magnitude and the phase angle of the total current drawn by the two branches.

$$[31\,\text{A}; 18.8°\text{ lagging}]$$

**6.20** Three circuits X, Y and Z are connected in series to a 200 V supply. If the voltage across circuit X is 100 V leading the supply voltage by 30°, and the voltage across circuit Y is 50 V lagging the supply voltage by 45°, determine the magnitude and phase angle of the voltage across circuit Z.

$$[79.4\ \text{V}; 10.6°\text{ lagging}]$$

# 7 Single-phase a.c. circuits and power

**7.1 Circuit notation**

This chapter deals with circuits having a sinusoidal supply. As with d.c. circuits, the current arrow on the circuit is drawn in the *assumed direction of current flow*. Again, as for d.c. circuits, voltage and p.d. arrows are drawn so that the arrowhead points towards the node which is *assumed to have the more positive potential*. In the case of both voltage and current, the value given is the r.m.s. value (unless otherwise stated).

**7.2 A circuit containing a pure resistor**

If an alternating voltage $v = V_m \sin \omega t$ is applied to a pure resistor [see Fig. 7.1(a)], by Ohm's law the current in the circuit is

$$i = \frac{v}{R} = \frac{V_m}{R} \sin \omega t = I_m \sin \omega t \qquad [7.1]$$

where $R$ is the resistance of the circuit in ohms, $V_m$ the maximum value of the applied voltage, $I_m$ the maximum current, and $\omega$ the angular frequency in rad/s. The voltage and current waveforms in Fig. 7.1(b) reflect the above expression. From Ohm's law and the waveforms we see that, for a circuit containing a pure resistor,

**the current flowing through the resistor is in phase with the voltage across it.**

The phasor diagram corresponding to the angle $\theta = 0°$ is shown in Fig. 7.1(c); remember, the phasors are scaled to $1/\sqrt{2}$ of the corresponding maximum value.

As explained in Chapter 6, the angle at which the phasor diagram is drawn depends on the point which we choose in Fig. 7.1(b). If we select the point where $\theta = 60°$, the phasor diagram appears as shown in Fig. 7.1(c); once again, the voltage and current phasors are in phase with one another, and the length of the $I$ and $V$ phasors are the same in Fig. 7.1(c) as they are in Fig. 7.1(d).

From eqn [7.1]

$$\frac{V_m}{R} = I_m$$

The r.m.s. voltage, $V$, is $V_m/\sqrt{2}$, and the r.m.s. current is $I = I_m/\sqrt{2}$ hence, for a purely resistive circuit

$$I = V/R \quad \text{or} \quad V = IR \qquad [7.2]$$

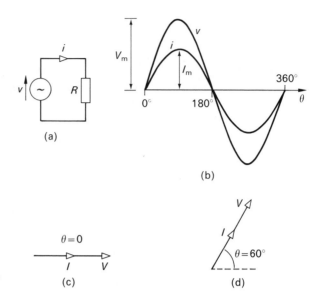

**Fig. 7.1** (a) An a.c. circuit containing a pure resistance, (b) waveforms in the circuit. Phasor diagram for (c) $\theta = 0$ and (d) $\theta = 60°$

*Worked example 7.1*   When a voltage $v = 30 \sin \omega t$ is applied to a resistor, the r.m.s. current is 10 A. Calculate the resistance of the resistor.

*Solution*   The maximum value of voltage is 30 V, hence the r.m.s. voltage is

$$V = 30/\sqrt{2} = 21.21 \text{ V}$$

From eqn [7.2] we see that

$$R = V/I = 21.21/10 = 2.121 \text{ } \Omega$$

## 7.3  Circuit containing a pure inductance

### Phase relationship between voltage and current

#### Qualitative approach

The circuit in Fig. 7.2(a) contains a pure (i.e. of zero resistance) inductor, which is excited by a sinusoidal supply. The reader will recall from the work in Chapter 5 that the voltage applied to the inductor is equal to the self-induced e.m.f. or 'back' e.m.f., produced by the flow of alternating current in the inductor. That is

$$v = L \times \text{rate of change of current} \qquad [7.3]$$

This situation is illustrated in Fig. 7.2(b), where the graph of the self-induced e.m.f. $(L\,di/dt)$ is superimposed on the voltage waveform. In Fig. 7.2(c) we separate out the rate of change (slope) of the current waveform $(di/dt)$ so that it can be considered in detail.

We see that at $\theta = 0°$ and $\theta = 180°$, $di/dt = 0$, that is the rate of change of current is zero; also it is has a positive value between the two points. Moreover, the slope of the current waveform is

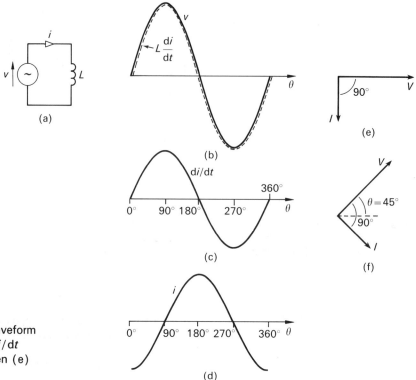

**Fig. 7.2** (a) An a.c. circuit containing a pure inductor. Waveform for (b) $v$ (and $L\,di/dt$), (c) $di/dt$ and (d) $i$. Phasor diagram when (e) $\theta = 0$ and (f) $\theta = 45°$

maximum at $\theta = 90°$. The corresponding current waveform between $\theta = 0°$ and $180°$ is therefore as shown in Fig. 7.2(d). The reader will find it an interesting exercise to show that, for the range of angles between $180°$ and $360°$, the current waveform is as shown in Fig. 7.2(d). That is, the current waveform has a $(-\text{cosine})$ shape.

If we look at the voltage and current waveforms, we conclude that the phasor diagram for the circuit at $\theta = 0°$ is as shown in Fig. 7.2(e). However, at $\theta = 45°$, the phasor diagram for the circuit is as shown in Fig. 7.2(f); in this case the phasor diagram has been rotated through $45°$ in a counter-clockwise direction. In general, from the waveform diagrams and the two phasor diagrams, we conclude:

> **In a circuit containing only a pure (of zero resistance) inductor, the current flowing through the inductor lags $90°$ behind the voltage across it.**

Alternatively, we may say that the voltage across the inductor leads the current by $90°$.

In most cases, engineers tend to think of the voltage as the reference phasor and we usually say that, in an inductive circuit, the current lags behind the voltage.

**Phase relationship by calculus**

Since the instantaneous applied voltage is equal to the back e.m.f., we may write

$$v = L\frac{d}{dt}(I_m \sin \omega t) = LI_m \frac{d}{dt}\sin \omega t$$

where $I_m$ is the maximum value of current through the inductor, and $\omega$ is the supply frequency in rad/s. The differential of $\sin \omega t$ is $(\omega \cos \omega t)$ and, since

$$\cos \omega t = \sin(\omega t + 90°),$$

then

$$v = \omega L I_m \sin(\omega t + 90°) \tag{7.4}$$

What interests us here is that, with a sinusoidal current $(I_m \sin \omega t)$, the voltage across the inductor is given by an equation containing $\sin(\omega t + 90°)$, i.e. the voltage leads the current by 90° or, alternatively, the current lags behind the voltage by 90°.

**7.4  Inductive reactance**

Despite the fact that a pure inductor has zero resistance, we find that when it is connected to an a.c. supply, the current is limited in value! The effective impedance to flow of alternating current is due to what is known as the *inductive reactance*, $X_L$, of the inductor, and has the dimensions of ohms. The reason for the inductive reactance is simply that the self-induced e.m.f. opposes the flow of current through the inductor. We can obtain the value of $X_L$ both quantitatively and by calculus as follows.

**Quantitative approach**

If we rewrite eqn [7.3] in the following form:

Average supply voltage = $L$ × average rate of change of current

and if we apply this to the first quarter-cycle of the waveform in Fig. 7.2, we get

$$\frac{V_m}{\pi/2} = L\frac{I_m}{T/4} = L\frac{I_m}{1/4f} = 4LI_m f$$

or

$$V_m = 2\pi f L I_m = \omega L I_m$$

where $T$ is the periodic time of the supply waveform, $f$ its frequency in Hz, $\omega$ is the angular frequency in rad/s, and $\pi/2$ rad corresponds to 90°. Now

$$X_L = \frac{V_m}{I_m} = \frac{V}{I} = \omega L = 2\pi f L \text{ ohms} \qquad [7.5]$$

**Solution by calculus**

From eqn [7.4] we see that the maximum voltage, $V_m$, is

$$V_m = \omega L I_m$$

hence

$$X_L = \frac{V_m}{I_m} = \frac{V}{I} = \omega L = 2\pi f L \text{ ohms}$$

*Worked example 7.2*    A 250 V r.m.s., 50 Hz sinusoidal supply is connected to a pure inductance of 0.75 H. Calculate the reactance of the inductor and the current flowing through it.

*Solution*    The inductive reactance is

$$X_L = 2\pi f L = 2\pi \times 50 \times 0.75 = 235.6 \ \Omega$$

hence the magnitude of the current is

$$I = V_S/X_L = 250/235.6 = 1.06 \text{ A}$$

Although, strictly speaking, this is all that was asked for, it is also both informative and useful to say that the current lags behind the supply voltage by 90°.

**7.5  Effect of frequency on inductive reactance and current magnitude**

Equation [7.5] shows that

$$X_L \propto f \qquad [7.6]$$

That is $X_L$ increases with frequency; its value is zero at zero frequency, rising to infinite value at infinite frequency. It follows that, by Ohm's law, the current in a circuit containing only a pure inductor diminishes in value as the frequency increases.

If the supply voltage is maintained constant, the current (theoretically) has an infinite value at zero frequency (d.c.), and has zero value at infinite frequency.

A graph showing how $X_L$ and $I$ vary with frequency is shown in Fig. 7.3.

*Worked example 7.3*    Calculate the reactance of a 100 mH pure inductor, and the current in the inductor when 100 V r.m.s. is applied to it at a frequency of (a) 50 Hz, (b) 1 kHz.

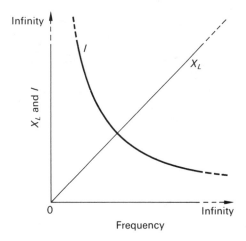

**Fig. 7.3** Effect of variation of frequency on $X_L$ and inductor current

*Solution*  (a) When $f = 50\,\text{Hz}$

$$X_L = 2\pi f L = 2\pi \times 50 \times 100 \times 10^{-3} = 31.42\,\Omega$$

and the current in the inductor is

$$I = V_S/X_L = 100/31.42 = 3.18\,\text{A}$$

(b) When $f = 1\,\text{kHz}$

$$X_L = 2\pi f L = 2\pi \times 1000 \times 100 \times 10^{-3} = 628.3\,\Omega$$

and the current in it is

$$I = V_S/X_L = 100/628.3 = 0.159\,\text{A}$$

## 7.6 Circuit containing a pure capacitor

In Chapter 3 it was shown that

Capacitor current, $i$

$$= C \times \text{rate of change of capacitor voltage} \qquad [7.7]$$

**Phase relationship between voltage and current**

**Qualitative approach**

Since the circuit [Fig. 7.4(a)] only contains a pure capacitor, the voltage is applied directly to the terminals of the capacitor, hence the waveform of $C \times$ (rate of change of capacitor voltage) is equal to the capacitor current waveform [see Fig. 7.4(b)].

Since the capacitor current is sinusoidal, the waveform for the rate of change of capacitor voltage is also sinusoidal [Fig. 7.4(c)]. This waveform has a positive value between $\theta = 0°$ and $180°$, so

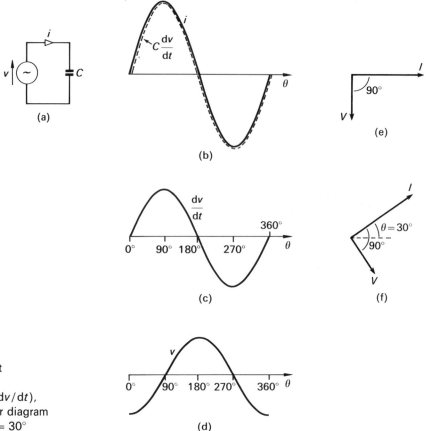

**Fig. 7.4** (a) An a.c. circuit containing a pure capacitor. Waveform for (b) $i$ (and $C \, dv/dt$), (c) $dv/dt$ and (d) $v$. Phasor diagram when (e) $\theta = 0°$ and (f) $\theta = 30°$

that the *slope of the voltage waveform* is positive in this range of angles. Also, the value of $dv/dt$ [Fig. 7.4(c)] – and therefore the slope of the voltage wave – is zero at $\theta = 0°$ and $180°$, and the voltage waveform has its maximum slope at $\theta = 90°$. Intuitively, we may say that the voltage waveform between $\theta = 0°$ and $180°$ is as shown in Fig. 7.4(d). It is left as an exercise for the reader to argue that the remainder of the voltage waveform in Fig. 7.4(d) is correct.

Clearly, when a sinusoidal alternating current flows through a pure capacitor, the voltage across the capacitor has a ($-$cosine) waveform. The corresponding phasor diagram for $\theta = 0°$ is shown in Fig. 7.4(e), and that for $\theta = 30°$ is in Fig. 7.4(f). We observe the following general relationship:

> **In a pure capacitor, the current flowing through the capacitor leads the voltage across it by 90°.**

Alternatively, we may say that the voltage across the capacitor lags behind the current flowing through it by 90°.

### Phase relationship by calculus

The mathematical relationship between the current through the capacitor and the voltage across it is $i = C \, dv_c/dt$, or

$$i = C \frac{d}{dt} V_m \sin \omega t = CV_m \frac{d}{dt} \sin \omega t = \omega CV_m \cos \omega t$$

$$= \omega CV_m \sin(\omega t + 90°) \tag{7.8}$$

Clearly, if the applied voltage is a sine wave, then the current follows a cosine wave; that is, the current leads the voltage across the capacitor by 90°.

## 7.7 Capacitive reactance

When a capacitor is connected to a sinusoidal voltage, the changing voltage causes the stored charge to alternate and, as it does, the charging current or displacement current also alternates. That is, the capacitor draws an alternating current from the supply. The magnitude of the current is limited by a quantity known as the *capacitive reactance*, $X_C$, which has the dimensions of ohms. We can obtain its value as follows.

We can rewrite eqn [7.7] as follows:

Average current

$$= C \times \text{average rate of change of capacitor voltage}$$

Taking values from the first quarter-cycle of Fig. 7.4, we get

$$\frac{I_m}{\pi/2} = C \frac{V_m}{T/4} = C \frac{V_m}{1/4f} = 4fCV_m$$

or

$$I_m = 2\pi f CV_m = \omega CV_m \tag{7.9}$$

where $T$ is the periodic time of the supply waveform, $f$ its frequency in Hz, $\omega$ the angular frequency in rad/s, and $\pi/2$ rad corresponds to 90°. Now

$$X_C = \frac{V_m}{I_m} = \frac{V}{I} = \frac{1}{2\pi f C} = \frac{1}{\omega C} \text{ ohms} \tag{7.10}$$

***Worked example 7.4***    A sinusoidal current of 2 mA r.m.s. flows through a 5 $\mu$F capacitor when it is connected to a 50 V, 1 kHz supply. Calculate the capacitance of the capacitor.

*Solution*   The reactance of the capacitor is

$$X_C = \frac{V}{I} = \frac{50}{2 \times 10^{-3}} = 25\,000\,\Omega$$

Now $X_C = 1/2\pi f C$, hence

$$C = 1/2\pi f X_C = 1/(2\pi \times 1000 \times 25\,000)$$

$$= 6.37 \times 10^{-9}\,\text{F} \quad \text{or} \quad 6.37\,\text{nF}$$

## 7.8 Effect of frequency on capacitive reactance and current

From eqn [7.10] we see that

$$X_C \propto \frac{1}{f} \tag{7.11}$$

That is as the frequency increases, the capacitive reactance reduces. Also, when $f = 0$ then $X_C = \infty$, and when $f = \infty$ then $X_C = 0$. This effect is shown in Fig. 7.5, together with the variation in the magnitude of the capacitor current if the r.m.s. value of the supply voltage is maintained constant.

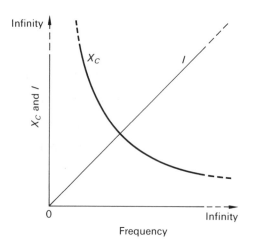

**Fig. 7.5** Effect of variation of frequency on $X_C$ and capacitor current

*Worked example 7.5*   At a frequency of (a) 50 Hz, (b) 50 kHz, calculate the reactance of a 0.1 µF capacitor, and the current through it if the supply voltage is 5 V r.m.s.

*Solution*   (a)

$$X_C = 1/2\pi f C = 1/(2\pi \times 50 \times 0.1 \times 10^{-6})$$

$$= 31.83\,\text{k}\Omega$$

and

$$I = V_S/X_C = 5/(31.83 \times 10^3) = 0.157\,\text{mA}$$

(b)

$$X_C = 1/2\pi fC = 1/(2\pi \times 50\,000 \times 0.1 \times 10^{-6})$$

$$= 31.83\,\Omega$$

hence

$$I = V_S/X_C = 5/31.83 = 0.157\ \text{A}$$

**7.9  CIVIL – a useful mnemonic**

A useful aid enabling the foregoing phase relationships to be remembered is

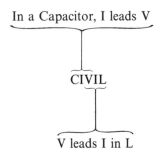

In a Capacitor, I leads V

CIVIL

V leads I in L

**7.10  Series circuit containing R and L**

In the circuit in Fig. 7.6(a), the current flows through both $R$ and $L$, hence the supply voltage, $V_S$, is given by

$$V_S = \textit{phasor sum of } V_R \textit{ and } V_L \qquad\qquad [7.12]$$

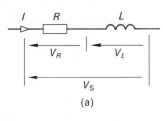

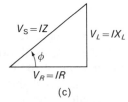

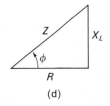

**Fig. 7.6**  (a) $R$–$L$ series circuit, (b) a typical phasor diagram, (c) the voltage triangle and (d) the corresponding impedance triangle

The phasor voltages $V_S$, $V_R$ and $V_L$ have both magnitude and 'direction' or phase angle (which could, in some cases, be zero!).

Readers interested in pursuing the study of electrical and electronic circuits to a higher level (which is necessary if the reader wishes to further his or her career) should read section 7.15, when the subject of *complex numbers* is introduced. This is an area of study where the boundaries of technology overlap with mathematics and, in particular, where we can describe a phasor as a mathematical quantity. Readers should not concern themselves about the word 'complex', because it is simply a way of saying that a value has a direction as well as a magnitude.

In a series circuit, the current is common to all the components, and it is usually the case that we use the current as the *reference phasor*. Consequently, we draw it in the *reference direction*, i.e. horizontally, pointing to the right of zero.

Referring to the phasor diagram in Fig. 7.6(b), the p.d., $V_R$, across the resistor *is in phase with the current*, $I$ (see also section 7.2). In the case of the inductor, the voltage, $V_L$, across it leads the current by 90° (see also section 7.3). Consequently, the phasor $V_L$ points vertically upwards in the phasor diagram [Fig. 7.6(b)].

From the phasor diagram in Fig. 7.6(b), we see that the magnitude of the supply voltage, $V_S$, is calculated by Pythagoras' theorem as follows:

$$V_S = \sqrt{[V_R^2 + V_L^2]} = \sqrt{[(IR)^2 + (IX_L)^2]}$$
$$= I\sqrt{(R^2 + X_L^2)} = IZ \qquad [7.13]$$

where $Z$ is the magnitude of the total **impedance** of the circuit to flow of alternating current, where

$$Z = \sqrt{(R^2 + X_L^2)} \qquad [7.14]$$

and has the dimensions of ohms.

The phasor diagram in Fig. 7.6(b) shows that, in a series $R$–$L$ circuit, *the current lags behind the voltage by angle $\phi$*, this angle being known as the *phase angle* of the circuit. The phase angle can either be expressed in degrees or radians, and may be calculated from any one of the following [see also Fig. 7.6(b)]:

$$\left.\begin{array}{l}
\cos \phi = V_R/V_S = IR/IZ = R/Z = R/\sqrt{(R^2 + X_L^2)} \\
\tan \phi = V_L/V_R = IX_L/IR = X_L/R = \omega L/R \\
\sin \phi = V_L/V_S = IX_L/IZ = X_L/Z = X_L/\sqrt{(R^2 + X_L^2)}
\end{array}\right\}$$
$$[7.15]$$

If we separate out the voltage components from the phasor diagram, as shown in Fig. 7.6(c), we are left with the *voltage triangle* of the circuit. Furthermore, if each side of the voltage triangle is divided by the magnitude of the current, $I$, the net result is the *impedance triangle* in Fig. 7.6(d).

### 7.10.1 Effective resistance or a.c. resistance of an $R$–$L$ series circuit

The reader will recall we showed, in Chapter 4, that hysteresis power loss occurs in iron-cored coils when the winding carries an alternating current. This power loss can be thought of as being produced by an extra resistance in the circuit. That is to say, the d.c. resistance (as measured by an ohmmeter) is less than the *effective resistance* or *a.c. resistance* of the coil (see also Problem 7.10).

The reader will also have noted in Chapter 4 that d.c. power loss in the coil due to its electrical resistance is called the copper loss, and the power loss due to hysteresis is known as the iron loss, the total power loss in the coil being

$$\text{Total power loss} = \text{copper loss} + \text{iron loss}$$

If $R_{dc}$ is the electrical resistance of the coil which produces the copper loss, and $R_{iron}$ is the electrical resistance which can be thought of as producing the iron loss, then

$$\text{Total power loss} = I^2(R_{dc} + R_{iron})$$

*Worked example 7.6*    A series $R$–$L$ circuit containing a $200\,\Omega$ resistor and a $0.5\,H$ inductor is energized by a $250\,V$, $50\,Hz$ supply. Calculate (a) the reactance of the inductor, (b) the impedance of the circuit, (c) the current in the circuit, (d) the phase angle between the current and the voltage and (e) the p.d. across the resistor and across the inductor. Draw the phasor diagram of the circuit.

*Solution*    (a) The inductive reactance is

$$X_L = 2\pi f L = 2\pi \times 50 \times 0.5 = 157.1\,\Omega$$

(b) From eqn [7.13]

$$Z = \sqrt{(R^2 + X_L^2)} = \sqrt{(200^2 + 157.1^2)} = 254.3\,\Omega$$

(c) By Ohm's law, the magnitude of the current is

$$I = V_S/Z = 250/254.3 = 0.983\,A$$

(d) From eqn [7.15]

$$\cos \phi = R/Z = 200/254.3 = 0.7865$$

hence

$$\phi = \cos^{-1}(0.7865) = 38.14° \text{ (or } 0.6657 \text{ rad)}$$

Since we are dealing with an $R$–$L$ circuit, the current lags behind the supply voltage by $38.14°$.

(e) Figure 7.6(b) shows us that

$$V_R = IR = 0.983 \times 200 = 196.6\,V$$

and

$$V_L = IX_L = 0.983 \times 157.1 = 154.4\,V$$

*Note*: From eqn [7.13]

$$V_S = \sqrt{[V_R^2 + V_L^2]} = \sqrt{[196.6^2 + 154.4^2]} = 249.98\,V$$

The very small difference between the specified value of $V_S$ (250 V) and the value calculated here is due to 'rounding' errors performed elsewhere in the calculation.

The phasor diagram is drawn in Fig. 7.7.

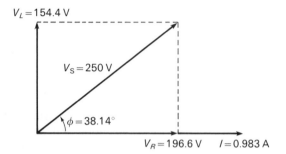

**Fig. 7.7** Phasor diagram for Worked Example 7.6

### 7.11 Admittance of an a.c. circuit

The magnitude of the *admittance*, $Y$, of an a.c. circuit is the reciprocal of the magnitude of its impedance. That is

$$Y = 1/Z$$

and has the dimensions of siemens (S). The *phase angle of the admittance* is $(-1) \times$ the phase angle of the impedance. In the case of Worked Example 7.6

$$Y = 1/Z = 1/254.3 = 3.93 \times 10^{-3} \text{ S} \quad \text{or} \quad 3.93 \text{ mS}$$

and its phase angle is $-38.14°$. Admittance of a.c. circuits is valuable in parallel circuit calculations.

### 7.12 Series circuit containing R and C

A typical a.c. series $R-C$ circuit is shown in Fig. 7.8(a). Once again, being a series circuit the current, $I$, is common to all the elements in the circuit. Consequently in the phasor diagram (b), the current is used as the reference phasor.

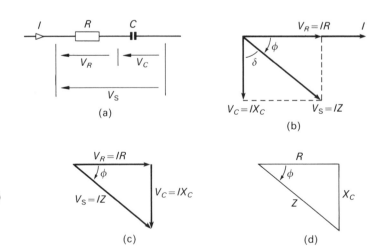

**Fig. 7.8** (a) Series $R-C$ circuit, (b) a typical phasor diagram, (c) the voltage triangle and (d) the corresponding impedance triangle

Since the voltage, $V_R$, across the resistor is in phase with the current, we also show it in the reference direction. Also, it was shown in section 7.6 that the voltage $V_C$ across the capacitor lags behind the current by 90°, and is shown pointing vertically downwards in the phasor diagram. From Pythagoras' theorem, the magnitude of $V_S$ is

$$V_S = \sqrt{[V_R^2 + V_C^2]} = \sqrt{[(IR)^2 + (IX_C)^2]}$$
$$= I\sqrt{[R^2 + X_C^2]} = IZ \qquad [7.16]$$

where

$$Z = \sqrt{[R^2 + X_C^2]} \qquad [7.17]$$

and is the *impedance* of the R–C series circuit.

It can be seen from the phasor diagram of the circuit, that *the current leads the supply voltage by angle φ*; the value of φ can be calculated from any of the following:

$$\left. \begin{array}{l} \cos\phi = V_R/V_S = IR/IZ = R/Z \\[2mm] \tan\phi = V_C/V_R = IX_C/IR = X_C/R = 1/\omega CR = 1/2\pi fCR \\[2mm] \sin\phi = V_C/V_S = IX_C/IZ = X_C/Z \end{array} \right\}$$
$$[7.18]$$

The *voltage triangle* for the series R–C circuit is shown in Fig. 7.8(c), and the corresponding *impedance triangle* in Fig. 7.8(d).

Unfortunately, no capacitor is ideal, and every practical capacitor has an electrical power loss when in an alternating current circuit. The main reasons for the power loss are *dielectric leakage* (which is due to the fact that the dielectric does not have infinite resistance), and *dielectric absorption* (which is a gradual change in the capacitor charge, even when a constant voltage is maintained between its terminals). For an individual capacitor, these effects can be regarded as being due to a low value of resistance in series with the capacitor, as shown in Fig. 7.8(a), and can be accounted for by the *loss angle, δ*, shown in Fig. 7.8(b); in a 'good' capacitor its value is normally less than about 2° or 0.04 rad [its value is exaggerated in Fig. 7.8(b)]. Its value can be calculated from

$$\delta = \tan^{-1}\frac{R}{1/\omega C} = \tan^{-1}\omega CR$$

Since δ is usually very small, then tan δ is very nearly equal to δ (in radians), so that

$$\delta\ (\text{radians}) \approx \omega CR$$

***Worked example 7.7***    A current of 5 A r.m.s. flows through a series R–C circuit which is energized from a 250 V, 50 Hz sinusoidal supply. If the circuit resistance is $20\,\Omega$, calculate (a) the circuit impedance, (b) the capacitive reactance, (c) the phase angle of the circuit, (d) the capacitance of the capacitor and (e) the voltage across each element in the circuit. Draw the phasor diagram of the circuit.

*Solution*   (a) By Ohm's law

$$Z = V_S/I = 250/5 = 50\,\Omega$$

(b) From eqn [7.17]

$$Z = \sqrt{[R^2 + X_C^2]}$$

or

$$X_C = \sqrt{[Z^2 - R^2]} = \sqrt{[50^2 - 20^2]} = \sqrt{2100} = 45.83\,\Omega$$

(c) From eqn [7.18]

$$\cos \phi = R/Z = 20/50 = 0.4$$

hence

$$\phi = \cos^{-1} 0.4 = 66.42° \text{ (or 1.16 rad)}$$

with $I$ leading $V_S$.

(d) Since $X_C = 1/2\pi f C$, then

$$C = 1/2\pi f X_C = 1/(2\pi \times 50 \times 45.83)$$

$$= 69.45 \times 10^{-6}\,F \quad \text{or} \quad 69.45\,\mu F$$

(e) The r.m.s. voltage across the resistor is

$$V_R = IR = 5 \times 20 = 100\,V$$

and that across the capacitor is

$$V_C = IX_C = 5 \times 45.83 = 229.15\,V$$

*Note*: $V_S = \sqrt{(V_R^2 + V_C^2)} = \sqrt{(100^2 + 229.15^2)} = 250\,V$
The phasor diagram for the circuit is shown in Fig. 7.9.

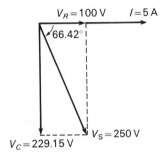

**Fig. 7.9** Phasor diagram for Worked Example 7.7

## 7.13 Series circuit containing R, L and C

Since the circuit contains both inductance and capacitance, any one of the three following conditions can prevail, namely:

1. $X_L > X_C$;
2. $X_L < X_C$;
3. $X_L = X_C$.

Case 3 gives rise to *series resonance*, and is dealt with comprehensively in Chapter 8. We will look at the other two cases here.

**Case 1:** $X_L > X_C$

Here, the voltage across the inductor ($V_L = IX_L$) is greater than that across the capacitor ($V_C = IX_C$), so that, overall, the circuit

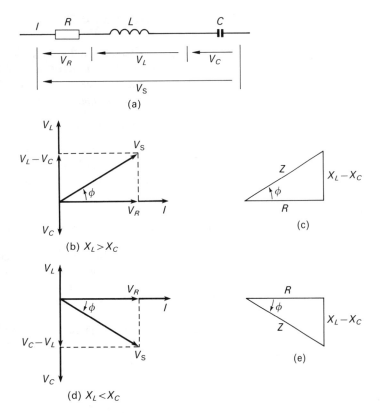

**Fig. 7.10** (a) Series $R$–$L$–$C$ circuit, (b) typical phasor diagram for $X_L > X_C$, and (c) the corresponding impedance triangle. (d) Phasor diagram for $X_L < X_C$, and (e) the corresponding impedance triangle

'looks' to the supply source as though it has a *net inductive reactance*, i.e. *the current lags behind the supply voltage* [see Fig. 7.10(b)]. The *net reactive voltage* is $(IX_L - IX_C) = I(X_L - X_C)$, which leads the current by $90°$; the reader will observe that, actually, we are *adding the phasors* $V_L$ and $V_C$ but, since $V_C$ is in phase opposition to $V_L$, we are effectively subtracting the magnitude of $V_C$ from $V_L$. From the phasor diagram in Fig. 7.10(b), the magnitude of the supply voltage is

$$V_S = \sqrt{[V_R^2 + (V_L - V_C)^2]} = \sqrt{[(IR)^2 + (IX_L - IX_C)^2]}$$
$$= I\sqrt{[R^2 + (X_L - X_C)^2]} = IZ \qquad [7.19]$$

where $Z$ is the impedance of the circuit, and is given by

$$Z = \sqrt{[R^2 + (X_L - X_C)^2]} = \sqrt{[R^2 + (\omega L - 1/\omega C)^2]} \quad [7.20]$$

The phase angle $\phi$ ($I$ lagging behind $V_S$) can be calculated from any of the following:

$$\left.\begin{aligned}
\cos\phi &= V_R/V_S = IR/IZ = R/Z \\
\tan\phi &= (V_L - V_C)/V_R = (X_L - X_C)/R \\
&= (\omega L - 1/\omega C)/R \\
\sin\phi &= (V_L - V_C)/V_S = (X_L - X_C)/Z
\end{aligned}\right\} \qquad [7.21]$$

**Case 2:** $X_L < X_C$

In this case, the voltage across the capacitor is greater than that across the inductor ($IX_L < IX_C$), and the circuit has an overall capacitive reactance, with the result that *the current leads the supply voltage*. In this case the net reactive voltage is

$$(IX_C - IX_L) = I(X_C - X_L)$$

which lags behind the current by 90° [see Fig. 7.10(d)]. Equations [7.19]–[7.21] satisfy the circuit, but the numerical values of the reactive components will differ from case 1 above.

*Worked example 7.8*   A series *R–L–C* circuit carries a current of 2 mA at a frequency of 150 kHz, the circuit containing a 5 mH inductor and a 0.5 nF capacitor. If the circuit resistance is 1 kΩ, calculate (a) the magnitude of the voltage applied to the circuit and its phase angle, and (b) the voltage across each element in the circuit and its phase angle with respect to the current. Draw the phasor diagram of the circuit.

*Solution*   The circuit diagram is shown in Fig. 7.11(a).
(a) The inductive reactance of the circuit is

$$X_L = 2\pi f L = 2\pi \times 150 \times 10^3 \times 5 \times 10^{-3} = 4712\,\Omega$$

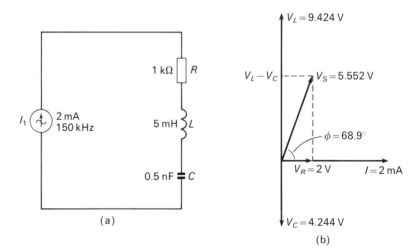

**Fig. 7.11** Figure for Worked Example 7.8

and the capacitive reactance is

$$X_C = 1/2\pi f C = 1/(2\pi \times 150 \times 10^3 \times 0.5 \times 10^{-9})$$

$$= 2122\,\Omega$$

hence the net circuit reactance is

$$X_L - X_C = 4712 - 2122 = 2590\,\Omega$$

i.e. a net inductive reactance of $2590\,\Omega$, which satisfies case 1 above; that is, the current lags behind the supply voltage. However, since the circuit is energized by a current, we should use the current as the reference phasor [see the phasor diagram in Fig. 7.11(b)].

The magnitude of the circuit impedance is

$$Z = \sqrt{(R^2 + (X_L - X_C)^2)} = \sqrt{(1000^2 + 2590^2)} = 2776\,\Omega$$

and the phase angle of the circuit is calculated from

$$\phi = \tan^{-1}((X_L - X_C)/R) = \tan^{-1}(2590/1000) = 68.9°$$

That is the current lags behind the supply voltage by $68.9°$ (or the supply voltage leads the current by $68.9°$). Also, the magnitude of the supply voltage is

$$V_S = IZ = 2 \times 10^{-3} \times 2776 = 5.552\,\text{V}$$

(b) The voltage across the resistor is

$$V_R = IR = 2 \times 10^{-3} \times 1000 = 2\,\text{V}$$

and is in phase with $I$. The voltage across the inductance is

$$V_L = IX_L = 2 \times 10^{-3} \times 4712 = 9.424\,\text{V}$$

leading $I$ by $90°$ (or leading $V_S$ by $90° - 68.9° = 21.1°$), and the voltage across the capacitor is

$$V_C = IX_C = 2 \times 10^{-3} \times 2122 = 4.244\,\text{V}$$

lagging behind $I$ by $90°$ (or lagging $V_S$ by $90° + 68.9° = 158.9°$).

## 7.14 Series connection of impedances

The total impedance of a number of elements connected in series is obtained by:

1. adding resistance values together;
2. adding inductive reactance values together;
3. adding capacitive reactance values together.

The net reactance of the circuit is

$$X = \text{total } X_L - \text{total } X_C$$

If $R$ is the total resistance of the circuit, the magnitude of the complete circuit impedance is

$$Z = \sqrt{(R^2 + X^2)}$$

and the phase angle of the circuit is

$$\phi = \tan^{-1}\frac{X}{R}$$

***Worked example 7.9***  Three impedances are connected in series, two of which are

$$R_1 = 10\,\Omega \qquad X_{L1} = 20\,\Omega$$

$$R_2 = 5\,\Omega \qquad L_2 = 31.83\,\text{mH}$$

If the supply frequency is 50 Hz, calculate (a) the resistance and reactance of the third impedance, given that the magnitude of the total impedance is 27.46 Ω, and its phase angle is 33.11° (lagging). State (b) the type of reactive element in the third impedance, and (c) its value.

*Solution*   The impedance triangle of the complete circuit is shown in Fig. 7.12, from which we can determine the total resistance of the complete circuit as follows:

$$R = Z \cos \phi = 27.46 \cos 33.11° = 23 \, \Omega$$

and the net reactance is

$$X = Z \sin \phi = 27.46 \sin 33.11° = 15 \, \Omega$$

(a) For the complete circuit

$$R = R_1 + R_2 + R_3$$

hence the resistance of the third impedance is

$$R_3 = R - (R_1 + R_2) = 23 - (10 + 5) = 8 \, \Omega$$

The reactance of the second impedance is

$$X_2 = 2\pi f L_2 = 2\pi \times 50 \times 31.83 \times 10^{-3} = 10 \, \Omega$$

and since

$$X = X_1 + X_2 + X_3$$

then the reactance of the third impedance is

$$X_3 = X - (X_1 + X_2) = 15 - (20 + 10) = -15 \, \Omega$$

(b) Since $X_3$ has a negative value of reactance, i.e. it has the opposite effect to an inductive reactance, then $X_3$ is a *capacitive reactance* of 15 Ω.
(c) Capacitive reactance is given by $X_C = 1/2\pi f C$, hence the capacitance $C_3$ in the third impedance is

$$C_3 = 1/2\pi f X_C = 1/(2\pi \times 50 \times 15) = 212.2 \, \mu F$$

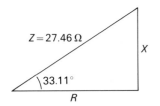

**Fig. 7.12** Impedance triangle for Worked Example 7.9

## 7.15 Introduction to complex values in a.c. circuits

This section is of particular relevance to the reader who hopes to proceed to more advanced studies in electrical and electronic engineering, who will need to have an understanding of 'complex' numbers and how they are used in circuit analysis.

The word 'complex' should not be confused with the word 'complicated', because engineers use 'complex' to describe a value that not only has magnitude, but also has a direction or an angle associated with it. Such a value may be a voltage, or a current, or an impedance, etc.

In electrical studies there are two principal forms of complex number, namely rectangular complex values and polar complex values.

A **rectangular complex value** tells us information about the magnitude and 'direction' of, say, an impedance, by stating the length of the 'horizontal' and the 'vertical' sides of the impedance

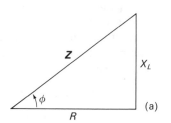
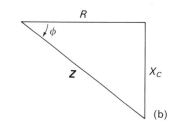

**Fig. 7.13** The impedance triangle for (a) an *R–L* circuit, (b) an *R–C* circuit

triangle (see Fig. 7.13). Much as we distinguish between horizontal values which extend to the right of the origin with a '+' sign, and those to the left with a '−' sign, we also distinguish between those which are 'upwards' and 'downwards' respectively, with '+' and '−' signs.

To indicate that a value is in the vertical plane, we multiply it by the **complex operator** 'j'. For example, in Fig. 7.13(a) we describe $X_L$ as the complex value $jX_L$, and in Fig. 7.13(b) we describe $X_C$ as $-jX_C$. That is, the **complex impedance, Z**, in Fig. 7.13(a) can be written in the form

$$\mathbf{Z} = R + jX_L \text{ ohms}$$

Once again, the reader is reminded that in this book we show complex numbers in **bold typeface**. The complex impedance in Fig. 7.13(b) can be written in the form

$$\mathbf{Z} = R - jX_C \text{ ohms}$$

Moreover, we describe any value in the horizontal or reference direction as a **real value**, and any value which has a 'j' associated with it as an **imaginary value**.

For example, in Worked Example 7.9 we can express the three impedance values as

$$\mathbf{Z}_1 = 10 + j20 \, \Omega$$

$$\mathbf{Z}_2 = \phantom{0}5 + j10 \, \Omega$$

$$\mathbf{Z}_3 = \phantom{0}8 - j15 \, \Omega$$

The process of adding complex impedance values together is illustrated here with the above values. The total impedance of the circuit is

$$\mathbf{Z} = \mathbf{Z}_1 + \mathbf{Z}_2 + \mathbf{Z}_3$$

$$= (10 + j20) + (5 + j10) + (8 - j15)$$

$$= (\text{sum of 'real' values}) + j(\text{sum of 'imaginary' values})$$

$$= (10 + 5 + 8) + j(20 + 10 - 15) = 23 + j15 \, \Omega$$

That is, we add the 'real' values together and the 'imaginary' values together.

A **polar complex value** provides generally similar information, but in a slightly different form. It gives information about the magnitude and the phase angle of the complex number. For the complex impedance in Fig. 7.13(a), we express the polar complex impedance in the form

$$\mathbf{Z} = \text{magnitude} \angle \text{phase angle} = Z \angle \phi$$

where '$\angle$' means the angle with respect to the reference direction.

The $\mathbf{Z}$ on the left-hand side of the equation is a complex value involving both magnitude and phase angle; the $Z$ on the right-hand side of the equation (shown as an *italic* character) is simply the magnitude or modulus of the impedance. The complex impedance in Fig. 7.13(b) is expressed in the form

$$\mathbf{Z} = Z \angle -\phi$$

For example, the total complex impedance of the complete circuit in Worked Example 7.9 can be expressed in the polar form

$$\mathbf{Z} = 27.46 \angle 33.11° \, \Omega$$

Using Pythagoras' theorem, we can relate polar and rectangular complex numbers as shown below. For the impedance in Fig. 7.13(a), the modulus of the impedance is

$$Z = \sqrt{(R^2 + X^2)}$$

and the phase angle can be calculated from

$$\phi = \tan^{-1}(R/X) = \cos^{-1}(R/Z) = \sin^{-1}(X/Z)$$

For example, the impedance $\mathbf{Z}_1$ in Worked Example 7.9 can be expressed in the polar form

$$\mathbf{Z}_1 = 10 + j20 = \sqrt{(10^2 + 20^2)} \angle \tan^{-1}(20/10)$$
$$= 22.36 \angle 63.44° \, \Omega$$

and

$$\mathbf{Z}_3 = 8 - j15 = \sqrt{(8^2 + 15^2)} \angle \tan^{-1}(-15/8)$$
$$= 17 \angle -61.95° \, \Omega$$

Converting a rectangular complex value to a polar complex value (or vice versa) is made very simple by the use of a calculator having the conversion facilities on it.

## 7.16 Parallel circuit containing R and L

A parallel circuit containing a resistor in one branch and a pure (of zero resistance) inductor in the other is shown in Fig. 7.14(a). *The applied voltage $V_S$ is common to both branches, and is used as the reference phasor.*

Since the current $I_R$ in the resistor is in phase with the voltage across $R$, then $I_R$ is also drawn in the reference direction [see

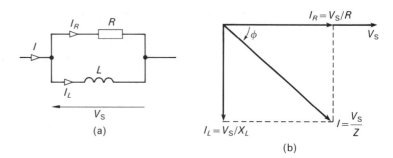

**Fig. 7.14** (a) Parallel *R–L* circuit and (b) a typical phasor diagram

Fig. 7.14(b)]. The magnitude of the current in the resistor is

$$I_R = V_S/R$$

The current $I_L$ in the inductor lags behind the voltage across $L$ by 90°, the $I_L$ phasor is drawn vertically downwards, and its magnitude is

$$I_L = V_S/X_L = V_S/\omega L = V_S/2\pi f L$$

Now, since we are dealing with phasor quantities, the current drawn from the supply is

$$I = phasor\ sum\ of\ I_R\ and\ I_L$$

and we can see from the phasor diagram in Fig. 7.14(b) that

$$\text{Magnitude of } I = \sqrt{(I_R^2 + I_L^2)}$$

and the magnitude of the circuit impedance is

$$Z = V_S/I$$

Since the circuit contains an inductance, the current drawn from the supply will lag behind the supply voltage. The way in which the lagging angle can be calculated is deduced from the phasor diagram, allowing us to use any of the following equations:

$$\cos \phi = \frac{I_R}{I} = \frac{V_S/R}{V_S/Z} = \frac{Z}{R}$$

$$\tan \phi = \frac{I_L}{I_R} = \frac{V_S/X_L}{V_S/R} = \frac{R}{X_L}$$

$$\sin \phi = \frac{I_L}{I} = \frac{V_S/X_L}{V_S/Z} = \frac{Z}{X_L}$$

*Worked example 7.10*  A 0.1 H inductor and a 40 Ω resistor are connected in parallel to a 200 V, 50 Hz supply. Calculate (a) the current in each branch of the circuit, (b) the total current and its phase angle, and (c) the magnitude of the impedance of the parallel circuit.

*Solution*　(a) The current in the resistive branch is

$$I_R = V_S/R = 200/40 = 5\text{ A in phase with } V_S$$

The reactance of the inductor is

$$X_L = 2\pi f L = 2\pi \times 50 \times 0.1 = 31.42\,\Omega$$

giving a current in the inductive branch of

$$I_L = V_S/X_L = 200/31.42 = 6.37\text{ A}$$

lagging behind $V_S$ by 90°.

(b) The magnitude of the total current is

$$I = \sqrt{(I_R^2 + I_L^2)} = \sqrt{(5^2 + 6.37^2)} = 8.1\text{ A}$$

which lags behind $V_S$ by [see also Fig. 7.14(b)]

$$\phi = \tan^{-1}(I_L/I_R) = \tan^{-1}(6.37/5) = \tan^{-1} 1.274$$
$$= 51.87°$$

(c) By Ohm's law, the magnitude of the circuit impedance is

$$Z = V_S/I = 200/8.1 = 24.69\,\Omega$$

## 7.17　Circuit containing *R* and *C* in parallel

The circuit diagram and a typical phasor diagram are shown in Fig. 7.15. Once again, the supply voltage is common to both branches, and is used as the reference phasor. Since the circuit contains a capacitor, the net reactance of the circuit is capacitive, and the current drawn from the supply will lead the voltage applied to the circuit.

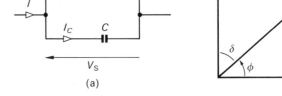

**Fig. 7.15**　(a) A parallel circuit containing *R* and *C* and (b) a typical phasor diagram

The magnitude of the current in the resistive branch is

$$I_R = \frac{V_S}{R}$$

and is in phase with $V_S$, and that in the capacitive branch is

$$I_C = \frac{V_S}{X_C} = \frac{V_S}{1/\omega C} = \omega C V_S = 2\pi f C V_S$$

which leads $V_S$ by 90°. From an inspection of the phasor diagram in Fig. 7.15(b) we see that

Total current, $I = $ *phasor sum* of $I_R$ and $I_C$

and its magnitude is

$$I = \sqrt{(I_R^2 + I_C^2)}$$

which leads $V_S$ by $\phi$. The value of $\phi$ can be calculated from any one of the following:

$$\cos \phi = I_R/I = Z/R$$

$$\tan \phi = I_C/I_R = R/X_C$$

$$\sin \phi = I_C/I = Z/X_C$$

and the circuit impedance is

$$Z = V_S/I$$

As discussed in section 7.12, capacitors are not ideal elements, and they exhibit a small power loss when operating in an a.c. circuit. One method of accounting for this power loss is to shunt the capacitor with a high value of resistance, as shown in Fig. 7.15(a). The value of the *loss angle*, $\delta$, can be calculated as follows:

$$\delta = \tan^{-1} \frac{I_R}{I_C} = \frac{V_S/R}{V_S \omega C} = \frac{1}{\omega CR}$$

The loss angle in a 'good' capacitor is very small (usually 2° or 0.04 rad), so that $\tan \delta$ is approximately equal to $\delta$ (in radians), hence

$$\delta \text{ (radians)} \approx 1/\omega CR$$

*Worked example 7.11*   A parallel $R$–$C$ circuit draws a current of 20 mA from a 20 V, 1 kHz supply. If the resistance of the resistive arm is 4 kΩ, calculate (a) the magnitude of the impedance of the complete circuit, (b) the current in the capacitor, (c) the reactance and capacitance of the capacitor and (d) the phase angle of the circuit.

*Solution*   (a) The magnitude of the impedance of the complete circuit is

$$Z = V_S/I = 20/20 \times 10^{-3} = 1000 \, \Omega$$

(b) The magnitude of the supply current is the phasor sum of $I_R$ and $I_C$. Since $I_R = V_S/R = 20/4000 = 5$ mA, the total current is

$$I = \text{phasor sum of 5 mA (in the reference direction) and } I_C$$

Hence the magnitude of $I_C$ is

$$I_C = \sqrt{[(20 \times 10^{-3})^2 - (5 \times 10^{-3})^2]} = 19.36 \text{ mA}$$

(c) The capacitive reactance can be calculated as follows:

$$X_C = V_S/I_C = 20/19.36 \times 10^{-3} = 1033 \, \Omega$$

and, since $X_C = 1/2\pi f C$, then

$$C = 1/2\pi f X_C = 1/(2\pi \times 1000 \times 1033)$$

$$= 0.154 \times 10^{-6}\,\text{F} \quad \text{or} \quad 0.154\,\mu\text{F}$$

(d) From the phasor diagram in Fig. 7.15(b), we see that

$$\phi = \tan^{-1}(I_C/I_R) = \tan^{-1}(19.36 \times 10^{-3}/5 \times 10^{-3})$$

$$= 75.5°$$

that is, $I$ leads $V_S$ by 75.5°.

## 7.18 Circuit containing *C* in parallel with *R–L*

This is a situation which occurs where a coil (which contains resistance and inductance) is in parallel with a capacitor – see Fig. 7.16 (see also section 7.21 on power factor correction).

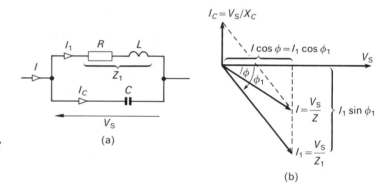

**Fig. 7.16** (a) A circuit containing *C* in parallel with a coil (*R–L*) and (b) a typical phasor diagram

The magnitude or modulus of the impedance of the inductive branch is $Z_1$, and its phase angle is $\phi_1$ [$= \tan^{-1}(X_L/R)$, with $I_1$ lagging behind $V_S$ by $\phi_1$]. A typical phasor diagram for the condition $I_C < I_1 \sin\phi_1$, i.e. when the capacitor current is less than the quadrature component of current in the inductive branch, is shown in Fig. 7.16(b).

Depending on the value of $I_C$, the circuit can assume any one of three operating conditions, namely

1. $I_C > I_1 \sin\phi_1$, when $I$ leads $V_S$;
2. $I_C = I_1 \sin\phi_1$, when $I$ is in phase with $V_S$;
3. $I_C < I_1 \sin\phi_1$, when $I$ lags $V_S$ [see Fig. 7.16(b)].

Condition 2 is known as *parallel resonance*, and is fully described in Chapter 8. The general equations of the circuit are

$$Z_1 = \sqrt{(R^2 + X_L^2)}$$

$$I_1 = V_S/Z_1$$

$$I_C = V_S/X_C = \omega C V_S = 2\pi f C V_S$$

Since the capacitor does not have a component of current which is in phase with $V_S$, the only in-phase component of current is $I_1 \cos \phi_1$. That is $I \cos \phi = I_1 \cos \phi_1$. The component of current which is perpendicular to $V_S$ is $(I_1 \sin \phi_1 - I_C)$, hence the magnitude of the current drawn from the supply is

$$I = \sqrt{[(I_1 \cos \phi_1)^2 + (I_1 \sin \phi_1 - I_C)^2]} \qquad [7.22]$$

and the overall circuit impedance is

$$Z = V_S/I$$

The phase angle $\phi_1$ of the inductive branch can be calculated from the values associated with that branch, and the following equation is typical:

$$\phi_1 = \tan^{-1}(X_L/R)$$

For the complete circuit, the following equations apply:

$$\phi = \cos^{-1}(I_1 \cos \phi_1/I)$$
$$= \tan^{-1}[(I_1 \sin \phi_1 - I_C)/(I_1 \cos \phi_1)] \qquad [7.23]$$

***Worked example 7.12***  A capacitor of capacitance 0.2 $\mu$F is connected in parallel with a coil of resistance 100 $\Omega$ and inductance 0.1 H, the combination being energized by a 10 V, 1 kHz supply. Calculate (a) the impedance and phase angle of the coil, (b) the reactance of the capacitor, (c) the current in each branch and its phase angle, and (d) the current drawn from the supply and its phase angle.

*Solution*  (a) The inductive reactance of the coil is

$$X_L = 2\pi f L = 2\pi \times 1000 \times 0.1 = 628.3 \, \Omega$$

Since $R = 100 \, \Omega$, the impedance of the coil is

$$Z_1 = \sqrt{(R^2 + X_L^2)} = \sqrt{(100^2 + 628.3^2)} = 636.2 \, \Omega$$

Since $R$ and $X_L$ are in series in the coil, the phase angle of the coil is

$$\phi_1 = \cos^{-1}(R/Z_1) = \cos^{-1}(100/636.2) = 80.96°$$

(b) The capacitive reactance is

$$X_C = 1/2\pi f C = 1/(2\pi \times 1000 \times 0.2 \times 10^{-6})$$
$$= 795.8 \, \Omega$$

(c) The magnitude of the current in the coil is

$$I_1 = V_S/Z_1 = 10/636.2 \, \text{A} = 15.72 \, \text{mA}$$

which lags behind $V_S$ by 80.96°, and

$$I_C = V_S/X_C = 10/795.8 \, \text{A} = 12.57 \, \text{mA}$$

which leads $V_S$ by 90°.
(d) The magnitude of the supply current is calculated using eqn [7.22]

as follows:

$$I = \sqrt{[(I_1 \cos \phi_1)^2 + (I_1 \sin \phi_1 - I_C)^2]}$$

$$= \sqrt{[(15.72 \times 10^{-3} \cos 80.96°)^2}$$

$$+ ((15.72 \times 10^{-3} \sin 80.96°) - 12.57 \times 10^{-3})^2] \, A}$$

$$= 3.85 \text{ mA}$$

It is interesting to note that the magnitude of the total current, $I$, is *less than* the magnitude of the current in either branch!

We calculate the overall phase angle of the circuitry from eqn [7.23] as follows [*Note*: since $(I_1 \sin \phi_1 - I_C)$ is positive, then the circuit is predominantly inductive and the current drawn from the supply lags behind the supply voltage; the phasor diagram is generally as shown in Fig. 7.16(b)]:

$$\phi = \cos^{-1}(I_1 \cos \phi_1 / I)$$

$$= \cos^{-1}(15.72 \times 10^{-3} \cos 80.86° / (3.85 \times 10^{-3}))$$

$$= 50.1° \text{ lagging}$$

## 7.19 Power in a.c. circuits

Initially we look at circuits with simple elements in them, namely either resistive or reactive elements. Later we will study the power consumed in circuits containing a combination of resistance and reactance.

### 7.19.1 Circuit containing a pure resistor

In a resistive circuit, the voltage and current waveforms are in phase with one another, as shown in Fig. 7.17. The *instantaneous power* waveform, $p = vi$, is the product of the instantaneous voltage and current waveforms, and is shown in Fig. 7.17. However, since the voltage and current waveforms are either both positive simultaneously, or are both negative simultaneously then, in a resistor, the instantaneous power *is always positive*. Power is the average value of the $v$–$i$ product wave, hence *power is always*

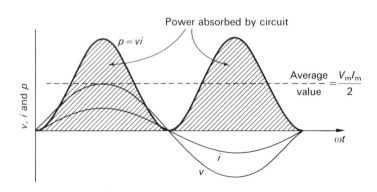

**Fig. 7.17** Waveforms of voltage, current and instantaneous power in a pure resistive circuit

*consumed when a resistor is connected to the supply*. Let us look at this mathematically; the expression for the instantaneous power is

$$p = vi = V_m \sin \omega t \times I_m \sin \omega t = V_m I_m \sin^2 \omega t$$
$$= \tfrac{1}{2} V_m I_m (1 - \cos 2\omega t) \qquad [7.24]$$

In eqn [7.24] we see that there is a term $\cos 2\omega t$ within the bracket. This, like any other sine wave, has *zero mathematical average value* when taken over one cycle. That is, the *average power, P,* consumed is

$$P = \tfrac{1}{2} V_m I_m$$

Also, since we are dealing with sine waves, we can rewrite this in the form

$$P = \tfrac{1}{2} V_m I_m = \frac{V_m}{\sqrt{2}} \frac{I_m}{\sqrt{2}} = VI = I^2 R = \frac{V^2}{R} \qquad [7.25]$$

where $V$ and $I$ are r.m.s. values. It is of interest to note that, when referring to the power consumed in a.c. circuits, we often use the expressions *active power* and *real power*.

### 7.19.2  Power absorbed in a pure inductor

When a circuit contains a pure (of zero resistance) inductor, the current lags 90° behind the voltage across the inductor. As before, the equation for the instantaneous power is $p = vi$ but, as we see from Fig. 7.18, the instantaneous power waveform is a sine wave of twice the supply frequency. Since this is the case, the instantaneous power waveform has *zero average value over any complete cycle*. That is, *the power absorbed by a pure inductance in an a.c. circuit is zero*.

Let us look at this mathematically. The mathematical expression for the current waveform is

$$i = I_m \sin \omega t$$

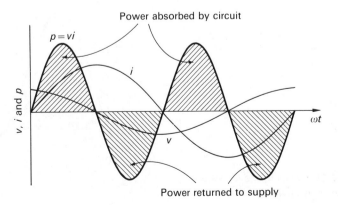

**Fig. 7.18** Waveforms of voltage current and power in a pure inductive circuit

and for the voltage waveform is

$$v = V_\mathrm{m} \sin(\omega t + 90°) = V_\mathrm{m} \cos \omega t$$

hence

$$p = vi = V_\mathrm{m} I_\mathrm{m} \sin \omega t \cos \omega t = \tfrac{1}{2} V_\mathrm{m} I_\mathrm{m} \sin 2\omega t$$

The mathematical expression shows that the instantaneous power waveform is a sinusoid of twice the supply frequency, which has zero mathematical average value.

The instantaneous power waveform in Fig. 7.18 highlights an interesting situation, in that power is *absorbed by the circuit* during part of the cycle (when the area under the graph is positive), and is *returned to the supply* during another part of the cycle (when the area under the graph is negative). The reason is that power is absorbed while the magnetic field is being built up, and is returned to the supply when the field collapses.

### 7.19.3 Power absorbed in an ideal capacitance

Once again, we have a situation where the current is 90° out of phase with the voltage (in this case, however, the current leads the voltage). The reader should draw the instantaneous $v$, $i$ and $p$ waves, and it will be seen that the instantaneous power wave is a sine wave of twice the supply frequency. That is, *no power is absorbed by an ideal capacitor in an a.c. circuit*.

### 7.19.4 Power consumed by a circuit containing resistance and reactance

Assume for the moment that the circuit contains a resistor and a pure inductor, so that the applied voltage leads the current in the circuit by angle $\phi$, as shown in Fig. 7.19. A plot of the $v$–$i$

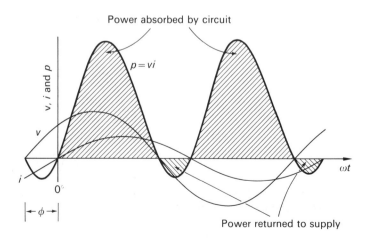

**Fig. 7.19** Waveforms of voltage current and power in an *R–L* circuit

product shows that the waveform is sinusoidal, and there are both positive and negative areas under the wave, the net positive area being greater than the negative area. That is, the average value of the power consumption is positive, i.e. *the circuit has a net power consumption.* as explained in section 7.19.2, the negative regions under the instantaneous power curve are due to the return of energy to the supply source. The mathematical expressions for the waves are

$$v = V_m \sin(\omega t + \phi) \qquad i = I_m \sin \omega t$$

and

$$p = vi = V_m I_m \sin(\omega t + \phi)\sin \omega t$$
$$= \tfrac{1}{2}V_m I_m[\cos \phi - \cos(2\omega t + \phi)]$$

where $\phi$ is the phase angle between the voltage and current waveforms. That is, $\phi$ is a fixed numerical value, so that $\cos \phi$ is a simple number and is independent of time. The term $\cos(2\omega t + \phi)$ is a function of time, consequently its average value taken over any complete cycle is zero. Hence, the average power consumed by the circuit is

$$P = \tfrac{1}{2}V_m I_m \cos \phi = VI \cos \phi \qquad\qquad [7.26]$$

where $V$ and $I$ are the respective r.m.s. values of voltage and current.

Using a similar argument for a series circuit containing a resistor and an ideal capacitor, it can be shown that the power consumed is, once more, given by eqn [7.26].

Since $\cos \phi$ is positive for both positive and negative values of $\phi$, it does not matter whether the current leads or lags the supply voltage. However, from an engineering point of view, it is useful to state whether the phase angle is a leading angle or a lagging angle.

## 7.20 Apparent power, reactive power and power factor

The voltage triangle for an *R–L* series circuit is redrawn in Fig. 7.20(a). If we multiply each side of the voltage triangle by the magnitude of the current, *I*, we are left with the **power triangle** of the circuit [see Fig. 7.20(b)], which is geometrically similar to the voltage triangle. It has already been shown that the average

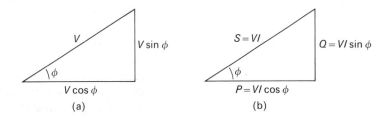

**Fig. 7.20** (a) Voltage triangle and (b) power triangle for an *R–L* series circuit

power consumed is

$$P = VI \cos \phi \tag{7.27}$$

If $V$ is in volts and $I$ in amperes, then the average power is in watts (W).

The product

$$S = VI \text{ volt-amperes} \tag{7.28}$$

is known as the **apparent power** in volt-amperes (VA). The **active power** or **real power**, $P$, is the 'in-phase' component of the apparent power. When we refer to 'power' we mean the active power, $P$, which is the ability of the circuit or element to produce heat or to do work.

The value

$$Q = VI \sin \phi \tag{7.29}$$

is known as the **reactive power** or **quadrature power**, and has units of volt-amperes reactive (VAr).

Alternatively, if we multiply each side of the impedance triangle for the series circuit [Fig. 7.21(a)] by $I^2$, we obtain the alternative power triangle in Fig. 7.21(b). Here we see that

$$S = I^2 Z \text{ volt-amperes}$$

where $Z$ is the magnitude of the impedance of the circuit. Also

$$P = I^2 R \text{ watts}$$

and

$$Q = I^2 X = \frac{V_X}{X} \frac{V_X}{X} X = \frac{V_X^2}{X} \text{ volt-amperes reactive}$$

where $V_X$ is the voltage across the circuit reactance.

The ratio of active power to apparent power is known as the **power factor** of the circuit or element. That is

$$\text{Power factor} = \frac{\text{active power}}{\text{apparent power}} = \frac{VI \cos \phi}{VI}$$

$$= \cos \phi \tag{7.30}$$

The above relationships have been derived from the phasor diagram for an $R$–$L$ circuit, but there is one point to which some caution must be applied, as follows. When the voltage triangle is drawn for an $R$–$C$ circuit, the applied voltage lags behind the current, and the net result is that *the corresponding reactive power is negative.*

It has been agreed internationally that reactive power in an inductor is assumed to be positive, and capacitive reactive power is negative. Alternatively, we can state reactive power as a positive value in both cases, but we must state whether it is inductive (lagging) or capacitive (leading).

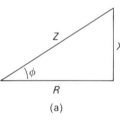

(a)

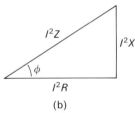

(b)

**Fig. 7.21**   Alternative power triangle for a series circuit

**7.21  Power factor correction**

If a single-phase alternator is rated to provide 100 A at voltage of 500 V, then the maximum *apparent power* it can supply is

$$S = VI = 500 \times 100 = 50\,000 \text{ VA}$$

If the current is in phase with the voltage ($\phi = 0°$), then the alternator can supply a power of

$$P = VI \cos \phi = 50\,000 \times \cos 0° = 50\,000 \text{ W}$$

On the other hand, if the load connected to the alternator has a power factor of 0.5 then the alternator can still supply 100 A, but the maximum power it can provide is

$$P = VI \cos \phi = 50\,000 \times 0.5 = 25\,000 \text{ W}$$

Worse still, if the load is a pure inductor (or capacitor), then $\phi = 90°$ ($\cos \phi = 0$) and the maximum power that can be supplied is

$$P = VI \cos \phi = 50\,000 \times 0 = 0 \text{ W}$$

Generally speaking, the customer pays for the amount of electricity he consumes, irrespective of the generating capacity he ties up. To ensure efficient use of generating plant, supply authorities impose a tariff on large consumers of electricity (usually industry) which penalizes them if the power factor of their installation has a low value.

Domestic equipment is generally associated with the production of heat, light and work, and the power factor of these items of plant have a high value (usually approaching unity). On the other hand, many items of industrial equipment (such as magnets and certain types of a.c. motor) have a low lagging power factor, and it is these which cause supply authorities some concern.

As mentioned above, it is industrial loads with a low lagging power factor which need attention. A number of methods of power factor improvement or correction are available, and are briefly outlined below.

A simple localized method of power factor correction is to connect a capacitor bank to the terminals of each item of load which has a poor power factor (see Worked Example 7.13). Alternatively, global power factor correction can be applied to some large industrial installations by the use of special three-phase machines such as *synchronous motors* and *synchronous induction motors*, which can be operated so that they draw a leading current from the supply, i.e. the supply 'sees' them as though they have a 'capacitor' action. The reader should refer to books on electrical machines for information on motors of this kind.

*Worked example 7.13*  A single-phase motor draws a current of 55 A at a power factor of 0.5 lagging from a 500 V, 50 Hz supply. (a) What value of capacitor must be connected to the terminals of the motor to raise the overall power factor to 0.9 lagging? (b) What value of current is taken by the motor and capacitor combination? (c) Calculate the kVAr rating of the capacitor bank.

*Solution*  The circuit arrangement is shown in Fig. 7.22(a). The motor draws a current $I_1$ from the supply, and $C$ is the power factor correction capacitor (which is connected as close as possible to the terminals of the motor). As will be seen from the numerical results later, the current $I_2$ drawn by the parallel combination of the capacitor and the motor is less than $I_1$, resulting (in addition to the power factor improvement) in reduced power loss in the supply cable. Referring to the phasor diagram, Fig. 7.22(b), we have

$$I_1 = \text{motor current}$$

$$\phi_1 = \text{motor phase angle}$$

$$I_2 = \text{new value of supply current}$$

$$\phi_2 = \text{new value of phase angle of the combined load}$$

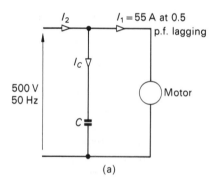

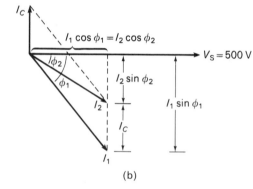

**Fig. 7.22**  (a) Circuit for Worked Example 7.13 and (b) its phasor diagram

The reader should note that it is assumed we are using an ideal capacitor, which does not consume power. That is, the connection of the capacitor does not increase the power consumption of the combination, and it does not increase the component of current which is in phase with the supply voltage. Hence

$$I_1 \cos \phi_1 = I_2 \cos \phi_2$$

(a) The in-phase component of the motor current is

$$I_1 \cos \phi_1 = 55 \times 0.5 = 27.5\,\text{A}$$

Now $\phi_1 = \cos^{-1} 0.5 = 60°$, hence

$$\tan \phi_1 = 1.732 = \frac{I_1 \sin \phi_1}{I_1 \cos \phi_1} = \frac{I_1 \sin \phi_1}{27.5}$$

or

$$I_1 \sin \phi_1 = 1.732 \times 27.5 = 47.63\,\text{A}$$

Now since $I_2 \cos \phi_2 = I_1 \cos \phi_1 = 27.5$ A, and

$$\tan \phi_2 = (I_2 \sin \phi_2)/(I_2 \cos \phi_2)$$

then

$$I_2 \sin \phi_2 = I_2 \cos \phi_2 \times \tan \phi_2$$

$$= 27.5 \times \tan(\cos^{-1} 0.9)$$

$$= 27.5 \tan 25.84° = 13.32 \text{ A}$$

From the phasor diagram in Fig. 7.22(b) we see that

$$I_C = I_1 \sin \phi_1 - I_2 \sin \phi_2 = 47.63 - 13.32 = 34.31 \text{ A}$$

But

$$X_C = V_S/I_C = 500/34.31 = 14.57 = 1/2\pi f C$$

or

$$C = 1/2\pi f X_C = 1/(2\pi \times 50 \times 14.57) \text{F} = 218.5 \ \mu\text{F}$$

(b) The magnitude of the current $I_2$ drawn by the parallel combination is

$$I_2 = \sqrt{[(I_2 \cos \phi_2)^2 + (I_2 \sin \phi_2)^2]}$$

$$= \sqrt{[27.5^2 + 13.32^2]} = 30.56 \text{ A}$$

which lags behind $V_S$ by 25.85° (corresponding to a lagging power factor of 0.9). The reader will note that, although the motor takes a current of 55 A, the parallel combination of the motor and the capacitor only draws 30.56 A. The addition of the capacitor therefore has two benefits, namely:

1. the overall power factor has improved from 0.5 lagging to 0.9 lagging;
2. the combination draws less current than before, with the consequence that it may be possible to use a smaller supply cable (with resulting improvement in economy).

(c) In industrial installations, power factor correction capacitors are large not only in terms of the capacitance, but also physically. Their volume may be $1 \text{ m}^3$ or larger, and are often housed in a steel tank, and immersed in oil not only because of the high voltage involved but also for cooling purposes. Their rating is expressed in VAr or, more likely, in kVAr. The kVAr rating of the capacitor in this problem is

$$\text{Rating} = V_S I_C/1000 = 500 \times 34.31/1000 = 17.16 \text{ kVAr}$$

## Problems

**7.1** A series $R–L$ circuit has a resistance of $10 \ \Omega$ and an impedance of $15 \ \Omega$. Calculate the inductive reactance and the phase angle of the circuit.

[$11.18 \ \Omega$; 48.19° (lagging)]

**7.2** A coil with a resistance of $5 \ \Omega$ and inductive reactance $10 \ \Omega$ is connected in series with another coil of resistance $8 \ \Omega$ and inductance 0.02 H. If the supply frequency is 50 Hz, calculate the total impedance of the circuit and its phase angle.

[$20.83 \ \Omega$; 51.39° (lagging)]

**7.3**   When a coil is energized by a 100 V, 100 Hz supply, the r.m.s. value of the current is 10 A. When supplied by 10 V d.c., the current is 2 A. Calculate the impedance of the coil, its resistance, its reactance, its inductance, and the phase angle of the coil at 100 Hz. The hysteresis loss may be ignored.

[10 Ω; 5 Ω; 8.66 Ω; 13.78 mH; 40.89° (lagging)]

**7.4**   An inductor of negligible resistance and a non-inductive resistor of 200 Ω resistance are connected in series to a 250 V, 50 Hz supply. A current of 1.12 A flows in the circuit. What is the inductance of the inductor and the phase angle between the voltage and the current?

[0.315 H; 26.36° (lagging)]

**7.5**   A coil of resistance 1 kΩ is connected in series with an ammeter of zero resistance to a 250 V a.c. supply. If the current is 0.1 A when the supply frequency is 50 Hz, calculate (a) the inductance of the coil, and (b) the current in the coil and its phase angle when the supply frequency is raised to 100 Hz.

[(a) 7.29 H; (b) 0.053 A (lagging by 77.68°)]

**7.6**   A pure resistor is connected in series with a capacitor of reactance 20 Ω. If the current in the circuit is 0.75 A when the applied voltage is 20 V r.m.s., calculate the resistance of the circuit.

[17.64 Ω]

**7.7**   A 50 Ω resistance is connected in series with two capacitors; $C_1$ has a capacitance of 0.4 μF and $C_2$ a capacitance of 0.6 μF. The circuit is supplied by a 20 V, 10 kHz supply. Determine (a) the net capacitive reactance of the circuit, (b) the modulus of the impedance of the circuit, and (c) the current in the circuit and its phase angle.

[(a) 66.32 Ω; (b) 83.06 Ω; (c) 0.24 A (leading by 53°)]

**7.8**   A series $R$–$C$ circuit having a resistance of 50 Ω and a capacitive reactance of 60 Ω, is connected to a 50 Hz sinusoidal a.c. source. If the current in the circuit is 3 A r.m.s., calculate (a) the impedance of the circuit, (b) the phase angle of the circuit, (c) the p.d. across each element in the circuit, (d) the supply voltage and (e) the capacitance of the capacitor.

[(a) 78.1 Ω; (b) 50.2° ($I$ leads $V_s$); (c) $V_R$ = 150 V, $V_C$ = 180 V; (d) 234.3 V (lagging behind $I$ by 50.2°); (e) 53.05 μF]

**7.9**   A series $R$–$L$–$C$ circuit contains a 10 Ω resistance, a capacitor of capacitance 25 μF, and an inductor of reactance 50 Ω. If the circuit is energized by a 150 V, 200 Hz supply, calculate (a) the capacitive reactance, (b) the inductance of the inductor, (c) the magnitude of the circuit impedance, (d) the current, (e) the phase angle of the circuit, (f) the voltage across each element in the circuit, (g) the power absorbed by each circuit element, and the total power; (h) the VA and VAr absorbed, and (i) the power factor of the circuit.

[(a) 31.83 Ω; (b) 39.8 mH; (c) 20.74 Ω; (d) 7.23 A; (e) 61.2° (lagging); (f) $V_R$ = 72.3 V, $V_C$ = 230.1 V, $V_L$ = 361.5 V (*Note*: both $V_C$ and $V_L$ are both in excess of the supply voltage); (g) Total power = power in resistor = 522.7 W, $P_L = P_C = 0$; (h) 1.085 kVA, 0.95 kVAr; (i) 0.482 (lagging)]

**7.10**   An iron-cored coil absorbs 6 kW when connected to a 415 V, 50 Hz sinusoidal supply, the current being 17.32 A r.m.s. When connected to a

12 V d.c. supply the current is 1 A. Calculate (a) the d.c. resistance of the coil, (b) its effective a.c. resistance, (c) its impedance, (d) the copper loss in the coil, (e) the iron loss, (f) the inductance of the coil and (g) its power factor. Explain the reasons for the difference between solutions (a) and (b).

[(a) $12\,\Omega$; (b) $20\,\Omega$; (c) $24\,\Omega$; (d) $3.6\,kW$; (e) $2.4\,kW$; (f) $42.2\,mH$; (g) $0.833$ lagging]

**7.11** A resistor of $50\,\Omega$ is connected in parallel with a $10\,\mu F$ capacitor, the combination being supplied by a $50\,V$, $318.3\,Hz$ a.c. supply. Calculate (a) the current in each branch, (b) the total current, (c) the impedance of the circuit, (d) the power factor of the circuit and (e) the power consumed by the circuit.

[(a) $I_R = 1\,A$, $I_C = 1\,A$; (b) $1.414\,A$; (c) $35.36\,\Omega$; (d) $0.707$ ($I$ leading $V_S$); (e) $50\,W$]

**7.12** Two series circuits, each having the same resistance, are connected in parallel with one another. The power factor of one circuit is 0.85 lagging, and that of the other is 0.6 leading. What is the power factor of the combination?

[Unity]

**7.13** A coil of resistance $20\,\Omega$ and inductance $50\,mH$ is connected in parallel with a non-inductive resistor of $25\,\Omega$. If a supply of $250\,V$, $50\,Hz$ is connected to the circuit, calculate (a) the current in each branch, (b) the total current drawn from the supply, (c) the phase angle of the circuit and (d) the apparent power, active power and VAr consumed.

[(a) $I_R = 10\,A$, $I_{COIL} = 9.83\,A$ (lagging $V_R$ by $38.13°$); (b) $18.74\,A$; (c) $18.9°$ lagging; (d) $4685\,VA$, $4432\,W$, $1517\,VAr$]

**7.14** A single-phase $240\,V$ induction motor has a $150\,\Omega$ resistor connected in parallel with it. The motor draws a current of $2.5\,A$, and the total current taken from the supply is $3.8\,A$. Determine the power consumed and the power factor of (a) the complete circuit and (b) the motor.

[(a) $807.5\,W$, $0.885$ lagging; (b) $424.3\,W$, $0.707$ lagging]

**7.15** A $450\,W$ gas discharge lamp draws a current of $3.5\,A$ at unity power factor. Determine the inductance of the choke (an inductor), connected in series with the lamp, which enables the lamp to operate from a $250\,V$, $50\,Hz$ supply. What value of capacitance must be connected across the lamp and choke combination in order to make the overall power factor unity?

[$0.195\,H$; $38.2\,\mu F$]

# 8 Resonance and filter circuits

## 8.1 Introduction

Resonance is a feature which occurs in many types of system, and results in a magnification of a voltage or a current because of the effects of energy storage elements within the circuit. At the resonant frequency, the forcing function (which may be a voltage or a current) is magnified in value by the circuit and, in this chapter, we will investigate the response of circuits.

A filter circuit is one which selectively transmits (or rejects) frequencies within certain bands. The bands which it transmits are known as *pass bands*, and those it rejects are known as *attenuation bands* or *stop bands*. We will look at a range of simple filters later in the chapter.

## 8.2 Series *R–L–C* circuits

The reader will recall from Chapter 7 that the reactance of inductors and capacitors alters in value with frequency. If the *R–L–C* series circuit in Fig. 8.1(a) is excited by a constant-voltage, variable-frequency a.c. source, we discover a marked variation in circuit performance as the frequency is changed.

At zero frequency (d.c.), the inductor has zero reactance and is, effectively, a short circuit; the capacitive reactance is infinite at this 'frequency' and represents an open circuit to the supply. Consequently, at zero frequency, the net impedance to current flow is infinite, and the current is zero.

As the supply frequency is increased, the inductive reactance increases and the capacitive reactance reduces, so that the net reactance of the circuit has a finite value [see Fig. 8.1(b)], and current begins to flow in the circuit.

At a frequency known as the *resonant frequency*, $\omega_0$ rad/s (or $f_0$ hertz), the value of $X_L$ and $X_C$ are equal in value [see Fig. 8.1(b)] and, since both carry the same current, the voltage across both the inductance and the capacitance have the same value. That is

$$V_L = V_C$$

or

$$I\omega_0 L = I/\omega_0 C$$

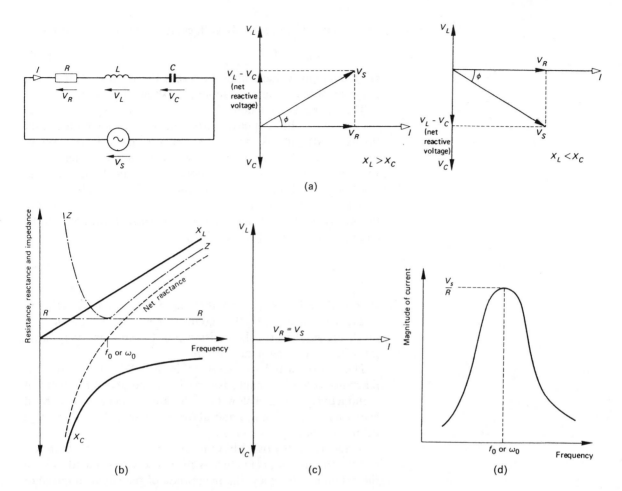

**Fig. 8.1** (a) $R$–$L$–$C$ series circuit, (b) variation of reactance and impedance with frequency, (c) phasor diagram for the circuit at resonance and (d) variation of current with frequency

hence

$$\omega_0 = \frac{1}{\sqrt{(LC)}} \text{ rad/s} \qquad [8.1]$$

that is

$$f_0 = \frac{1}{2\pi\sqrt{(LC)}} \text{ hertz} \qquad [8.2]$$

Since we are dealing with a series circuit, the magnitude of the impedance at resonance is

$$Z_0 = \sqrt{(R^2 + (X_L - X_C)^2)} = \sqrt{R^2} = R \qquad [8.3]$$

That is, *the impedance of the series circuit at resonance is purely resistive*, and the current is in phase with the voltage applied to the circuit. Also, the supply voltage at resonance is given by

$$V_S = IZ_0 = IR \qquad [8.4]$$

i.e. the supply voltage appears across the resistor. The phasor diagram of the circuit at resonance is shown in Fig. 8.1(c).

The reader should note that, since $X_L$ and $X_C$ are finite at resonance, a voltage appears across them, as shown in the phasor diagram. Also, it was shown in Chapter 7 that while the voltage across $L$ leads the current by $90°$, the voltage across $C$ lags behind the current by $90°$. Hence, at resonance, $V_L$ and $V_C$ are in phase opposition, and the net reactive voltage $I(X_L - X_C)$ is zero!

Also, at resonance, the circuit has its lowest value of impedance (which is equal to $R$), and *the maximum current*, $I_0$, flows in the circuit [see Fig. 8.1(d)]. Its value is

$$I_0 = \frac{V_S}{R} \qquad [8.5]$$

It should be borne in mind that the *a.c. resistance* or *effective resistance* of a coil (which accounts for hysteresis loss) may be much higher than the d.c. resistance of the coil, and may affect the value of the resonance current.

The reader will have observed in Fig. 8.1(b) that the net reactance below resonance is capacitive, and above resonance it is inductive. That is, below the resonant frequency, the current leads the supply voltage, and above resonance the current lags behind the voltage (see below).

As the frequency rises above resonance, $X_L$ is greater than $X_C$, and the circuit has a net inductive reactance. Consequently, above the resonant frequency, the magnitude of the circuit impedance rises in value and the current falls in value. At infinite frequency, the current falls to zero.

The series circuit is sometimes described as an *acceptor circuit*, since it accepts a high value of current at the resonant frequency.

In *power systems*, the circuit resistance is inherently low, and series resonance can give rise to a dangerously high current. Since the current flows in both $L$ and $C$, it can also result in an excessive voltage across these elements. In *electronic circuits*, the circuit resistance is usually very high, and the current at resonance is fairly small. Series resonance is sometimes employed in radio and telecommunications equipment to 'magnify' the voltage across $L$ and $C$ when compared with the applied voltage (see also section 8.3).

***Worked example 8.1*** A series *RLC* circuit contains an inductance of 10 mH and a capacitance of 1 $\mu$F. Determine the resonant frequency of the circuit. If the circuit resistance is 10 $\Omega$ and the supply voltage is 10 V, calculate at the resonant frequency (a) the current in the circuit, (b) the voltage across $L$ and (c) the voltage across $C$.

*Solution*     From eqn [8.1]

$$\omega_0 = 1/\sqrt{(LC)} = 1/\sqrt{(10 \times 10^{-3} \times 1 \times 10^{-6})}$$
$$= 10\,000\,\text{rad/s} \quad \text{or} \quad 1.592\,\text{kHz}$$

(a) At resonance

$$I_0 = V_S/R = 10/10 = 1\,\text{A}$$

(b) The reactance of the inductance at resonance is

$$X_L = \omega_0 L = 10\,000 \times 10 \times 10^{-3} = 100\,\Omega$$

and

$$V_L = I_0 X_L = 1 \times 100 = 100\,\text{V}$$

(c) The reactance of the capacitance at resonance is

$$X_C = 1/(\omega_0 C) = 1/(10\,000 \times 1 \times 10^{-6}) = 100\,\Omega$$

and

$$V_C = I_0 X_C = 1 \times 100 = 100\,\text{V}$$

The fact that $X_L = X_C$ confirms that resonance has occurred. The reader should note two points. The first is that the voltage across $L$ and across $C$ is the same and second, that the voltage across $L$ and $C$ is *ten times greater than the supply voltage*. The latter is discussed fully in section 8.3.

**8.3  *Q*-factor of a series resonant circuit**

The *quality factor* or *Q-factor* of a series circuit is a figure of merit used to describe the performance of the circuit at resonance, and gives the amount by which the circuit *magnifies the supply voltage*, the magnified voltage appearing across both $L$ and $C$. The reader should note that we refer here to a pure $L$ and *not to a coil* (which has some resistance). The $Q$-factor of the circuit is given either by

$$Q\text{-factor} = \frac{\text{voltage across } L \text{ at resonance}}{\text{supply voltage}} = \frac{I\omega_0 L}{IR}$$

$$= \frac{\omega_0 L}{R} \qquad\qquad\qquad [8.6]$$

or

$$Q\text{-factor} = \frac{\text{voltage across } C \text{ at resonance}}{\text{supply voltage}} = \frac{I/\omega_0 C}{IR}$$

$$= \frac{1}{\omega_0 CR} \qquad\qquad\qquad [8.7]$$

If we insert $\omega_0 = 1/\sqrt{(LC)}$ into eqn [8.6] we get

$$Q\text{-factor} = \frac{1}{\sqrt{(LC)}} \times \frac{L}{R} = \frac{1}{R}\sqrt{\left[\frac{L}{C}\right]} \qquad [8.8]$$

If we draw a graph over a range of frequencies showing the

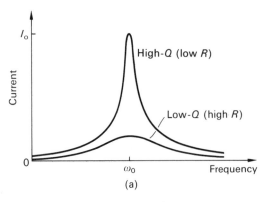

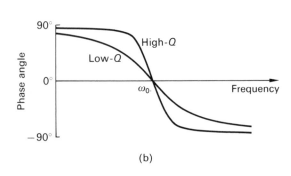

**Fig. 8.2** Frequency response of a series $R-L-C$ circuit showing the variation with frequency of (a) current magnitude and (b) phase angle

variation in the magnitude of current to a base of frequency we get, for $Q$-factors of 5 and 20, the curves in Fig. 8.2(a). It is seen that the circuit is more responsive or *frequency selective* when the $Q$-factor is high; power circuits rarely have a $Q$-factor greater than about 5.

The variation in phase shift with frequency is shown in Fig. 8.2(b); again, a high-$Q$ circuit is more frequency selective, and gives a much more rapid change in phase shift in the region of the resonant frequency.

The two graphs show the change in magnitude and change in phase shift to a base of frequency, and collectively give the *frequency response* of the circuit.

On occasions, engineers refer to the '$Q$-factor of a coil'; this should not be confused with the $Q$-factor of a resonant circuit. The $Q$-factor of a coil can be given *at any frequency*, and is given by

$$Q\text{-factor of coil} = \frac{\omega L}{R}$$

where $\omega$ is the angular frequency at which the $Q$-factor is calculated.

***Worked example 8.2*** A series $R-L-C$ circuit contains a pure inductor of inductance 0.253 mH, a 1 $\mu$F capacitor and a 0.76 $\Omega$ resistor. Calculate the resonant frequency of the circuit and its $Q$-factor. Determine also the current in the circuit if 1 V r.m.s. at the resonant frequency is applied to the circuit.

Determine also the magnitude of the current and its phase angle if 1 V at 8.4 kHz is applied to the circuit.

*Solution*  From eqn [8.1]

$$\omega_0 = 1/\sqrt{(LC)} = 1/\sqrt{(0.253 \times 10^{-3} \times 1 \times 10^{-6})}$$

$$= 62.87 \times 10^3 \text{ rad/s} \quad \text{or} \quad 10 \text{ kHz}$$

and from eqn [8.8]

$$Q\text{-factor} = \frac{1}{R}\sqrt{\frac{L}{C}} = \frac{1}{0.76}\sqrt{\left[\frac{0.253 \times 10^{-3}}{1 \times 10^{-6}}\right]} = 20.9$$

That is to say, at resonance, the voltage across $L$ and across $C$ is 20.9 times greater than the supply voltage.

The current which flows in the circuit at the resonant frequency is

$$I_0 = V_S/R = 1/0.76 = 1.316\,\text{A}$$

The frequency of 8.4 kHz is below the resonant frequency, and we would expect to find at this frequency (see also Fig. 8.2) that (i) the current in the circuit is less than that at resonance and (ii) the circuit has a leading phase angle. When $f = 8.4\,\text{kHz}$

$$X_L = 2\pi f L = 2\pi \times 8.4 \times 10^{-3} \times 0.263 \times 10^{-3}$$

$$= 13.35\,\Omega$$

$$X_C = 1/2\pi f C = 1/(2\pi \times 8.4 \times 10^{-3} \times 1 \times 10^{-6})$$

$$= 18.9\,\Omega$$

The magnitude of the circuit impedance at this frequency is

$$Z = \sqrt{(R^2 + (X_L - X_C)^2)}$$

$$= \sqrt{(0.76^2 + (13.35 - 18.9)^2} = 5.6\,\Omega$$

and the current in the circuit is

$$I = V_S/Z = 1/5.6 = 0.179\,\text{A}$$

The phase angle of the circuit at 8.4 kHz is

$$\phi = \cos^{-1}\frac{R}{Z} = \cos^{-1}\frac{0.76}{5.6} = 82°$$

Since $X_C > X_L$, the current leads the supply voltage.

## 8.4 Resonance in an ideal parallel circuit

Consider the parallel circuit in Fig. 8.3(a), which contains ideal $L$ and $C$ elements in parallel with one another. As the supply frequency increases the reactance of the inductor increases, and the current through it reduces in value. At the same time, an increase in frequency causes the capacitive reactance to fall in value, and the current through it increases.

At the resonant frequency of the circuit, $\omega_0$ rad/s or $f_0$ hertz, the reactance of each of the elements has the same value and, therefore, the current in each branch has the same numerical value. However, since the current in the capacitor leads the supply voltage by 90°, and the current through the inductor lags behind it by 90° then, at the resonant frequency, the two currents cancel out. The phasor diagram for the ideal parallel circuit at resonance is shown in Fig. 8.3(c). By KCL, the total current, $I$, drawn from the supply is

$$I = \textit{phasor sum of } I_L \text{ and } I_C$$

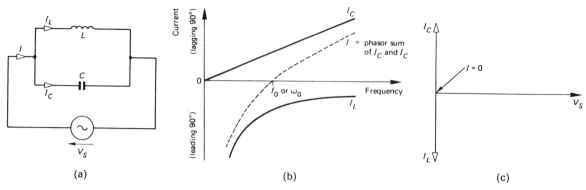

**Fig. 8.3** (a) Parallel *L–C* circuit containing ideal elements, (b) variation of current with frequency and (c) the phasor diagram at resonance

Since $I_L$ and $I_C$ both have the same value at resonance and are in phase opposition to one another, *the total current drawn from the supply at resonance is zero*. That is to say, the magnitude of the impedance of an ideal parallel *L–C* circuit at resonance is infinity! The impedance at resonance is known as the *dynamic impedance*, $Z_D$, or since the phase angle of the current is zero, it is sometimes called the *dynamic resistance*, $R_D$. Hence, for the circuit in Fig. 8.3(a)

$$Z_D = \infty$$

Let us briefly look at the reason why the current drawn from the supply at resonance is zero. Since both elements in the circuit have zero resistance, the average energy consumed by both of them is zero and, at resonance, the source does not supply any current. In fact, the energy stored by one element (say *L*) is equal to the energy returned to the circuit by the other (i.e. *C*). At a frequency other than resonance, the reactance values of *L* and *C* are unequal, so that there is an imbalance in the current in the two branches, and the circuit draws a current from the supply.

As mentioned above, resonance occurs in an ideal parallel *L–C* circuit when $X_L = X_C$, or when

$$\omega_0 L = \frac{1}{\omega_0 C}$$

or

$$\omega_0 = \frac{1}{\sqrt{(LC)}} \text{ rad/s} \qquad [8.9]$$

that is

$$f_0 = \frac{1}{2\pi\sqrt{(LC)}} \text{ rad/s} \qquad [8.10]$$

***Worked example 8.3*** Calculate the resonant frequency of a parallel circuit containing an ideal inductor of 0.2 mH in one branch and a pure capacitor of 1 $\mu$F in the other branch. If the supply voltage at the resonant frequency is 10 V, calculate the current in each branch of the circuit.

*Solution* From eqn [8.9]

$$\omega_0 = 1/\sqrt{(LC)} = 1/\sqrt{(0.2 \times 10^{-3} \times 1 \times 10^{-6})}$$

$$= 70\,711 \text{ rad/s}$$

or

$$f_0 = \omega_0/2\pi = 11\,254 \text{ Hz}$$

At the resonant frequency

$$X_L = \omega_0 L = 70\,711 \times 0.2 \times 10^{-3} = 14.14\,\Omega$$

and

$$X_C = 1/\omega_0 C = 1/(70\,711 \times 1 \times 10^{-6}) = 14.14\,\Omega$$

and the magnitude of the current in each branch is

$$I = V_S/\text{reactance} = 10/14.14 = 0.707 \text{ A}$$

Since the currents in the two branches are in phase opposition with one another [see Fig. 8.3(c)], the current drawn from the supply is zero.

## 8.5 Parallel resonance of a practical circuit

In practice, an inductor or coil has some resistance, $R$, so that the circuit diagram of a practical parallel circuit is as shown in Fig. 8.4(a) (a practical capacitor has very little effective resistance in series with it).

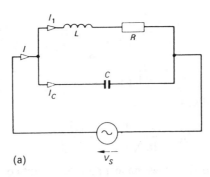

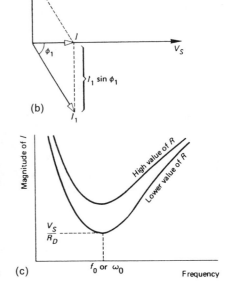

**Fig. 8.4** (a) Practical parallel circuit, (b) its phasor diagram at resonance and (c) variation of supply current with frequency

Since the circuit has some resistance, it will draw some current from the supply at resonance (this current supplies the energy loss in $R$ at resonance). We say that resonance occurs in this circuit when the current, $I$, drawn from the supply is in phase with $V_S$. This occurs when the quadrature component of the current in the coil ($I_1 \sin \phi_1$) is equal to the capacitor current. That is when

$$I_C = I_1 \sin \phi_1 \qquad [8.11]$$

where $\phi_1$ is the phase angle of the coil. Now

$$\sin \phi_1 = \frac{X_L}{Z_{coil}} = \frac{\omega_0 L}{\sqrt{(R^2 + (\omega_0 L)^2)}}$$

Hence, from eqn [8.11]

$$V_S \omega_0 C = \frac{V_S}{\sqrt{(R^2 + (\omega_0 L)^2)}} \times \frac{\omega_0 L}{\sqrt{(R^2 + (\omega_0 L)^2)}}$$

That is

$$C = \frac{L}{R^2 + (\omega_0 L)^2}$$

An equation for $\omega_0$ is obtained in the following steps:

$$R^2 + (\omega_0 L)^2 = \frac{L}{C}$$

$$(\omega_0 L)^2 = \frac{L}{C} - R^2$$

$$\omega_0^2 = \frac{1}{LC} - \frac{R^2}{L^2}$$

$$\omega_0 = \sqrt{\left[ \frac{1}{LC} - \frac{R^2}{L^2} \right]} \qquad [8.12]$$

that is

$$f_0 = \frac{1}{2\pi} \sqrt{\left[ \frac{1}{LC} - \frac{R^2}{L^2} \right]} \qquad [8.13]$$

If $R = 0$, we have a parallel circuit containing ideal elements [see Fig. 8.3(a)], whose resonant frequency is $\omega_0 = 1/2\pi\sqrt{(LC)}$. A small value of $R$ results in a resonant frequency which is slightly lower than that of the ideal circuit.

As the resistance of the coil increases in value, the phase angle $\phi_1$ (see Fig. 8.4) becomes less than 90°, with the net result that current is drawn from the supply at the resonant frequency (see Worked Example 8.4). The larger the value of $R$, *the greater the value of the current drawn from the supply at resonance!* This is illustrated in the graph in Fig. 8.5(a).

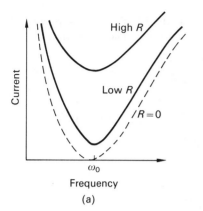

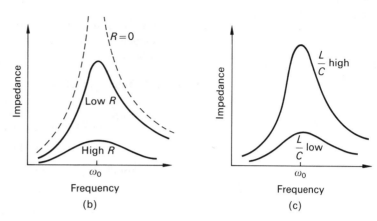

**Fig. 8.5** (a) The effect of variation of $R$ on current with $L/C$ constant, (b) effect of the variation of $R$ on $Z$ with $L/C$ constant and (c) the effect of variation of $L/C$ on $Z$ with $R$ constant

It follows that as the value of $R$ increases, the dynamic impedance of the circuit at resonance reduces [see Fig. 8.5(b)]. We can determine the dynamic impedance of the circuit as follows. From the phasor diagram in Fig. 8.4(b).

$$\tan \phi_1 = \frac{\text{quadrature current}}{\text{supply current}} = \frac{I_C}{I}$$

or

$$I = I_C / \tan \phi_1$$

The dynamic impedance of the circuit at resonance is

$$Z_D = \frac{V_S}{I} = \frac{V_S}{I_C} \tan \phi_1 = X_C \frac{X_L}{R} = \frac{1}{2\pi f_0 C} \frac{2\pi f_0 L}{R}$$

$$= \frac{L}{CR} \qquad\qquad [8.14]$$

Equation [8.14] tells us two things, namely:

1. With a fixed value of $L/C$, the lower the value of $R$, the greater the dynamic impedance [see Fig. 8.5(b)].
2. With a fixed value of $R$, the higher the value of $L/C$, the greater the dynamic impedance [see Fig. 8.5(c)].

Clearly, both the value of $R$ and the ratio $L/C$ affect the shape of the impedance/frequency graph (and also the current/frequency graph).

***Worked example 8.4*** A two-branch parallel circuit has a coil of effective resistance $20\,\Omega$ and inductance $0.2\,H$ in one branch, and a $0.1\,\mu F$ capacitor in the other branch. Calculate (a) the resonant frequency of the circuit, (b) the dynamic impedance of the circuit at resonance and (c) the current drawn from the supply if it is excited by a 10 V a.c. sinusoidal supply at the resonant frequency. Calculate also (d) the current flowing in the capacitor at resonance.

*Solution*   (a) From eqn [8.12]

$$\omega_0 = \sqrt{\left[\frac{1}{LC} - \frac{R^2}{L^2}\right]} = \sqrt{\left[\frac{1}{0.2 \times 0.1 \times 10^{-6}} - \frac{20^2}{0.2^2}\right]}$$

$$\approx 7070 \text{ rad/s (or 1125 Hz)}$$

(b) Also, from eqn [8.14]

$$Z_D = \frac{L}{CR} = \frac{0.2}{0.1 \times 10^{-6} \times 20} \Omega = 100 \text{ k}\Omega$$

(c) From Ohm's law

$$I = V_S/Z_D = 10/100 \times 10^3 \text{ A} = 0.1 \text{ mA}$$

(d) $X_C = 1/\omega_0 C = 1/(7071 \times 0.1 \times 10^{-6}) = 1414 \Omega$

hence

$$I_C = V_S/X_C = 10/1414 \text{ A} = 7.071 \text{ mA}$$

That is, the capacitor current at resonance is 70.71 times greater than the supply current! This is explained in section 8.6.

## 8.6   Q-factor of a parallel resonant circuit

The *Q-factor of a parallel circuit at resonance* is given by the ratio of the current in $C$ to the current drawn by the circuit, i.e. it is the *current magnification at resonance*. It is therefore the ratio of the quadrature current [$I_C$ or ($I_{coil} \sin \phi_1$)] to the current drawn from the supply, as follows:

$$Q = \frac{I_C}{I} = \frac{I_{coil} \sin \phi_1}{I} = \frac{\sin \phi_1}{I/I_{coil}} = \frac{\sin \phi_1}{\cos \phi_1}$$

$$= \tan \phi_1 = \frac{\omega_0 L}{R} \qquad \qquad [8.15]$$

*Worked example 8.5*   Determine the Q-factor of the parallel circuit in Worked Example 8.4(a) using the results from Worked Example 8.4 and (b) from eqn [8.15].

*Solution*   (a) From the results in Worked Example 8.4

$$I_C = 7.071 \text{ mA} \qquad I = 0.1 \text{ mA}$$

hence

$$Q = I_C/I = 7.071/0.1 = 70.71$$

(b) From eqn [8.15]

$$Q = \frac{\omega_0 L}{R} = \frac{7071 \times 0.2}{20} = 70.71$$

**8.7 Resonance with a complex wave**

As mentioned in Chapter 6, a complex wave can be thought of as being the sum of many sine waves of different (integral multiple) frequencies. Consequently, if a complex wave contains frequencies of, say, 50 Hz, 1 kHz and 5 kHz, then a circuit whose natural resonant frequency is 1 kHz, will selectively resonate with the 1 kHz component of the complex wave. The circuit will present its 50 Hz impedance to the 50 Hz frequency, and its 5 kHz impedance to the 5 kHz frequency. Depending on whether it is a series or parallel circuit, the net result will give exaggerated effects to the 1 kHz component of the complex wave.

This may cause particular difficulties when trying to measure the voltage, current and power in the circuit. It is possible for analysis of the circuit to be carried out, but it is beyond the scope of this book.

**8.8 Filter circuits**

A filter circuit is an electrical network which will transmit signals within certain designated ranges (*pass bands*), and suppress signals in other frequency ranges (*attenuation bands* or *stop bands*). A frequency which separates a pass band and a stop band is known as a *cut-off frequency*, and there may be several of these. The four main classifications of filter circuits are:

(a) low-pass;
(b) high-pass;
(c) band-pass;
(d) band-stop.

Figure 8.6 shows typical ideal (broken line) and practical (full line) frequency response characteristics for all four types.

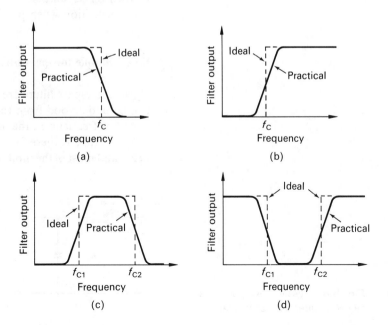

**Fig. 8.6** Filter characteristics: (a) low-pass, (b) high-pass, (c) band-pass and (d) band-stop

A *low-pass filter* is one which passes low frequencies to the output terminals but, beyond the cut-off frequency, $f_C$, the output rapidly reduces. A *high-pass filter* is one which will only pass signals at a frequency above the cut-off frequency, and rejects those at lower frequencies. Where we need to select a band of frequencies from within a complex signal, we can use a *band-pass filter*, which passes signals within the pass band from $f_{C1}$ to $f_{C2}$, and prevents other frequencies from reaching the output; radio, television and telecommunications equipment contain filters with this kind of characteristic. A *band-stop filter* is one which severely attenuates all signals in the range $f_{C1}-f_{C2}$, and passes other frequencies, ideally with no attenuation. The *bandwidth* of both the band-pass and band-stop filters is $f_{C2}-f_{C1}$ (in the former it is known as the *pass band*, and as the *stop band* in the latter).

Filter circuits can be constructed either from *passive circuit elements* (e.g. *R*, *L* and *C*), or from *active circuits* (operational amplifiers together with *R* and *C*). The latter are described in books on electronic circuits.

Figure 8.7 shows the general circuit for many types of passive filter circuit. One of the two elements, $Z_1$ and $Z_2$, is frequency selective, the other being a resistor. The impedance of the frequency-selective element varies in such a way that it gives the required filter characteristic. We shall refer to Fig. 8.7 in each of the following filter circuits.

We have not discussed the effect of the elements in the circuit on the phase shift in the circuit. In fact phase shift does occur, so that when a given frequency is transmitted through the filter it will be subject to a phase shift. Moreover, each frequency has its own phase shift, so that the input signal will probably suffer some distortion when passing through the filter.

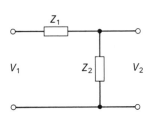

**Fig. 8.7**   Principle of a passive filter circuit

### 8.8.1   Simple low-pass filter circuit

Typical low-pass filters are shown in Fig. 8.8. The basic principle can be understood from the generalized circuit in Fig. 8.7.

At low frequencies, the impedance of $Z_1$ is much less than that of $Z_2$, so that at these frequencies there is very little voltage drop in $Z_1$, and most of the input signal is transmitted to the output.

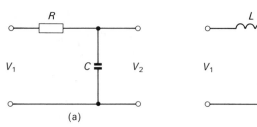

**Fig. 8.8**   Passive low-pass filter: (a) *R–C* filter, (b) *L–R* filter

At high frequencies, the impedance of $Z_1$ is much higher than $Z_2$, so that only a small proportion of the input signal appears between the output terminals.

In the case of the $R$–$C$ filter [Fig. 8.8(a)], capacitor $C$ is in the $Z_2$ position of the filter and, above the cut-off frequency, its reactance has a value much less than the resistance of $R$. Consequently, the output terminals can be regarded as though they are short-circuited at high frequencies. In the case of the $L$–$R$ filter, Fig. 8.8(b), inductor $L$ is in the $Z_1$ position of the filter. At frequencies above the cut-off frequency, the inductive reactance is much higher than that of the resistance of $R$; consequently, above the cut-off frequency it practically presents an open circuit to the flow of current in the filter. Once again, the output voltage is practically zero at high frequency.

We will perform a simple analysis on the $R$–$C$ filter. The cut-off frequency is defined as that frequency at which the magnitude of $Z_1$ (see Fig. 8.7) is equal to the magnitude of $Z_2$. That is, in Fig. 8.8(a)

$$R = X_C = 1/\omega_C C$$

where $\omega_C$ is the cut-off frequency of the filter. Hence

$$\omega_C = 1/RC$$

or

$$f_C = 1/2\pi RC$$

Applying a similar analysis to the $L$–$R$ filter yields

$$\omega_C = R/L$$

or

$$f_C = R/2\pi L$$

### 8.8.2 Simple high-pass filter

The generalized passive filter circuit is shown in Fig. 8.7. For it to perform as a high-pass filter, the impedance of $Z_1$ must, at high frequency, have a low value when compared with the magnitude of $Z_2$ (this allows easy propagation of the input signal); also, at low frequency, the magnitude of $Z_1$ must be high when compared with the magnitude of $Z_2$ (to prevent the propagation of the input signal).

In the case of the $C$–$R$ filter in Fig. 8.9(a), the reactance of the capacitor is very high at low frequency, giving very little output from the circuit. As the frequency increases the capacitive reactance falls in value, and a greater proportion of the input signal is transmitted to the output. In the case of the $R$–$L$ circuit,

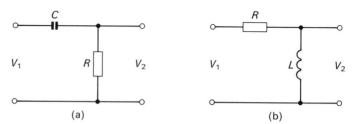

**Fig. 8.9** Passive high-pass filter: (a) $C$–$R$ filter, (b) $R$–$L$ filter

the reactance of $L$ is very low at low frequency, and it practically short-circuits the output. As the frequency rises, the inductive reactance rises, being very high indeed at high frequency. Consequently, a greater proportion of the input signal appears at the output terminals at high frequency.

At this point we will carry out a simple analysis on the $C$–$R$ filter in Fig. 8.9(a). At the cut-off frequency we have

$$R = X_C = 1/\omega_c C$$

where $\omega_c$ is the cut-off frequency of the filter. Hence

$$\omega_c = 1/RC$$

or

$$f_c = 1/2\pi RC$$

Applying a similar analysis to the $L$–$R$ filter yields

$$\omega_c = R/L$$

or

$$f_c = R/2\pi L$$

***Worked example 8.6*** An $R$–$C$ combination is used as (i) a low-pass filter, (ii) a high-pass filter. If the cut-off frequency in both cases is 10 kHz and $R = 1\,\mathrm{k\Omega}$, calculate the value of $C$ for the two filters. Calculate for the low-pass filter and an input volage of 1 V, the output voltage if the signal frequency is (a) 1 kHz, (b) 100 kHz.

*Solution* If $\omega_c$ is the cut-off frequency we have, for both circuits

$$\omega_c = 1/RC$$

or

$$C = 1/\omega_c R = 1/(2\pi \times 10\,000 \times 1000)$$

$$= 1.592 \times 10^{-8}\,\mathrm{F} \quad \text{or} \quad 15.92\,\mathrm{nF}$$

(a) The capacitive reactance at 1 kHz is

$$X_C = 1/2\pi f C = 1/(2\pi \times 1000 \times 1.592 \times 10^{-8})$$

$$= 9997\,\Omega$$

and the input impedance of the circuit at this frequency is

$$Z = \sqrt{(R^2 + X_C^2)} = \sqrt{(1000^2 + 9997^2)} = 10\,047\,\Omega$$

The current flowing through the circuit is

$$I = V_S/Z = 1/10\,047 = 99.5 \times 10^{-6}\,\text{A} \quad \text{or} \quad 99.5\,\mu\text{A}$$

Assuming that very little electrical load is connected to the output terminals, the output voltage is equal to the voltage across $C$ as follows:

$$V_{\text{out}} = IX_C = 99.5 \times 10^{-6} \times 9997 = 0.995\,\text{V}$$

which represents only a 0.5 per cent loss in voltage.
(b) The capacitive reactance at 100 kHz is

$$X_C = 1/2\pi f C = 1/(2\pi \times 100\,000 \times 1.592 \times 10^{-8})$$

$$= 99.97\,\Omega$$

and the input impedance is

$$Z = \sqrt{(R^2 + X_C^2)} = \sqrt{(1000^2 + 99.97^2)} = 1005\,\Omega$$

The current flowing through the circuit is

$$I = V_S/Z = 1/1005 = 0.995 \times 10^{-3}\,\text{A} \quad \text{or} \quad 0.995\,\text{mA}$$

The output voltage is

$$V_{\text{out}} = IX_C = 0.995 \times 10^{-3} \times 99.97 \approx 0.1\,\text{V}$$

### 8.8.3  Simple band-pass filter

We can still use the simple basic filter arrangement of Fig. 8.7 but, in this case, we use a series $L$–$C$ circuit (see Fig. 8.10) as the impedance $Z_1$ in Fig. 8.7.

Since the reactance of $C$ is high at low frequency, and the reactance of $L$ is high at high frequency, the $L$–$C$ section in Fig. 8.10 presents a high impedance at both low frequency and at high frequency, so that very little of the input signal is transmitted to the output at low and high frequency.

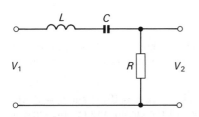

**Fig. 8.10** Simple band-pass filter

At the resonant frequency, when $X_L = X_C$, the impedance of the $L$–$C$ section is very low (ideally zero), and practically all of the input voltage is transmitted to the output.

That is, the value of the ratio $V_2/V_1$ is zero (or nearly so) at low frequency, rising to unity at the resonant frequency of the $L$–$C$ circuit $(1/2\pi\sqrt{(LC)})$, and falls to zero at high frequency [see Fig. 8.6(c)].

The bandwidth of the filter depends on the $Q$-factor of the $L$–$C$ circuit; a high-$Q$ circuit has a narrower bandwidth than does a low-$Q$ circuit.

The reader will find it interesting to show that it is possible to use a parallel $L$–$C$ circuit as part of a band-pass filter.

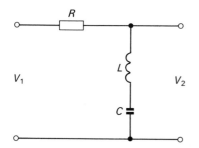

**Fig. 8.11**   Simple band-stop filter

### 8.8.4   Simple band-stop filter

In this circuit the $R$ element and the $L$–$C$ section are interchanged (see Fig. 8.11) when compared with the band-pass circuit in Fig. 8.10.

Both at low and high frequency, the impedance of the $L$–$C$ section is very high, and is an effective open circuit. Providing that the value of $R$ is not too high, most of the input signal is transmitted to the output at these frequencies [see Fig. 8.6(d)].

In the region of the resonant frequency of the $L$–$C$ circuit $(1/2\pi\sqrt{(LC)})$, the impedance of the $L$–$C$ section is very low, and it practically short-circuits the output terminals. Consequently, in the region of the resonant frequency, the output voltage is very low.

That is, the ratio $V_2/V_1$ for the circuit has a value of about unity at low frequency, falling to nearly zero at the resonant frequency of the $L$–$C$ section, and rises to nearly unity at high frequency.

As with other circuits of the type described here, the output voltage also depends on the value of the impedance of the load connected to the output terminals.

Once again, it is possible to construct a band-stop filter using a parallel-tuned circuit as one of the sections of the filter. It is left as an exercise for the reader to devise a suitable circuit using a parallel-tuned circuit.

### Problems

**8.1**   A parallel circuit contains a coil of resistance $R$ and inductance $L$ in parallel with a capacitor $C$. Show that the impedance of the circuit at resonance is $L/CR$.

**8.2**   In the parallel circuit in problem 8.1, $R = 1\,\Omega$, $L = 10\,\mu H$ and $C = 10\,nF$. Calculate (a) the resonant frequency of the circuit and (b) the current drawn from a 10 V supply at the resonant frequency.

[503 kHz; 0.01 A]

**8.3**   For the circuit in problem 8.2, determine at the resonant frequency (a) the current in the coil, (b) the current in the capacitor and (c) the $Q$-factor of the circuit.

[(a) 0.316 A; (b) 0.316 A; (c) 31.6]

**8.4**   A series $R$–$L$–$C$ circuit contains $R = 2\,\Omega$, $L = 0.03\,H$ and $C = 3.38\,\mu F$. Calculate the resonant frequency of the circuit. If the circuit is energized at the resonant frequency by a 20 V supply, calculate (a) the current in the circuit, (b) the voltage across each element in the circuit and (c) the $Q$-factor of the circuit.

[500 Hz; (a) 10 A; (b) $V_L = 942$ V, $V_C = 942$ V, $V_R = 20$ V; (c) $Q = 47.1$]

**8.5**  A series circuit of resistance $10\,\Omega$, inductance $0.15\,\mathrm{H}$ and capacitance $50\,\mu\mathrm{F}$ is supplied at $200\,\mathrm{V}$ at the resonant frequency. Calculate (a) the resonant frequency, (b) the current drawn by the circuit, (c) the $Q$-factor of the circuit and (d) the power consumed.

[(a) $58.11\,\mathrm{Hz}$; (b) $20\,\mathrm{A}$; (c) $5.48$; (d) $4\,\mathrm{kW}$]

**8.6**  Resonance occurs in a series $R–L–C$ circuit at a frequency of $200\,\mathrm{Hz}$. If the capacitance in the circuit is $5\,\mu\mathrm{F}$, calculate the inductance in the circuit. If a second capacitance of $1\,\mu\mathrm{F}$ is connected in parallel with the first one, determine the new value of the resonant frequency.

[$0.127\,\mathrm{H}$; $489\,\mathrm{Hz}$]

**8.7**  An iron-cored coil absorbs $6\,\mathrm{kW}$ when connected to an a.c. supply of $415\,\mathrm{V}$, $50\,\mathrm{Hz}$, the current being $17.32\,\mathrm{A}$. When the a.c. supply is replaced by a $12\,\mathrm{V}$ d.c. supply, the current is $1\,\mathrm{A}$. Calculate for the coil (a) its d.c. resistance, (b) its effective 'a.c. resistance', (c) its impedance, (d) the copper loss, (e) the iron loss, (f) its inductance and (g) its power factor. Explain the reason for the difference between the answers for (a) and (b).

[(a) $12\,\Omega$; (b) $20\,\Omega$; (c) $24\,\Omega$; (d) $3.6\,\mathrm{kW}$; (e) $2.4\,\mathrm{kW}$; (f) $42.2\,\mathrm{mH}$; (g) $0.833$]

**8.8**  A resistor, a capacitor and a variable inductor of zero resistance are connected in series to a $200\,\mathrm{V}$, $50\,\mathrm{Hz}$ supply. When the inductance is varied, the maximum r.m.s. current in the circuit is $0.314\,\mathrm{A}$, at which time the voltage across the capacitor is $300\,\mathrm{V}$. Calculate the value of $R$, $L$ and $C$. Draw to scale the phasor diagram of the circuit at resonance.

[$636.9\,\Omega$; $3.04\,\mathrm{H}$; $3.33\,\mu\mathrm{F}$]

**8.9**  A pure inductor of $0.12\,\mathrm{H}$ is connected in parallel with a $60\,\mu\mathrm{F}$ capacitor. Calculate the resonant frequency of the circuit. What current circulates in the capacitor and in the inductor if an a.c. supply of $20\,\mathrm{V}$ at the resonant frequency is connected to the circuit?

[$59.3\,\mathrm{Hz}$; $0.447\,\mathrm{A}$]

**8.10**  If a non-inductive resistance of $20\,\Omega$ is connected in series with the inductor in problem 8.9, determine the new value of the resonant frequency. What current flows in the capacitor when a source of $20\,\mathrm{V}$ at the new resonant frequency is applied to the circuit? Determine also the dynamic impedance of the circuit and the $Q$-factor at resonance.

[$53.05\,\mathrm{Hz}$; $0.4\,\mathrm{A}$; $100\,\Omega$; $2$]

**8.11**  The following test figures were obtained from a single-phase inductive load: power $1.6\,\mathrm{kW}$, voltage $240\,\mathrm{V}$, frequency $71.25\,\mathrm{Hz}$, current $10\,\mathrm{A}$. Determine the resistance and inductance of the circuit (assume it to be series-connected). What capacitance must be connected in parallel with the load in order to raise the overall power factor to unity? What is then the $Q$-factor of the parallel circuit?

[$16\,\Omega$; $40\,\mathrm{mH}$; $124.8\,\mu\mathrm{F}$; $1.12$]

# 9 Three-phase circuits

## 9.1 Introduction

Three-phase systems can be shown to have a number of advantages over single-phase systems and include the following:

1. power is transmitted more efficiently;
2. the total volume of conductor material in the transmission system is less than in the single-phase case, and is therefore more cost-effective;
3. motors and control gear are simpler, more efficient and less costly.

Additionally, industry makes use of 6-, 12- and 24-phase systems, and these can be produced from a 3-phase supply.

## 9.2 Generating a three-phase supply

Figure 9.1 shows a simple three-phase a.c. generator or *alternator*, which has three separate windings on its fixed part or *stator*. each winding has an e.m.f. induced in it by a permanent magnet on its rotating part or *rotor*. The rotor is driven round by a *prime mover*, which may be a turbine, or a diesel engine, etc.

Each coil on the stator is described as a *phase winding*, in which the *phase voltage* is induced and, since the phase windings are physically displaced 120° apart around the periphery of the stator, the e.m.f. induced in each phase differs from the next phase by 120° [see Fig. 9.1(b)]. In order to differentiate between the windings we give them (in the UK) the names *red phase* (R-phase),

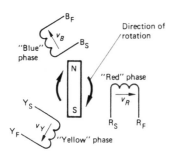

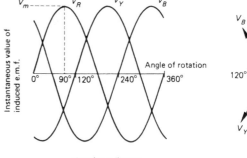

Waveform diagram

(b)

Phasor diagram

(c)

**Fig. 9.1** (a) Simple three-phase alternator, (b) voltage waveform diagrams and (c) typical phasor diagram

*yellow phase* (Y-phase) and *blue phase* (B-phase); in the USA they are called the A-, B- and C-phases, respectively.

As with single-phase circuits, we can draw the phasor diagram corresponding to any particular point in time by 'freezing' the phasors at that time (the phasor being scaled to represent the r.m.s. value of the voltage or current). If we take the R-phase to be the reference phase, then the phasor diagram in Fig. 9.1(c) would correspond to the waveform diagram at $t = 0$ in Fig. 9.1(b). Strictly speaking, the phasor diagram in Fig. 9.1(c) is only correct if we have connected the winding in a particular way, which we discuss in section 9.5.

## 9.3 Phase sequence

The *phase sequence* of a three-phase system is the sequence or order in which the sine waves reach their maximum voltage (or current). The normal order in which this occurs is called the *positive phase sequence* (or PPS). In the case of the waveforms in Fig. 9.1, we see that it is the sequence red, yellow, blue (RYB). Technically speaking, a phasor does not 'rotate' since it is fixed in time, but the line whose tip draws out the corresponding sine wave can be thought of as rotating in an anticlockwise direction. If we observe the way in which the lines draw out the waveforms in Fig. 9.1(b), we see that they pass a given point in the sequence RYB. When one of the phasors is in the horizontal direction, as is $V_R$ in Fig. 9.1(c), we say that it points in the *reference direction*.

The opposite or reverse phase sequence is known as *negative phase sequence* (or NPS), i.e. the sequence red, blue, yellow (RBY), as shown in Fig. 9.2. Negative phase sequence voltage and currents can occur in what is normally a PPS system as a result of an electrical fault. In fact certain electrical protection systems depend on the detection of non-PPS voltages and currents in order to detect certain faults (an earth fault detection system is an example).

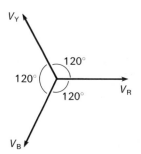

**Fig. 9.2** Negative phase sequence

## 9.4 Balanced and unbalanced systems

A *balanced* or *symmetrical* three-phase supply has phase voltages (or currents) which are:

1. equal in magnitude; and
2. phase displaced from one another by 120°.

If either of these conditions do not apply, the supply is said to be *unbalanced*. A balanced system can become unbalanced either when certain types of fault occur or, in some cases, when an unbalanced load is connected.

A load is said to be *balanced* if the impedance in each phase of the load:

1. has the same magnitude; and
2. has the same phase angle.

If either condition is not satisfied, the load is said to be *unbalanced*.

### 9.5 Star connection of balanced three-phase voltages

If the 'start' points $R_S$, $Y_S$ and $B_S$ of the windings in Fig. 9.1(a) are connected together, we get the *star connection*[†] in Fig. 9.3(a). The common point is known as the *star point*; the star point of an alternator is usually connected to earth, and is therefore at 'neutral' or zero potential, and is known as the *neutral point* (N) as shown in Fig. 9.3(a).

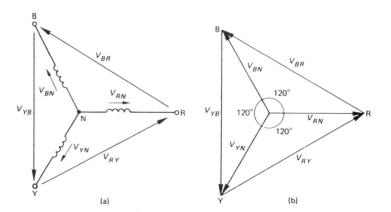

**Fig. 9.3** (a) Star connection of three-phase windings and (b) its phasor diagram for balanced voltages

The 'finish' connection of each alternator winding, i.e. $R_F$, $Y_F$ and $B_F$, is connected to the red, yellow and blue lines, respectively, of the supply system. The voltage between the star point and a line is known as the *phase voltage*; they are:

$V_{RN}$ = voltage of the R-line with respect to the neutral point

$V_{YN}$ = voltage of the Y-line with respect to the neutral point

$V_{BN}$ = voltage of the B-line with respect to the neutral point

Since the neutral point is at zero potential, we sometimes drop the suffix N and describe $V_{RN}$ as $V_R$, $V_{YN}$ as $V_Y$, and $V_{BN}$ as $V_B$.

The voltage between a pair of lines is known as the *line-to-line voltage* or, more simply, as the *line voltage*. The line voltages of a three-phase system are:

$V_{BR}$ = voltage of the B-line with respect to the R-line

$\quad = V_{BN} - V_{RN}$

$V_{YB}$ = voltage of the Y-line with respect to the B-line

$\quad = V_{YN} - V_{BN}$

$V_{RY}$ = voltage of the R-line with respect to the Y-line

$\quad = V_{RN} - V_{YN}$

---

[†] In the USA, the star connection is known as the *wye connection* or the *Y-connection*.

In a balanced system, the magnitude of each phase voltage is the same. If we call this $V_P$, then

$$V_P = \text{magnitude of } V_R = \text{magnitude of } V_Y$$

$$= \text{magnitude of } V_B$$

Moreover, if the phase voltages are balanced, then the line voltages are also balanced. If the line voltage is $V_L$, we have, in a balanced system

$$V_L = \text{magnitude of } V_{BR} = \text{magnitude of } V_{YB}$$

$$= \text{magnitude of } V_{RY}$$

The relationship between the line and phase voltages in a balanced star-connected system can be determined as follows. Looking at the line voltage $V_{RY}$ in Fig. 9.4(b) (shown as $V_L$), we have

$$V_L = \text{phasor } V_R - \text{phasor } V_Y$$

$$= \text{phasor } V_R + (-\text{phasor } V_Y)$$

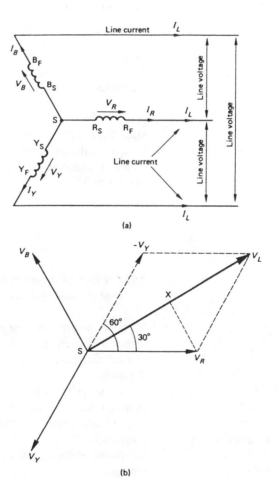

**Fig. 9.4** Relationship between the phase and line voltages in a balanced star-connected system

The subtraction is carried out by reversing the phasor $V_Y$ (i.e. rotating it through $180°$) and then adding it to $V_R$. This gives

Magnitude of $V_L = 2 \times$ SX   [see Fig. 9.4(b)]

$$= 2 \times V_R \cos 30° = 2 \times V_P \cos 30°$$

$$= 2 \times V_P \times \frac{\sqrt{3}}{2} = \sqrt{(3)} V_P$$

That is, in a balanced star-connected system

**Line voltage, $V_L = \sqrt{3} \times$ phase voltage**　　　　　　　　[9.1]

The reader should note that when an engineer refers to the 'voltage' of a three-phase system, he always means the line voltage. However, a 440 V three-phase system means that the line voltage is 440 V, and its phase voltage is

$$V_P = V_L/\sqrt{3} = 440/\sqrt{3} = 254 \text{ V}$$

Looking at Fig. 9.4 it will be seen that, since each phase winding of the alternator is directly connected to a line, then

**Phase current = line current**

Also, *if the supply voltage is balanced and the load is balanced,* then the magnitude of the current in each phase will have the same numerical value. In this circumstance, for all lines

$$I_P = I_L　　　　　　　　　　　　　　　　　[9.2]$$

Since the supply system in Fig. 9.4 has only three lines connected to it, it is described as a *three-phase, three-wire, star-connected supply*. A three-phase, four-wire, star-connected system is described in section 9.6.

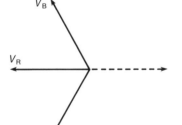

**Fig. 9.5** Effect of incorrect connection of phase R in a star-connected system

### 9.5.1 Effect of incorrect connection of one phase in a star-connected system

If the connections to one phase of an alternator are accidentally reversed, the effect is to reverse that particular phasor on the phasor diagram, as shown in Fig. 9.5 for the reversal of the R-phase.

The net effect is that, even if all the phase voltages are equal in magnitude, the line voltages are unbalanced. If the phasors in Fig. 9.5 are drawn carefully to scale, it will be found that the magnitude of $V_{BR}$ and $V_{RY}$ are both equal to the magnitude of the phase voltage, and $V_{YB}$ is $\sqrt{3}$ times greater than the phase voltage.

### 9.6 Star connection of unbalanced three-phase voltages

In certain circumstances, the phase voltages can be unbalanced as shown, for example, in Fig. 9.6. When this occurs, the relationship between the line and phase voltages described in section 9.5 for a balanced system does not hold good. We can determine the value of the line voltages either graphically or by calculation.

The graphical procedure is fairly straightforward, and it is merely necessary to draw the phase voltages to scale as shown in Fig. 9.6: the line voltages are determined by measuring them on the diagram. However, to ensure reasonable accuracy, the phasor diagram should be drawn as large as possible.

To calculate the magnitude of a line voltage, it is necessary to determine the horizontal and vertical components of the associated phase voltages, and then add (or subtract) them in the manner described in section 9.5, and in Worked Example 9.1 below.

*Worked example 9.1*  A star-connected set of unbalanced three-phase voltages is as follows:

$$V_{RN} = 100 \text{ V at } 20° \text{ to the reference direction}$$

$$V_{YN} = 150 \text{ V at } -110° \text{ to the reference direction}$$

$$V_{BN} = 200 \text{ V at } 100° \text{ to the reference direction}$$

Calculate the value of the line voltages.

*Solution*  For $V_{RN}$

$$\text{Horizontal component} = V_{RN} \cos 20° = 100 \cos 20°$$
$$= 93.97 \text{ V}$$

$$\text{Vertical component} = V_{RN} \sin 20° = 100 \sin 20°$$
$$= 34.2 \text{ V}$$

For $V_{YN}$

$$\text{Horizontal component} = V_{YN} \cos(-110°)$$
$$= 150 \cos(-110°) = -51.3 \text{ V}$$

$$\text{Vertical component} = V_{YN} \sin(-110°)$$
$$= 150 \sin(-110°) = -141 \text{ V}$$

For $V_{BN}$

$$\text{Horizontal component} = V_{BN} \cos 100° = 200 \cos 100°$$
$$= -34.7 \text{ V}$$

$$\text{Vertical component} = V_{BN} \sin 100° = 200 \sin 100°$$
$$= 197 \text{ V}$$

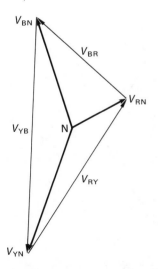

**Fig. 9.6** Star connection of unbalanced voltages

Now $V_{BR}$ = phasor difference $(V_{BN} - V_{RN})$, hence horizontal component of $V_{BR}$ is

[Horizontal component of $V_{BN}$] − [horizontal component of $V_{RN}$]

$$= -34.7 - 93.97 = -131.4 \text{ V}$$

The vertical component of $V_{BR}$ is

[Vertical component of $V_{BN}$] − [vertical component of $V_{RN}$]

$$= 197 - 34.2 = 162.8 \text{ V}$$

Hence

Magnitude of $V_{BR} = \sqrt{((-131.4)^2 + 162.8^2)} = 209.2 \text{ V}$

Applying a similar method to $V_{YB}$ and $V_{RY}$ using the equations $V_{YB} = V_{YN} - V_{BN}$ and $V_{RY} = V_{RN} - V_{YN}$ we get

Magnitude of $V_{YB} = \sqrt{((-16.6)^2 + (-338)^2)} = 338.4 \text{ V}$

and

Magnitude of $V_{RY} = \sqrt{(145.3^2 + 175.2^2)} = 227.6 \text{ V}$

## 9.7 Three-phase, four-wire, star-connected system

The usual way of distributing electricity in villages, towns and small industries is the three-phase, four-wire system, illustrated in Fig. 9.7. The supply source is star-connected and, in the case of towns and villages, each house is connected to a different phase; the neutral point of the source is connected to the star point of the load by means of a *neutral wire* (ideally having zero resistance). We can therefore assume for the moment that the star point of the load is at the same electrical potential as the neutral point of the supply, so that the terms 'star point' and 'neutral point' are synonymous.

**Fig. 9.7** A three-phase, four-wire, star-connected system

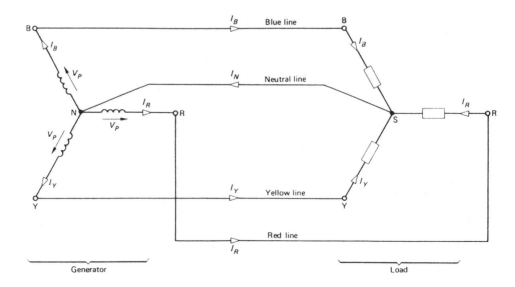

Generally speaking, the supply voltage is balanced, but the load may be unbalanced. The reason for the latter will be understood from the following. If the load is a housing estate then, for practical reasons, groups of houses are connected to different phases of the supply. Since each house has a different electrical loading requirement, the electrical load connected to the supply is rarely balanced.

For the purpose of calculation in a three-phase, four-wire system, we can regard the load as comprising three single-phase loads, each being supplied by one phase of the generator. That is, the impedance in the R-phase of the load is regarded as being energized by the R-phase of the alternator, the impedance in the Y-phase of the load is regarded as being energized by the Y-phase of the alternator, etc.

Applying Kirchhoff's current law (KCL) either to the neutral point of the source or to the star point of the load, we see that the current in the neutral wire is

$$I_N = phasor\ sum\ (I_R + I_Y + I_B) \tag{9.3}$$

The reader should carefully note that it is the *phasor sum* and *not the numerical sum* of the currents (see also Worked Example 9.3).

***Worked example 9.2*** Calculate the current flowing in each line and in the neutral wire of a 440 V, three-phase, four-wire system, having a balanced load consisting of a resistance of $3\,\Omega$ in series with an inductive reactance of $4\,\Omega$ per phase. Draw the phasor diagram showing the phase voltages and currents.

*Solution* The magnitude of the impedance in each phase of the load is

$$Z_P = \sqrt{(R^2 + X_L^2)} = \sqrt{(3^2 + 4^2)} = 5\,\Omega$$

and its phase angle is

$$\phi = \tan^{-1}(X_L/R) = \tan^{-1}(4/3) = 53.13°\ \text{(lagging)}$$

Since the supply is balanced, the phase voltage is $440/\sqrt{3} = 254\,\text{V}$. Moreover, since we are dealing with a balanced load, we need only calculate the magnitude of the current in one phase which is

$$I_P = V_P/Z_P = 254/5 = 50.8\,\text{A}$$

which is also equal to the magnitude of the line current, that is

$$I_R = I_Y = I_B = 50.8\,\text{A}$$

Since the phase angle of the impedance in each phase is 53.13° lagging, the phase current lags behind the appropriate phase voltage by 53.13° as follows:

$I_R = 50.8$ A at an angle of $-53.13°$

$I_Y = 50.8$ A at an angle of $(-120° - 53.13°)$ or $-173.13°$

$I_R = 50.8$ A at an angle of $(120° - 53.13°)$ or $66.87°$

The phasor diagram showing the phase voltages and currents is in Fig. 9.8. Finally, we use eqn [9.3] to determine the current in the neutral wire as follows:

$$I_N = \text{phasor sum } (I_R + I_Y + I_B) = 0$$

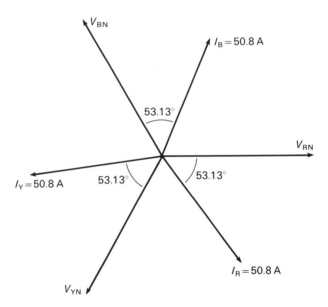

**Fig. 9.8** Graphical solution of Worked Example 9.2

That is, *in a balanced load with a balanced supply, the current in the neutral wire is zero.*

This condition is the usual case for a large industrial load, when we can dispense with the neutral wire completely.

***Worked example 9.3*** A balanced 433 V, star-connected, three-phase, four-wire supply is connected to the following four-wire, star-connected load:

R-phase: a 10 Ω resistor;

Y-phase: an 8 Ω resistor in series with a 2 Ω inductive reactance;

B-phase: a 4 Ω resistor in series with a 5 Ω capacitive reactance.

Determine the current in each phase of the load and in the neutral wire.

*Solution* The phase voltage is

$$V_P = V_L/\sqrt{3} = 433/\sqrt{3} = 250 \text{ V}$$

The magnitude and phase angle of the impedance in each phase are

$$Z_R = R_R = 10\,\Omega \qquad \phi_R = 0$$
$$Z_Y = \sqrt{(R_Y^2 + X_Y^2)} = \sqrt{(8^2 + 2^2)} = 8.25\,\Omega$$

$$\phi_Y = \tan^{-1}(X_Y/R_Y) = \tan^{-1}(2/8) = 14° \text{ lagging}$$

$$Z_B = \sqrt{(R_B^2 + X_B^2)} = \sqrt{(4^2 + 5^2)} = 6.4\,\Omega$$

$$\phi_B = \tan^{-1}(X_B/R_B) = \tan^{-1}(5/4) = 51.3° \text{ leading}$$

The current in each phase is

$$I_R = V_P/Z_R = 250/10 = 25\,\text{A}, \text{ and is in phase with } V_{RN}$$

$$I_Y = V_P/Z_Y = 250/8.25 = 30.3\,\text{A}, \text{ and lags behind } V_{YN} \text{ by } 14°$$

$$I_B = V_P/Z_B = 250/6.4 = 39.1\,\text{A}, \text{ and leads } V_{BN} \text{ by } 51.3°$$

$I_N$ is the phasor sum of $I_R$, $I_Y$ and $I_B$, and can be determined graphically or by calculation. We will do both.

*Graphical solution*
The phasor diagram of the voltages and the currents is drawn to scale in Fig. 9.9. To calculate $I_N$ we must add the phasor currents together, which we do in two stages. We commence by adding any two of the currents; we select $I_R$ and $I_Y$, as shown in the figure. Next we add $I_B$ to the sum of these to give $I_N$, which is estimated to be 38 A lagging behind the reference direction by $-153°$.

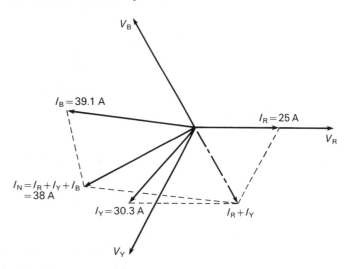

**Fig. 9.9** Graphical solution of Worked Example 9.3

While this method is straightforward, it is not as accurate as the answer obtained by calculation (see below).

*Solution by calculation*
In this case we resolve each current into its horizontal and vertical components. By adding all the horizontal components together, we obtain the net horizontal components of $I_N$, and the sum of the vertical components gives the net vertical component of $I_N$. From these values we calculate the magnitude and the phase angle of $I_N$.

For $I_R$

$$\text{Horizontal component} = I_R \cos 0° = I_R = 25\,\text{A}$$

$$\text{Vertical component} = I_R \sin 0° = 0$$

For $I_Y$

Horizontal component $= I_Y \cos(-120° - 14°)$

$= 30.3 \cos(-134°) = -21.05 \, \text{A}$

Vertical component $= I_Y \sin(-120° - 14°)$

$= 30.3 \sin(-134°) = -21.8 \, \text{A}$

For $I_B$

Horizontal component $= I_B \cos(120° + 51.3°)$

$= 39.1 \cos 171.3° = -38.65 \, \text{A}$

Vertical component $= I_B \sin(120° + 51.3°)$

$= 39.1 \sin 171.3° = 5.91 \, \text{A}$

Hence

Net horizontal component of $I_N = 25 - 21.05 - 38.65$

$= -34.7 \, \text{A}$

Net vertical component of $I_N = 0 - 21.8 + 5.91$

$= -15.89 \, \text{A}$

and

Magnitude of $I_N$

$= \sqrt{[(\text{horizontal component})^2 + (\text{vertical component})^2]}$

$= \sqrt{[(-34.7)^2 + (-15.89)^2]} = 38.17 \, \text{A}$

and

$$\phi_N = \tan^{-1}\left[\frac{\text{vertical component}}{\text{horizontal component}}\right]$$

$$= \tan^{-1}\left[\frac{-15.89}{-34.7}\right] = -153.4°$$

*The reader should be very cautious when using an electronic calculator to determine a phase angle*, since it generally gives the *principal angle* in the range $\pm 90°$ which satisfies the calculation. In this case it is 24.6°. The reader should study the mathematical sign associated with the horizontal and vertical component before deciding what the answer should be.

## 9.8 Three-phase, three-wire, star-connected system

In the case of a system with a balanced supply and a balanced load, the neutral wire current is zero, and *the neutral wire can be removed* without causing any disturbance to the system (see Worked Example 9.2), leaving a three-wire system.

The majority of industrial systems are balanced, and a three-phase, three-wire supply can be used.

If either the supply or the load is unbalanced, the calculation is quite complex, and is beyond the scope of this book. However, it can be solved by the SPICE software package described in Chapter 18.

### 9.9 The delta connection or mesh connection

If, in the alternator in Fig. 9.1(a), the 'start' connection of one phase is connected to the 'finish' connection of the next phase, the *delta connection* or *mesh connection* is formed. This is shown in Fig. 9.10.

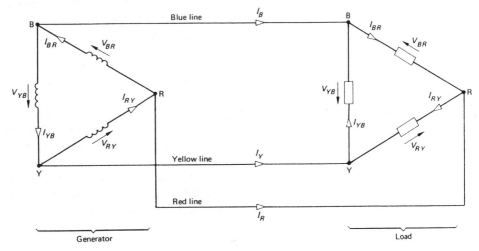

**Fig. 9.10** A delta or mesh connected system

In this case, each of the alternator windings is connected between a pair of the supply lines and, *if the phase voltages are balanced*, then

Phase voltage = line voltage

or

$$V_P = V_L \qquad [9.4]$$

The phasor diagram for the line (and therefore the phase) voltages of a balanced delta-connected system is shown in Fig. 9.11. In such a system, the phasor diagram is a 'closed' triangle, so that no voltage exists between the 'start' of the first voltage phasor and the 'end' of the third voltage phasor (see also section 9.9.1). *This implies that no current circulates around the delta when it is 'closed'.*

Current does, of course, flow in the generator windings when a load is connected. By the nature of the delta connection, it provides a three-phase, three-wire supply.

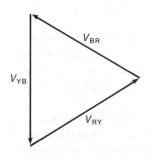

**Fig. 9.11** Line voltage phasors for a balanced delta-connected system

We can now explain the directions of the current arrows in Fig. 9.10. In the case of the generator, current flows (inside the windings) in the direction of the induced voltage in each winding. That is, *the voltage and current arrows associated with each phase of the generator act in the same direction.*

Since each phase of the generator is directly connected to one phase of the load, then the potential arrow across that phase of the load points in the same direction as the e.m.f. arrow in the corresponding generator phase. For example, $V_{BR}$ in the generator points in the same direction as $V_{BR}$ across the load.

Also, current in the load flows from a point of higher potential to one of lower potential. That is, *the current arrow in each phase of the load points in the opposite direction to the potential arrow across that phase of the load.*

This follows the notation explained for the d.c. system, that is *the voltage and current arrows in a power source point in the same direction, and the current and voltage arrows in a load point in the opposite direction.*

We can see from Fig. 9.10 that each phase of the supply is directly connected to one phase of the load and, for the purposes of calculation, we can regard the system as comprising three separate single-phase loads, each supplied by its own single-phase source. That is, the impedance connected between lines B and R is energized by $V_{BR}$, etc. However, each 'line' wire carries two currents in opposite directions.

### 9.9.1 Effect of incorrect connection of one phase in a delta-connected system

If the connections to one phase of a delta-connected generator are accidentally reversed, the effect is to reverse that phasor on the phasor diagram. For example, if the phase involved is the YB-phase, the net phasor diagram is as shown in Fig. 9.12. It will be seen that the voltage between Y and Y′ on the diagram is *twice the line voltage.*

In a correctly connected alternator, the p.d. between these two points is zero. Consequently, if one phase is incorrectly connected, then twice the line voltage appears between the final pair of terminals to be connected! This would be disastrous for both man and machine.

A simple method of testing for this condition is to leave the final link in the delta open until the voltage between the ends has been tested. If the voltage is zero, it will be safe to close the delta, otherwise it will be necessary to find the reason for the voltage between the ends.

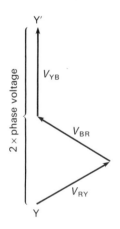

**Fig. 9.12** Effect of incorrect connection of one phase of a delta-connected system

### 9.10 Phase and line currents in a delta-connected system

To deduce the relationship between the line and phase currents in a delta-connected system, we apply KCL to each node in turn of Fig. 9.10.

At the R-node we see that

$$I_R + I_{BR} = I_{RY}$$

or

$$I_R = I_{RY} - I_{BR} \tag{9.5}$$

The reader will note that *this applies both to the generator and the load*.

Applying KCL to the Y-node yields

$$I_Y + I_{RY} = I_{YB}$$

or

$$I_Y = I_{YB} - I_{RY} \tag{9.6}$$

and at the B-node

$$I_B + I_{YB} = I_{BR}$$

or

$$I_B = I_{BR} - I_{YB} \tag{9.7}$$

The above relationships hold good for either balanced or unbalanced conditions.

**Balanced system**

As explained earlier, we can regard each phase of the load as if it were energized by its own single-phase supply. That is, we can draw a phasor diagram for each phase of the load. Assuming that the load is resistive, i.e. the current in each phase of the load is in phase with the voltage across it, the phasor diagrams for the separate phases are as shown in Fig. 9.13.

From eqn [9.5], the line current $I_R$ is given by the expression $I_R = I_{RY} - I_{BR}$ so that at node R we can draw the phasors for $I_{RY}$ and $I_{BR}$ from a common point, as shown in Fig. 9.14. We can

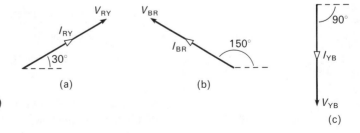

**Fig. 9.13** Phasor diagram for a balanced delta-connected load for (a) phase RY, (b) phase BR and (c) phase YB

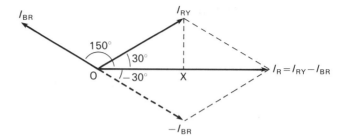

**Fig. 9.14** Relationship between the phase and line currents in a balanced delta-connected system

deduce the relationship between the line and phase current in a balanced system as follows.

Since the load is balanced, the current in each phase has the same numerical value, and we will call this $I_P$. That is

Magnitude of $I_{RY} = I_P$

Magnitude of $I_{BR} = I_P$

Magnitude of $-I_{BR} = I_P$

and the line currents will all have the same magnitude, which we shall call $I_L$; that is $I_R = I_L$, $I_Y = I_L$ and $I_B = I_L$. We can see from the phasor diagram in Fig. 9.14 that

$$I_R = I_L = 2 \times OX = 2 \times I_P \cos 30° = 2 \times I_P \times \frac{\sqrt{3}}{2}$$

$$= \sqrt{(3)}I_P = 1.732 I_P \qquad [9.8]$$

That is, in a balanced delta-connected system

**Line current $= \sqrt{3} \times$ phase current**

*Worked example 9.4*   Three $10\,\Omega$ resistors are connected to a balanced $440\,V$ three-phase supply. If the resistors are connected (a) in star, (b) in delta, calculate in each case the phase current and the line current.

*Solution*   (a) *Star connection*
The phase voltage is

$$V_P = V_L/\sqrt{3} = 440/\sqrt{3} = 254\,V$$

and the current in each phase is

$$I_P = V_P/Z_P = 254/10 = 25.4\,A$$

In a star-connected system, $I_P = I_L$, hence

$$I_L = 25.4\,A$$

Moreover, since the load is resistive, the phase current is in phase with the phase voltage, and the line current is in phase with the line voltage.

(b) *Delta connection*

The voltage applied to each phase of the load is, in this case, equal to the line voltage; that is $V_P = V_L = 440$ V. The current in each phase of the load is

$$I_P = V_P/Z_P = 440/10 = 44 \text{ A}$$

and

$$I_L = \sqrt{(3)}I_P = \sqrt{3} \times 44 = 76.2 \text{ A}$$

### 9.11 Summary of balanced three-phase systems

Table 9.1 gives the relationships between voltages and currents in balanced three-phase systems.

**Table 9.1**

|  | Phase current | Line current | Phase voltage | Line voltage |
|---|---|---|---|---|
| Star | $I_P = I_L$ | $I_L = I_P$ | $V_P = V_L/\sqrt{3}$ | $V_L = \sqrt{(3)}V_P$ |
| Delta | $I_P = I_L/\sqrt{3}$ | $I_L = \sqrt{(3)}I_P$ | $V_P = V_P$ | $V_L = V_P$ |

### 9.12 VA, power and VAr consumed by a balanced load

The VA consumed per phase by a balanced load is $V_P I_P$, hence

Total VA consumed, $S = 3 \times$ VA consumed per phase

$$= 3V_P I_P$$

The power consumed per phase is $V_P I_P \cos \phi$, where $\cos \phi$ is the power factor of the load, hence

Total power consumed, $P = 3 \times$ power per phase

$$= 3V_P I_P \cos \phi$$

and the VAr consumed per phase is $V_P I_P \sin \phi$, hence

Total VAr consumed, $Q = 3 \times$ VAr per phase

$$= 3V_P I_P \sin \phi$$

**Star-connected system**

In this case $I_P = I_L$ and $V_P = V_L/\sqrt{3}$, hence

$$3V_P I_P = 3 \times \frac{V_L}{\sqrt{3}} \times I_L = \sqrt{(3)}V_L I_L$$

### Delta-connected system

In this case $I_P = I_L/\sqrt{3}$ and $V_P = V_L$, hence

$$3V_P I_P = 3 \times V_L \times \frac{I_L}{\sqrt{3}} = \sqrt{(3)} V_L I_L$$

Substituting these values in the above equation gives (for either star or delta connection)

$$\text{Total VA consumed, } S = \sqrt{(3)} V_L I_L \qquad\qquad [9.9]$$

$$\text{Total power consumed, } P = \sqrt{(3)} V_L I_L \cos \phi \qquad\qquad [9.10]$$

$$\text{Total VAr consumed, } Q = \sqrt{(3)} V_L I_L \sin \phi \qquad\qquad [9.11]$$

and

$$S = \sqrt{(P^2 + Q^2)} \qquad\qquad [9.12]$$

***Worked example 9.5***   A three-phase, 440 V a.c. motor provides a full-load mechanical output of 10 kW when its power factor is 0.8 (lagging) and its efficiency is 90 per cent. Calculate at full load (a) the power consumed by the motor, (b) its line current and (c) the VA and VAr consumed.

*Solution*   (a) The input power to the motor is

$$P = \text{output power/efficiency} = 10/0.9 \text{ kW}$$
$$= 11.1 \text{ kW}$$

(b) Since the input power is equal to $\sqrt{(3)} V_L I_L \cos \phi$, then

$$I_L = P/(\sqrt{(3)} V_L \cos \phi) = 11.1 \times 10^3/(\sqrt{3} \times 440 \times 0.8)$$
$$= 18.22 \text{ A}$$

(c) The total VA consumed is

$$S = \sqrt{(3)} V_L I_L = \sqrt{3} \times 440 \times 18.22 = 13.9 \text{ kVA}$$

Since the power factor is 0.8, then

$$\phi = \cos^{-1} 0.8 = 36.87° \text{ (lagging)}$$

hence $\sin \phi = 0.6$. The VAr consumed is therefore

$$Q = \sqrt{(3)} V_L I_L \sin \phi = 13.9 \sin \phi \text{ kVAr}$$
$$= 13.9 \times 0.6 = 8.33 \text{ kVAr}$$

*Note:*
$$\sqrt{(P^2 + Q^2)} = \sqrt{[(11.1 \times 10^3)^2 + (8.33 \times 10^3)^2]}$$
$$= 13.9 \text{ } kVA = S$$

***Worked example 9.6***   Three identical coils are connected in star to a 440 V, three-phase, three-wire supply, and consume 1.5 kW when the line current is 2.5 A.

Calculate (a) the phase voltage applied to the load, (b) the power factor of the load, (c) the magnitude of the impedance of each coil, (d) the total VA, power and VAr consumed. If the coils are reconnected in delta calculate (e) the VA, power and VAr consumed.

*Solution*  (a) The phase voltage is

$$V_P = V_L/\sqrt{3} = 440/\sqrt{3} = 254 \text{ V}$$

(b) Since the load is balanced, the power consumed is given by $P = \sqrt{(3)} V_L I_L \cos \phi$, hence

$$\cos \phi = P/(\sqrt{(3)} V_L I_L) = 1500/(\sqrt{3} \times 440 \times 2.5)$$

$$= 0.7873$$

Moreover, since we are dealing with an inductive load, the phase angle is lagging.

(c) From Ohm's law, $V_P = I_P Z_P$, where $Z_P$ is the magnitude of the impedance of the load. Hence

$$Z_P = V_P/I_P = 254/2.5 = 101.6 \, \Omega$$

(d) The total VA consumed when connected in star is

$$S = \sqrt{(3)} V_L I_L = \sqrt{3} \times 440 \times 2.5 = 1905 \text{ VA}$$

The total power consumed is

$$P = \sqrt{(3)} V_L I_L \cos \phi = 1905 \times 0.7873 = 1500 \text{ W}$$

The phase angle of the load is $\cos^{-1}(0.7873) = 38.07°$, hence the VAr consumed is

$$Q = \sqrt{(3)} V_L I_L \cos \phi = 1905 \sin 38.07° = 1905 \times 0.6166$$

$$= 1175 \text{ VAr}$$

*Note:* $\sqrt{(P^2 + Q^2)} = \sqrt{(1500^2 + 1175^2)} = 1905 \text{ VA} = S$.

(e) When the coils are connected in delta, the phase voltage is 440 V, and the phase current is

$$I_P = V_P/Z_P = 440/101.6 = 4.33 \text{ A}$$

and

$$I_L = \sqrt{(3)} I_P = 7.5 \text{ A}$$

The VA consumed by the delta-connected load is

$$S = \sqrt{(3)} V_L I_L = \sqrt{3} \times 440 \times 7.5 = 5715 \text{ VA}$$

and

$$P = \sqrt{(3)} V_L I_L \cos \phi = 5715 \times 0.7873 = 4500 \text{ W}$$

$$Q = \sqrt{(3)} V_L I_L \sin \phi = 5715 \times 0.6166 = 3525 \text{ VAr}$$

When we compare the values of VA, power and VAr consumed in the delta system with those consumed in the star system, we see that the values are *three times greater in the delta case than in the star case*.

**9.13 Power consumed by an unbalanced three-phase load**

The power consumed by an unbalanced load is the *sum of the power consumed in each phase of the load*. The relationships developed in Section 9.12 *do not hold good for an unbalanced load*. The following example illustrates the method of determining the power consumed.

*Worked example 9.7*

A three-phase, delta-connected load contains the following elements, the line voltage being 1100 V:

> Phase B–R: a $100\,\Omega$ resistor in series with a $100\,\Omega$ capacitive reactance;

> Phase Y–B: a $50\,\Omega$ resistance in series with a $50\,\Omega$ inductive reactance;

> Phase Y–R: a $200\,\Omega$ resistance.

Calculate (a) the current in each phase of the load, (b) the power consumed by each phase of the load and (c) the total power consumed.

*Solution*

(a) For phase B–R

$$Z_{BR} = \sqrt{(R^2 + X_C^2)} = \sqrt{(100^2 + 100^2)} = 141.4\,\Omega$$

and the phase current is

$$I_{BR} = V_L/Z_{BR} = 1100/141.4 = 7.78 \text{ A}$$

For phase Y–B

$$Z_{YB} = \sqrt{(R^2 + X_L^2)} = \sqrt{(50^2 + 50^2)} = 70.7\,\Omega$$

and the phase current is

$$I_{YB} = V_L/Z_{YB} = 1100/70.7 = 15.56 \text{ A}$$

For phase R–Y

$$Z_{RY} = 200\,\Omega$$

and the phase current is

$$I_{RY} = V_L/Z_{RY} = 1100/200 = 5.5 \text{ A}$$

(b) The power consumed by each phase of the load is

> Power = (phase current)$^2$ × phase resistance

For phase B–R

> Power = $7.78^2 \times 100 = 6053$ W

For phase Y–B

> Power = $15.56^2 \times 50 = 12\,106$ W

For phase R–Y

> Power = $5.5^2 \times 200 = 6050$ W

(c) Total power = sum of the power consumed in each phase

$$= 6053 + 12\,106 + 6050 = 24\,209 \text{ W}$$

## 9.14 Measurement of power in a balanced three-phase load

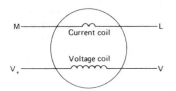

**Fig. 9.15** Wattmeter terminal markings

Electrical power is usually measured by means of an *electrodynamic wattmeter*, which is an analogue instrument. It contains two coils, as shown in Fig. 9.15; one coil is known as the *voltage coil* or *potential coil*, and is connected between the supply lines, and the other coil carries the load current, and is known as the *current coil*. The voltage coil has a high resistance, and very little current flows in it. The current coil has a very low resistance, and there is very little voltage drop across it.

The terminals are generally marked as shown in Fig. 9.13. The current coil terminal marked M is connected to the Mains, and terminal L is connected to the Load. The voltage coil terminal marked V+ is connected to a point in the circuit which has a higher potential than the point to which terminal V is connected; in most applications, V+ is connected to terminal M. The way in which a wattmeter is connected to measure power in a single-phase circuit is shown in Fig. 9.16.

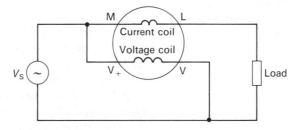

**Fig. 9.16** Wattmeter connections for a single-phase load

The *average deflecting torque* developed by a dynamometer wattmeter is

[Voltage across the voltage coil ($V$)]

$\times$ [current through current coil ($I$)]

$\times$ [cosine of the angle between $V$ and $I$]     [9.13]

A single wattmeter can be used to measure the power consumed by a *balanced three-phase load*, as shown in Fig. 9.17. The power in one phase of a balanced star-connected system is measured as shown in Fig. 9.17(a), and the total power consumed is three times the power indicated by the wattmeter.

Where the star point of the load is not available (as, for example, in a delta-connected load), an *artificial star point* can be created by connecting two impedances, $Z$, to the wattmeter as shown in Fig. 9.17(b); the value of the impedances must be equal to the impedance of the wattmeter voltage coil.

An extension to the method in Fig. 9.17(a) is to use *three wattmeters*, one connected in each phase. The total power consumed is the sum of the readings of the three meters.

Another alternative, in a three-phase three-wire system, is to use the *two-wattmeter method* of power measurement described

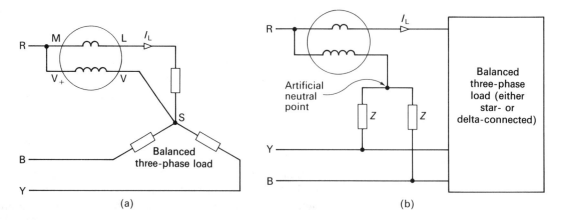

**Fig. 9.17** Measurement of balanced three-phase power using only one wattmeter (a) with a star-connected load in which the star point is available and (b) in a load whose star point is not available

in section 9.15. If two separate wattmeters are used, not only can the total power be measured (for either balanced or unbalanced loads) but also (if the load is balanced) we can determine the power factor of the load. Double-element wattmeters are manufactured, the two wattmeter movements having a common mechanical shaft; the net result is that the meter directly indicates the total power consumed.

*Worked example 9.8*   A wattmeter connected as shown in Fig. 9.17(a) is used to measure the total power consumed by a three-phase, star-connected system. The load is balanced and the impedance per phase comprises a $5\,\Omega$ resistance in series with a $10\,\Omega$ inductive reactance. If the line voltage is $415.7\,\text{V}$, determine (a) the power indicated by the wattmeter and (b) the total power consumed.

*Solution*   The magnitude of the impedance per phase is

$$Z_P = \sqrt{(R^2 + X_L^2)} = \sqrt{(5^2 + 10^2)} = 11.18\,\Omega$$

and its phase angle is

$$\phi = \tan^{-1}(X_L/R) = \tan^{-1}(10/5) = 63.43° \text{ (lagging)}$$

The magnitude of the current in each phase (and also in each line) is

$$I_L = V_P/Z_P = (415.7/\sqrt{3})/11.18 = 240/11.18 = 21.47\,\text{A}$$

(a) The power indicated by the wattmeter is calculated from eqn [9.13] as follows:

$$P = V_P I_P \cos\phi = 240 \times 21.47 \times \cos 63.43° = 2305\ \text{W}$$

(b) The total power consumed is

$$P_T = 3P = 6915 \text{ W}$$

*Note:*

$$\text{Total power} = \sqrt{(3)} V_L I_L \cos \phi$$

$$= \sqrt{3} \times 415.7 \times 21.47 \times \cos 63.43°$$

$$= 6915 \text{ W}$$

**9.15   The two-wattmeter method of measuring power**

Research has shown that we need $N$ wattmeters to measure the power consumed by a load supplied by $(N + 1)$ lines. That is, *we only need two wattmeters to measure the power supplied by a three-phase, three-wire system.* This applies whether the load is balanced or unbalanced.

One method of connecting the two wattmeters is shown in Fig. 9.18. Assuming that the load is star-connected, the *instantaneous power* consumed by the circuit is

$$P = v_{RS} i_R + v_{YS} i_Y + v_{BS} i_B \qquad [9.14]$$

We will now look at the instantaneous power acting on the two wattmeters in Fig. 9.18.

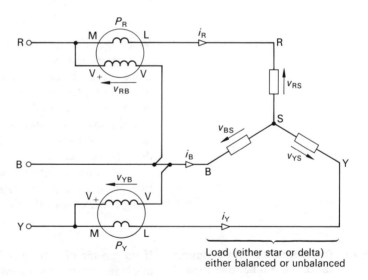

**Fig. 9.18**  Measurement of power in a three-phase, three-wire system using the two-wattmeter method

Load (either star or delta) either balanced or unbalanced

**For $P_R$**

The current in through the current coil is $i_R$, and the voltage of terminal V+ with respect to terminal V is $v_{RB} = v_{RS} - v_{BS}$. The instantaneous power acting on $P_R$ is

$$i_R v_{RB} = i_R(v_{RS} - v_{BS}) = i_R v_{RS} - i_R v_{BS}$$

**For $P_Y$**

The current in through the current coil is $i_R$, and the voltage of terminal V+ with respect to terminal V is $v_{RB} = v_{RS} - v_{BS}$. The instantaneous power acting on $P_Y$ is

$$i_Y v_{YB} = i_Y(v_{YS} - v_{BS}) = i_Y v_{YS} - i_Y v_{BS}$$

**Total instantaneous power**

This is given by

$$i_R v_{RB} + i_Y v_{YB} = i_R v_{RS} - i_R v_{BS} + i_Y v_{YS} - i_Y v_{BS}$$
$$= i_R v_{RS} + i_Y v_{YS} - v_{BS}(i_R + i_Y) \qquad [9.15]$$

If we apply KCL to the star point, S, of Fig. 9.18 we see that

$$i_R + i_Y + i_B = 0$$

or

$$i_B = -(i_R + i_Y) \qquad [9.16]$$

Substituting eqn [9.16] into eqn [9.15] tells us that the total instantaneous power indicated by the two wattmeters is

$$i_R v_{RS} + i_Y v_{YS} + i_B v_{BS}$$

When we compare this with eqn [9.14], we see that the two wattmeters indicate the total power consumed by the system. Since we have not indicated whether the load is balanced or unbalanced, the two-wattmeter method can be used to measure power in any three-phase, three-wire system.

The above proof was for a star-connected load. The reader will find it an interesting exercise to show that it is also true for a delta-connected load.

As a matter of interest, we can connect the wattmeters in any two phases, the low-voltage or 'V' terminal of the wattmeters being connected to the third phase.

**9.16 Determination of the power factor of a balanced load using two wattmeters**

If the power of a balanced inductive load of phase angle $\phi$ is measured by the two wattmeters connected as shown in Fig. 9.18, then the phasor diagram of the circuit for a phase angle less than 30° lagging is as shown in Fig. 9.19.

**For wattmeter $P_R$**

The r.m.s. current in the current coil is $I_R = I_L$, and the voltage across the voltage coil is $V_{RB} = V_L$. The phase angle between the

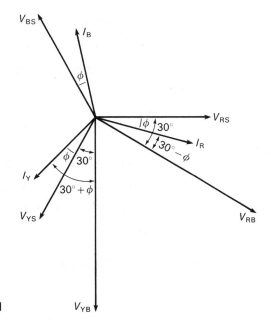

**Fig. 9.19** Phasor diagram for the measurement of power in a balanced load using the two-wattmeter method

current and the voltage is $(30° − \phi)$ (see Fig. 9.19). Hence

$$\text{Indication of } P_R = V_L I_L \cos(30° − \phi) \qquad [9.17]$$

**For wattmeter $P_Y$**

The r.m.s. current in the current coil is $I_Y = I_L$, and the voltage across the voltage coil is $V_{YB} = V_L$. The phase angle between the current and the voltage is $(30° + \phi)$ (see Fig. 9.19). Hence

$$\text{Indication of } P_Y = V_L I_L \cos(30° + \phi) \qquad [9.18]$$

**Total power**

The total power consumed by the load is

$$
\begin{aligned}
P_R + P_Y &= V_L I_L [\cos(30° − \phi) + \cos(30° + \phi)] \\
&= V_L I_L [(\cos 30° \cos \phi + \sin 30° \sin \phi) \\
&\quad + (\cos 30° \cos \phi − \sin 30° \sin \phi)] \\
&= \sqrt{(3)} V_L I_L \cos \phi \\
&= \text{total power in a balanced three-phase system}
\end{aligned}
$$

$$[9.19]$$

Also from eqns [9.17] and [9.18] we get

$$P_R - P_Y = V_L I_L [\cos(30° - \phi) - \cos(30° + \phi)]$$

$$= V_L I_L [(\cos 30° \cos \phi + \sin 30° \sin \phi)$$

$$- (\cos 30° \cos \phi - \sin 30° \sin \phi)]$$

$$= V_L I_L \sin \phi = \frac{\text{three-phase VAr}}{\sqrt{3}} \qquad [9.20]$$

Dividing eqn [9.20] by eqn [9.19] gives

$$\frac{P_R - P_Y}{P_R + P_Y} = \frac{V_L I_L \sin \phi}{\sqrt{(3)} V_L I_L \cos \phi} = \frac{\tan \phi}{\sqrt{3}} \qquad [9.21]$$

but

$$\tan^2 \phi = \sec^2 \phi - 1 = \frac{1}{\cos^2 \phi} - 1$$

or

$$\cos^2 \phi = \frac{1}{1 + \tan^2 \phi}$$

therefore

$$\cos \phi = \frac{1}{\sqrt{(1 + \tan^2 \phi)}} \qquad [9.22]$$

From eqn [9.21] we see that

$$\tan^2 \phi = 3 \left[ \frac{P_R - P_Y}{P_R + P_Y} \right]^2$$

substituting this into eqn [9.22] gives

$$\cos \phi = \frac{1}{\sqrt{\left( 1 + 3 \left[ \frac{P_R - P_Y}{P_R + P_Y} \right]^2 \right)}} \qquad [9.23]$$

Alternatively we may take a simpler route as follows. From the power triangle we may say that

$$\tan \phi = \frac{\text{total VAr}}{\text{total watts}} = \frac{\sqrt{3}(P_R - P_Y)}{P_R + P_Y}$$

from which we may evaluate $\phi$. Hence

$$\text{Power factor} = \cos \phi = \cos(\tan^{-1} \phi)$$

$$= \cos \left( \tan^{-1} \left[ \frac{\sqrt{3}(P_R - P_Y)}{P_R + P_Y} \right] \right)$$

***Worked example 9.9*** The power consumed by a balanced load is measured by two wattmeters connected as shown in Fig. 9.18. If $P_R$ indicates 80 kW and $P_Y$ indicates 40 kW determine (a) the total power consumed, (b) the total kVAr consumed and (c) the power factor of the load.

*Solution* (a) The total power consumed is

$$P_T = P_R + P_Y = 80 + 40 = 120\,\text{kW}$$

(b) From eqn [9.20], the total kVAr consumed is

$$Q_T = \sqrt{3}(P_R - P_Y) = \sqrt{3}(80 - 40) = 69.28\ \text{kVAr}$$

(c) From eqn [9.23], the power factor of the load is

$$\cos\phi = \cfrac{1}{\sqrt{\left(1 + 3\left[\dfrac{P_R - P_Y}{P_R + P_Y}\right]^2\right)}}$$

$$= \cfrac{1}{\sqrt{\left(1 + 3\left[\dfrac{(80 - 40) \times 10^3}{(80 + 40) \times 10^3}\right]^2\right)}} = 0.866$$

Alternatively

$$\tan\phi = \frac{\text{total kVAr}}{\text{total kW}} = \frac{69.28}{120} = 0.5773$$

or

$$\phi = \tan^{-1} 0.5773 = 30°$$

and

$$\text{Power factor} = \cos\phi = 0.866$$

**9.17 Effect of particular values of phase angle on the two-wattmeter method**

There are a few special values of phase angle which are of interest to engineers. These values are as follows (the reader should also refer to the phasor diagram in Fig. 9.19):

1. $\phi = 0°$ (*unity power factor*)

$$P_R = P_Y$$

2. $\phi = 60°$ *lag* (*power factor = 0.5 lagging*). $P_R$ indicates the total power consumed, and $P_Y$ indicates zero.
3. $\phi > 60°$ *lag* (*power factor < 0.5 lagging*). $P_R$ indicates a value greater than the total power, and $P_Y$ gives a negative indication.
4. $\phi = 60°$ *lead* (*power factor = 0.5 leading*). $P_R$ indicates zero and $P_Y$ indicates the total power consumed.
5. $\phi > 60°$ *lead* (*power factor < 0.5 lagging*). $P_R$ gives a negative indication and $P_Y$ indicates a value greater than the total power.

Even when one of the instruments gives a negative (reverse) reading, the total power consumed is still equal to the sum of the readings of the two instruments. When one of the instruments gives a negative reading, the correct magnitude reading is obtained either by reversing the connections to the current coil, or reversing the connections to the voltage coil (*but not both*). The resulting reading is then given a negative sign.

### Problems

**9.1** A balanced three-phase load takes a line current of 40 A at a line voltage of 11 kV. Determine the line and phase values of the voltage and current if the load is (a) star-connected, (b) delta-connected.

[(a) $I_L = 40$ A, $I_P = 40$ A, $V_L = 11$ kV, $V_P = 6.35$ kV; (b) $I_L = 40$ A, $I_P = 23.1$ A, $V_L = 11$ kV, $V_P = 11$ kV]

**9.2** If the phase voltage of a three-phase, star-connected alternator is 250 V, determine the line voltage. If the connections to one of the phases are reversed, what will be the value of the line voltages?

[433 V; 433 V, 250 V, 250 V]

**9.3** The line currents taken by a three-phase, four-wire load are as follows:

R-line: 25.4 A in the reference direction;
Y-line: 16.93 A lagging $I_R$ by $150°$;
B-line: 25.4 A leading $I_R$ by $150°$.

Determine the magnitude of the current in the neutral wire and its phase angle with respect to $I_R$.

[12.03 A; $159.4°$ leading]

**9.4** A three-phase alternator supplies a balanced 600 kVA load. if the line current is 80 A, calculate the line voltage.

[4.33 kV]

**9.5** A balanced three-phase load takes a line current of 100 A at a line voltage of 450 V. If the power factor of the load is 0.8 lagging, calculate the apparent power, the active power and the reactive power consumed.

[77.94 kVA; 62.35 kW; 46.76 kVAr]

**9.6** A balanced star-connected, three-phase load consumes 50 kW at a power factor of 0.8 lagging, the line voltage being 550 V. Calculate (a) the line current, (b) the phase voltage and (c) the kVA and kVAr consumed.

[(a) 65.6 A; (b) 317.5 V; (c) 62.49 kVA, 37.5 kVAr]

**9.7** A three-phase, four-wire, star-connected load comprises the following:

R-phase: a $10 \, \Omega$ resistor in series with an inductive reactance of $15 \, \Omega$;
Y-phase: a $25 \, \Omega$ resistor;
B-phase: a $30 \, \Omega$ resistor in series with a capacitive reactance of $50 \, \Omega$.

Determine (a) the magnitude of each phase current and its phase angle with respect to its own phase voltage, (b) the magnitude of the neutral current and its phase angle with respect to $V_{RN}$, (c) the power consumed by each phase of the load and the total power consumed. Draw the phasor diagram to scale.

[(a) $I_R = 17.61$ A lagging $V_{RN}$ by $56.3°$, $I_Y = 12.7$ A in phase with $V_{YN}$, $I_B = 5.45$ A leading $V_{BN}$ by $59°$; (b) $I_N = 25.63$ A lagging $V_{RN}$ by $94.52°$; (c) $P_R = 3.1$ kW, $P_Y = 4$ kW, $P_B = 0.89$ kW, $P_T = 7.99$ kW]

**9.8**   A three-phase, 440 V, delta-connected induction motor takes a line current of 40 A at a power factor of 0.8 lagging. Calculate (a) the input power to the motor, (b) the kVAr consumed and (c) the phase current.

[(a) 24.39 kW; (b) 18.29 kVAr; (c) 32.1 A]

**9.9**   (i) Three coils each of resistance $9\,\Omega$ and inductive reactance $12\,\Omega$ are connected in delta to a three-phase 440 V supply. Calculate (a) the line current, (b) the power factor of the load, (c) the kVA consumed and (d) the kW consumed.
(ii) If the coils are reconnected in star, determine the new values for (a) to (d) above.

[(i) (a) 50.7 A; (b) 0.6; (c) 38.6 kVA; (d) 23.16 kW; (ii) (a) 16.9 A; (b) 0.6; (c) 12.866 kVA; (d) 7.719 kW]

**9.10**  The power consumed by a partially loaded three-phase induction motor is measured by the two-wattmeter method. The readings are as follows:

Wattmeter $P_1$: 15 kW;
Wattmeter $P_2$: 3 kW.

If the efficiency of the motor is 60 per cent, calculate (a) the power output from the motor in kW and (b) the power factor of the motor.

[(a) 10.8 kW; (b) 0.65]

**9.11**  The power consumed by a balanced three-phase load is measured by the two-wattmeter method, and the readings of the two wattmeters are 2 kW (forward) and 0.33 kW (reverse), respectively. Calculate (a) the total power dissipated by the load and (b) the power factor of the load.

[(a) 1.67 kW; (b) 0.38]

# 10 The transformer

## 10.1 Introduction

The general principle of the transformer was briefly touched on when mutual inductance was described in Chapter 5. Broadly speaking, there are two types of transformer, namely the ideal transformer (or power transformer) and the linear transformer.

A **power transformer** or **ideal transformer** can be thought of as an a.c. 'machine' for transforming an alternating voltage (or current) of one level into a voltage (or current) of another level with minimum power loss. In these transformers the magnetic coupling between the windings can be regarded as near-perfect.

For example, electricity may be generated at, say, 11 kV. For a given amount of power, it is more economic to transmit it at a high voltage (say 275 or 400 kV) because the current needed at the higher voltage is less than is needed if the transmission voltage is lower.

The lower value of current needed to transmit the power brings with it two further advantages. Firstly, a smaller conductor is needed for the smaller current, with a consequent economic advantage. Secondly, since the current is smaller, the $I^2R$ power loss (the *copper loss*) in the cable is reduced.

Once the power has reached the consumer it is transformed to a lower voltage (say 3.3 kV or 415 V three-phase, and 240 or 110 V single-phase) for use by the consumer.

In addition to the conventional power transformer, there are other versions of the ideal transformer, including auto-transformers and instrument transformers, which we shall study.

A **linear transformer** has an air core, in which the magnetic coupling between the windings is not perfect. Typical applications include radio-frequency and communications circuits.

## 10.2 The ideal transformer or power transformer

An 'ideal' transformer is one which has no power loss, and all the magnetic flux which leaves the primary winding links with all windings on the magnetic core.

In fact such a transformer does not exist in practice, but the *power transformer* approaches it as nearly as possible. We devote the majority of this chapter to the power transformer.

The general construction of an ideal transformer is shown in Fig. 10.1, in which the two windings (namely the *primary winding* and the *secondary winding*) are wound on an *iron core* or *iron*

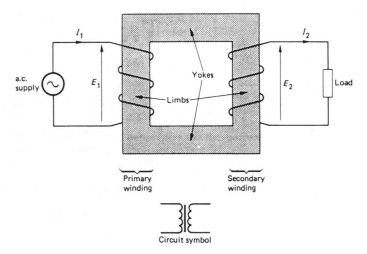

**Fig. 10.1** A two-winding transformer

circuit in which the magnetic flux is established. Since the transformer has two separate windings, we refer to it as a *two-winding* transformer.

The supply voltage is connected to the primary winding of the transformer, and the load is connected to the secondary winding (some transformers may, in fact, have more than one secondary winding). The windings are supported on the *limbs* of the transformer, which are linked together by the *yokes*.

As explained in Chapter 5, the current which flows in the primary winding establishes a magnetic flux in the iron circuit. It is this magnetic flux which induces an e.m.f. in the secondary winding, and causes current to flow in the load. It was also shown in Chapter 5 that the e.m.f. induced in a coil of $N$ turns on the magnetic circuit is given by

$$e = N \times \text{rate of change of flux} = N \times d\Phi/dt$$

or

$$\frac{d\Phi}{dt} = \frac{e}{N}$$

Since every coil in an ideal transformer is subject to the same magnetic flux, they are also subject to the same rate of change of flux. It follows that if a sinusoidal voltage of r.m.s. value $E_1$ is applied to the primary winding of $N_1$ turns then, for a two-winding transformer, the relationship between $E_1$ and $E_2$ is

$$\frac{E_1}{N_1} = \frac{E_2}{N_2} \qquad [10.1]$$

That is to say, *each winding on the transformer supports the same number of volts per turn.*

Equation [10.1] can be rewritten in the form

$$\frac{N_2}{N_1} = \frac{E_2}{E_1} \qquad [10.2]$$

If $E_2$ is greater than $E_1$, we say that the transformer has a *step-up voltage ratio*, and if $E_2$ is less than $E_1$, we say that the transformer has a *step-down voltage ratio*.

Since we are dealing with an ideal transformer we can, for the moment, assume that the power loss in the transformer is zero. That is

VA supplied to the primary winding

= VA supplied to the secondary winding

or

$$E_1 I_1 = E_2 I_2 \qquad [10.3]$$

hence

$$\frac{E_2}{E_1} = \frac{I_1}{I_2} \qquad [10.4]$$

Arising from eqn [10.3] we see that the volt-ampere product associated with both windings is the same. That is

the 'high' voltage winding carries a 'low' current, and the 'low' voltage winding carries a 'high' current

and from eqn [10.4] we see that

a transformer with a 'step-up' voltage ratio has a 'step-down' current ratio, and vice versa.

Combining eqns [10.2] and [10.4] yields

$$\frac{N_2}{N_1} = \frac{E_2}{E_1} = \frac{I_1}{I_2} \qquad [10.5]$$

and from this equation it can be seen that

$$N_1 I_1 = N_2 I_2 \qquad [10.6]$$

That is *each winding supports the same number of ampere-turns.*

*Worked example 10.1*  A single-phase power transformer reduces the voltage from 240 to 6 V. If the secondary winding has 10 turns of wire on it and the primary current is 2.5 A, determine (a) the number of turns on the primary winding and (b) the current in the secondary winding.

*Solution*  (a) Since each winding supports the same number of volts per turn then $E_1/N_1 = E_2/N_2$, hence

$$N_1 = \frac{E_1}{E_2} N_2 = \frac{240}{6} \times 10 = 400 \text{ turns}$$

(b) To maintain ampere-turn balance between the windings $N_1 I_1 = N_2 I_2$, or

$$I_2 = N_1 I_1 / N_2 = 2.5 \times 400/10 = 100 \text{ A}$$

## 10.3  Rating of transformers

The rating of electrical apparatus is set by its ability to dissipate the heat developed within it when it is operating. Heat is developed in a transformer arising from its iron loss and its copper loss. The iron loss is a 'constant' loss, and the copper loss depends on the load current handled by the transformer.

For this reason, the rating of a transformer is given in VA (or, in the case of a large transformer in kVA or MVA), rather than in watts. For example, the rating of the transformer in Worked Example 10.1 must be at least

$$240 \text{ V} \times 2.5 \text{ A} = 600 \text{ VA}$$

## 10.4  Impedance matching using a transformer

It is shown in Chapter 12 that maximum power is transferred from a power supply into a resistive load, when the internal resistance of the source is equal to the resistance of the load.

We will show here that an ideal (lossless) transformer can be used as a **matching device** in an electronic circuit, to convert the apparent resistance of a load (say a loudspeaker) into another value. In this way it is possible in electronic circuits to obtain maximum power transfer from a source to a load.

We will use the usual convention that the primary voltage and current are $V_1$ and $I_1$, the secondary voltage and current are $V_2$ and $I_2$, and the number of turns on the primary winding and secondary winding are $N_1$ and $N_2$; the load resistance is $R_L$. The current in the load is

$$I_2 = V_2 / R_L$$

or

$$R_L = V_2 / I_2$$

Now, for an ideal transformer

$$I_1 = \frac{N_2}{N_1} I_2$$

and

$$V_1 = \frac{N_1}{N_2} V_2$$

The apparent resistance, $R_{in}$, 'seen' when 'looking into' the primary winding terminals is

$$R_{in} = \frac{V_1}{I_1} = \left[\frac{N_1}{N_2}V_2\right] \times \left[\frac{N_1}{N_2 I_2}\right] = \left[\frac{N_1}{N_2}\right]^2 \frac{V_2}{I_2}$$

$$= \left[\frac{N_1}{N_2}\right]^2 R_L \hspace{2cm} [10.7]$$

That is, the apparent resistance 'seen' when 'looking into' the primary winding terminals of the transformer is $(N_1/N_2)^2$ times the value of the actual load resistance. Thus, by altering the value of the turns ratio it is possible to 'alter' the apparent resistance of the load as seen by the amplifier.

*Worked example 10.2*   An amplifier having an **output resistance** of $375\,\Omega$ is connected via a transformer to a $15\,\Omega$ loudspeaker. Calculate the turns ratio of the transformer needed to ensure maximum power transfer to the load.

*Solution*   From eqn [10.7]

$$R_{in} = \left[\frac{N_1}{N_2}\right]^2 R_L$$

or

$$\frac{N_1}{N_2} = \sqrt{(R_{in}/R_L)}$$

For maximum power transfer to take place, $R_{in}$ must be $375\,\Omega$, or

$$\frac{N_1}{N_2} = \sqrt{(375/15)} = 5$$

That is, we need a transformer with a step-down voltage ratio of 5:1.

## 10.5   The auto-transformer

The **auto-transformer** has a single winding, as shown in Fig. 10.2, the winding being 'tapped' to provide either the secondary or primary winding connection points [see diagrams (a) and (b), respectively], depending on whether a step-down or step-up voltage ratio is required.

The reader will be familiar with this type of transformer as a variable-voltage a.c. source used in laboratory experiments. In this case, the secondary voltage is obtained from a sliding connection on the winding.

Once again, the relationships for the ideal transformer hold good. That is, each winding supports the same number of volts per turn, and each winding supports the same number of ampere-turns, or

$$\frac{E_1}{N_1} = \frac{E_2}{N_2} \quad \text{and} \quad I_1 N_1 = I_2 N_2$$

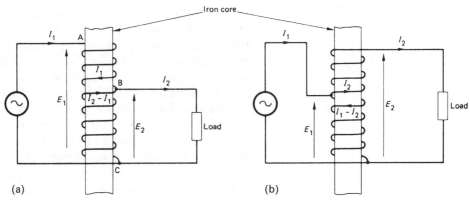

**Fig. 10.2**  An auto-transformer
giving (a) a step-down voltage ratio,
(b) a step-up voltage ratio

where $N_1$ is the number of turns between A and C in Fig. 10.2
and $N_2$ the number of turns between B and C in the same figure.
While we refer to the primary and secondary windings above, in
an auto-transformer the winding is common to both! If, in Fig.
10.2(a), the step-down voltage ratio is 1.2:1, and the current drawn
from the supply is 12 A, then

$$I_2 = I_1 \frac{N_1}{N_2} = 12 \times 1.2 = 14.4 \, \text{A}$$

The current in the upper part of Fig. 10.2(a) is 12 A, and in the
lower part is $(14.4 - 12) = 2.4 \, \text{A}$.

## 10.6  Instrument transformers

It is impractical to operate some instruments at, say, the very
high voltage associated with the grid distribution system, or the
very high current associated with an industrial electroplating
process. It is therefore necessary to use transformers which reduce
the voltage or current (or both in the case of wattmeters) to one
which is realistic.

To reduce the voltage, we use a two-winding **voltage transformer**
or **potential transformer** (abbreviated to VT or PT, respectively),
whose nominal secondary voltage is 110 V in the UK or 115 V
in the USA. These transformers are ideal transformers, and *should
only have one instrument connected to their secondary winding*.

To reduce the value of the current we use a two-winding **current
transformer** (abbreviated to CT), whose nominal full-load
secondary current is 5 A. Since the transformer is designed to
carry current, the windings have only a few turns on them and
have very little resistance. In fact, the primary 'winding' may
simply be the conductor which carries the main current, the
secondary winding being a few turns of wire wrapped around the
conductor.

Once again, the CT is an ideal transformer, and *there is an implied risk in its use* as follows. From earlier theory, an ideal transformer maintains ampere-turn balance between the windings and, whatever the value of the primary winding current, the secondary winding current is proportional to it.

If the ammeter connected to the secondary winding becomes disconnected for any reason, the transformer will try to force the secondary current through the disconnection point. This has two effects: firstly, it produces a very high voltage at the disconnection point and, secondly, the transformer temperature can rise to a dangerously high value with the implied risk of fire.

The secondary winding of a CT must therefore never be open-circuited when in use. *Always short-circuit the secondary winding of a CT when the ammeter is disconnected.*

## 10.7 Transformer efficiency

The efficiency, $\eta$, of a transformer is given by

$$\text{Efficiency, } \eta = \frac{\text{output power}}{\text{input power}}$$

Since the efficiency is the ratio of two values of the same type (namely power), it is dimensionless. In this case it is given as a *per unit* (p.u.) value; in other cases the ratio is multiplied by 100, and is called the per cent efficiency. Now, in a transformer

Input power = output power + power loss

where the power loss is given by

Power loss = iron loss + copper loss

where

Iron loss = hysteresis loss + eddy current loss

and

$$\text{Copper loss} = I_1^2 R_1 + I_2^2 R_2$$

Hence

$$\text{Efficiency} = \frac{\text{output power}}{\text{output power} + \text{iron loss} + \text{copper loss}}$$

Once again, this value is in p.u.; when multiplied by 100 it is the per cent efficiency.

*Worked example 10.3*    A 15 kVA single-phase transformer has an iron loss of 240 W. When the transformer is on full load at unity power factor, the efficiency is 97.5 per cent. Estimate the full-load copper loss.

*Solution*   At unity power factor the full-load output power is 15 kW, and the input power is

Input power = output/efficiency

$$= 15\,000/0.975 = 15\,384\,\text{W} \quad\text{or}\quad 15.384\,\text{kW}$$

Now

Total loss = input power − output power

$$= 15\,384 − 15\,000 = 384\,\text{W}$$

$$= \text{copper loss} + \text{iron loss}$$

hence

Copper loss = total loss − iron loss

$$= 384 − 240 = 144\,\text{W}$$

**10.8  The linear transformer or air-core transformer**

**Linear transformers** have an air core, and have a very low value of magnetic coupling coefficient between the windings; that is, not too much of the flux produced by the primary winding links with the secondary winding. It is this feature which makes the linear transformer of particular value in radio, TV and communications circuits. They are widely used in tuned coupled circuits, and the low value of magnetic coupling coefficient, in association with other components in the circuit, gives them desirable band-pass or band-stop filter characteristics.

**Problems**

**10.1**   A 600/240 V single-phase transformer has a rating of 15 kVA. Determine the full-load value of the primary current and the secondary current. Neglect the effects of power loss in the transformer.

[25 A; 62.5 A]

**10.2**   If the rated secondary voltage of a 25 kVA transformer is 3.3 kV, calculate the full-load value of the secondary current.

[7.58 A]

**10.3**   A 250/415 V single-phase transformer is supplied at 250 V and draws a current of 10 A from the supply. Neglecting the effects of power loss in the transformer, determine (a) the rating of the transformer in kVA and (b) the full-load secondary current.

[2.5 kVA; 6.02 A]

**10.4**   A single-phase transformer supplies a load current of 16 A. If the primary-to-secondary turns ratio is 160:20, determine the value of the primary current if the effects of power loss can be ignored.

[2 A]

**10.5**   The induced e.m.f. per turn on a two-winding transformer is 0.8 V. Determine the number of turns on each winding if the transformer has a step-down voltage ratio of 3.3 kV/240 V.

[4125; 300]

**10.6**   A 5 kVA transformer supplies a lighting load comprising parallel-connected 240 V, 100 W tungsten filament lamps. If the transformer is rated at 3.3 kV, 240 V, calculate (a) the number of lamps that may be connected without overloading the transformer, (b) the full-load secondary current and (c) the full-load primary current. Neglect the effects of power loss.

[(a) 50; (b) 20.83 A; (c) 1.52 A]

**10.7**   A single-phase transformer has an iron loss of 40 W at a supply frequency of 40 Hz, and an iron loss of 80 W at 60 Hz, the same voltage being applied in each case. Assuming the same value of peak flux density in both cases, calculate the hysteresis loss and the eddy current loss for a supply frequency of 50 Hz.

[16.6 W; 41.75 W]

**10.8**   The power loss in a transformer amounts to 5 kW when it supplies a load of 160 kW. Determine the efficiency of the transformer.

[97 per cent]

**10.9**   A 5 kVA single-phase transformer has an iron loss of 80 W and a full-load copper loss of 100 W. Determine the full-load efficiency if the power factor of the load is (a) unity, (b) 0.8 lagging.

[(a) 96.5 per cent; (b) 95.7 per cent]

**10.10**   The output power supplied by a single-phase transformer is 10 kW. If the copper loss at this load is 200 W and the efficiency is 97 per cent, determine the iron loss.

[109.3 W]

**10.11**   The primary and secondary voltages of an auto-transformer are 500 and 450 V, respectively. Determine the value of the current flowing in each section of the transformer winding if the secondary load current is 50 A. Neglect losses in the transformer.

[45 A; 5 A]

**10.12**   An auto-transformer is used to increase a voltage from 240 to 260 V. If the total number of turns on the transformer is 1300, determine (a) the position of the tapping point and (b) the current in each part of the winding when the connected load is 5 kVA. Neglect losses in the transformer.

[50 turns from one end; 19.23 A, 1.6 A]

# 11 Rectifier circuits and an introduction to power electronics

## 11.1 Introduction

A **rectifier** is a device which allows current to flow freely in one direction only; this introduces us to the concept of **power electronics**, with which we are all familiar.

The primary function of a rectifier circuit is to convert or to 'rectify' an alternating voltage into a direct voltage. Since the speed of d.c. machines can be easily controlled, *controlled rectifier* (**thyristor**) systems are widely used in industry.

Applications of power electronics range from simple battery-charging circuit, through the control of electric lights, to the control of large industrial systems.

The reverse of rectification is **inversion**, in which direct current is converted into alternating current. Invertors themselves incorporate controlled rectifiers.

## 11.1 The diode and the thyristor

In this section we will take a brief look at the operation of two very important semiconductor devices, namely the diode and the thyristor – see Fig. 11.1.

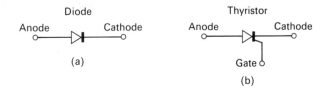

**Fig. 11.1** (a) A semiconductor diode and (b) a reverse blocking thyristor

A **diode** has two electrodes, namely its **anode** and its **cathode**, and allows *easy flow of current through it when the anode is positive with respect to the cathode*. However, when the anode is negative with respect to the cathode, no current (or, in practice, very little current) flows through it. That is, the diode is the electrical equivalent of a mechanical non-return valve.

When the anode of the diode is positive with respect to the cathode, we say that it is **forward biased**; when the anode is negative with respect to the cathode it is **reverse biased**, and it is in a **reverse blocking mode**.

The **thyristor** (also known as a *silicon controlled rectifier* or SCR) has three electrodes. Beside having an anode and a cathode, it has a **gate** electrode [see Fig. 11.1(b)] which is used to 'trigger' the thyristor into its **conducting mode**. *If the gate is not activated,* the thyristor blocks the flow of current through the device (whatever the polarity of the anode).

The thyristor can be triggered into operation when a pulse of current is injected into the gate *provided that the anode is positive with respect to the cathode*; the thyristor will not be triggered into operation if the anode is negative with respect to the cathode.

From the above, we see that *the diode acts as an uncontrolled rectifier*, because it conducts whenever the anode is positive with respect to the cathode. It also follows that *the thyristor can be thought of as a controlled rectifier*, because the gate electrode allows us to control the point where the thyristor is triggered or 'fired'.

## 11.3 Single-phase half-wave rectifier circuit

When a rectifier diode is connected in circuit, Fig. 11.2(a), current flows in the circuit when the anode of the diode is positive with respect to the cathode; that is, current flows in the positive half-cycles of the supply voltage waveform, as shown in Fig. 11.2(b). In the negative half-cycles, the diode blocks the flow of current in the circuit. Since current only flows during positive half-cycles, the circuit is described as a **single-phase, half-wave rectifier circuit**.

When an 'ideal' diode conducts it has no *forward resistance*, and no voltage is dropped across it. Similarly, when it is in its reverse blocking mode, it has infinite resistance, and no current flows through it. In the latter case, all of the negative half-cycle of the supply appears across the diode, and the *reverse blocking*

**Fig. 11.2** (a) A single-phase half-wave rectifier circuit and (b) the supply voltage, load voltage and load current waveforms for a resistive load

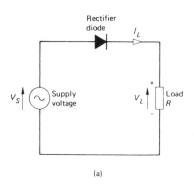

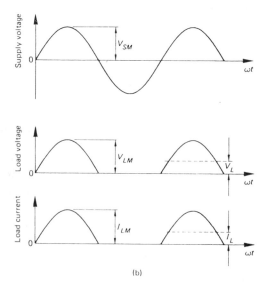

*voltage* rating of the diode must be at least equal to the peak supply voltage (in fact, it is usually much greater than this). The maximum voltage across the diode when it is blocking is known as the **peak inverse voltage**.

We may therefore think of a diode as being equivalent to a voltage-operated switch, which is 'closed' when the anode is positive with respect to the cathode, and is 'open' when the anode is negative. For all practical purposes, we can think of the load as being directly connected to the supply during positive half-cycles, and being disconnected from the supply during negative half-cycles. *The rectifier circuit therefore converts the alternating supply voltage into a unidirectional (d.c.) voltage across the load.* For a sinusoidal supply, the mean voltage across a resistive load can be estimated by the mid-ordinate rule as follows.

If we take ordinates at 20° intervals, the *mid-ordinates* during the first half-cycle for a sinusoidal supply of maximum value $V_M$ are given in Table 11.1; the sum of these is $5.756V_M$. Between 180° and 360°, the voltage across the load is zero, so that the sum of the voltage mid-ordinates over the complete cycle remains at $5.756V_M$. Consequently, using the mid-ordinate rule, we estimate the mean value of the load voltage to be $5.756V_M/18 = 0.319V_M$. To be strictly accurate, we should take an infinite number of mid-ordinates, i.e. we should integrate the wave with respect to time. If we integrate the rectified wave, we obtain an average load voltage of

**Table 11.1**

| Mid-ordinate | Value |
|---|---|
| 10° | $0.173V_M$ |
| 30° | $0.5V_M$ |
| 50° | $0.766V_M$ |
| 70° | $0.939V_M$ |
| 90° | $V_M$ |
| 110° | $0.939V_M$ |
| 130° | $0.766V_M$ |
| 150° | $0.5V_M$ |
| 170° | $0.173V_M$ |
| Total | $5.756V_M$ |

$$V_L = 0.318V_M = \frac{V_M}{\pi} = 0.45V_S \qquad [11.1]$$

where $V_S$ is the r.m.s. supply voltage. Also, the average value of the load current is

$$I_L = \frac{V_L}{R} = \frac{I_M}{\pi} = 0.318I_M$$

where $I_M$ is the maximum value of the load current.

In practice the 'ideal' diode does not exist. When conducting, a practical diode has some resistance, and there is a small voltage drop across diode. When it is in its reverse blocking mode, there is some leakage current, and a small value of negative voltage across the load. The net effect is that the mean voltage across the load is slightly less than one would expect, and the mean value of the load current is also less than one would expect.

*Worked example 11.1*   A 10 Ω resistor is supplied by a single-phase, half-wave rectifier circuit which uses an ideal diode. If the a.c. supply is 240 V r.m.s., calculate (a) the maximum value of the voltage across the load, (b) the average load voltage, (c) the direct current flowing in the load, (d) power dissipated by the load and (e) the peak inverse voltage across the diode.

*Solution*  (a) Since the diode is 'ideal', no voltage drop occurs in the diode and the maximum voltage across the load is equal to the maximum supply voltage, and is

$$V_{LM} = \sqrt{(2)} V_{SM} = \sqrt{2} \times 240 = 339.4 \text{ V}$$

(b) The mean voltage across the load is

$$V_L = 0.318 V_{SM} = 0.318 \times 339.4 = 108 \text{ V}$$

*Note*: From eqn [11.1]

$$V_L = 0.45 V_S = 0.45 \times 240 = 108 \text{ V}$$

(c) The load current is

$$I_L = V_L/R = 108/10 = 10.8 \text{ A}$$

(d) The power consumed by the load is

$$P = I_L^2 R = 10.8^2 \times 10 = 1166 \text{ W}$$

(e) The peak inverse voltage across the diode is equal to the peak negative supply voltage, that is

$$\text{Peak inverse voltage} = -V_{SM} = -339.4 \text{ V}$$

## 11.4 Single-phase, full-wave rectifier circuits

The half-wave rectifier circuit in Fig. 11.2 is not very efficient because it only uses one-half of the a.c. wave in the rectification process. The circuits in Figs 11.3 and 11.4 are much more efficient, since they use both half-cycles of the a.c. wave in the rectification process, and are known as **full-wave rectifier circuits**.

The **biphase rectifier circuit** in Fig. 11.3 is so named because it is energized by the *centre-tapped secondary winding* of a two-winding transformer. The two halves of the secondary winding have e.m.f.s induced in them which act in the same direction so that, for example, the 'top' connection of each half-winding has the same instantaneous potential.

Consequently in, say, the positive half-cycle of the waveform,

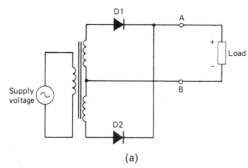

(a)

**Fig. 11.3** (a) Biphase rectifier circuit and (b) the output voltage waveform across a resistive load

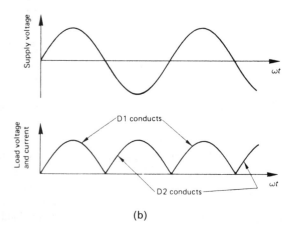

(b)

diode D1 is forward biased and conducts, while diode D2 is reverse biased and blocks the flow of current. In the negative half-cycle, the polarity of the induced voltage in the secondary winding reverses, so that D1 blocks the flow of current and D2 conducts. The waveforms in Fig. 11.3(b) illustrate the general result.

Since the biphase circuit needs a transformer, the d.c. output voltage can be selected by means of the turns ratio of the transformer. Moreover, since the transformer electrically isolates the secondary winding from the primary, the circuit is particularly useful where safety is important. While these features may be advantageous, it has the disadvantage of the cost, size and weight of the transformer. Moreover, the transformer will reduce the electrical efficiency of the circuit.

The **bridge rectifier circuit** in Fig. 11.4 does not need a transformer, but one may be necessary if both the d.c. and a.c. sides of the circuit must be separately earthed.

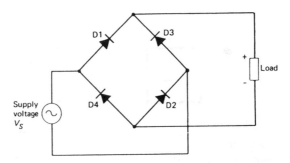

**Fig. 11.4** Single-phase bridge rectifier circuit

In operation, diagonally opposed pairs of rectifiers conduct. In, say, the positive half-cycle, diodes D1 and D2 conduct, while D3 and D4 block the flow of current. In the negative half-cycle, D1 and D2 are reverse biased while D3 and D4 are forward biased. The supply and output waveforms in Fig. 11.3(b) apply to the circuit in Fig. 11.4.

If we assume that the same a.c. supply voltage is used in the full-wave circuit as in the half-wave circuit in Fig. 11.2, and that the turns ratio of the transformer in Fig. 11.3(a) is 1:(1 + 1), then the d.c. output voltage produced by the full-wave circuits is twice that from the half-wave circuit. That is

$$V_L = \frac{2V_M}{\pi} = 0.636V_M = 0.9V_S \qquad [11.2]$$

and

$$I_L = \frac{2I_M}{\pi} = 0.636I_M$$

where the terms have the same meaning as before.

## 11.5 Single-phase, phase-controlled thyristor circuit

As mentioned earlier, the thyristor (or, to be strictly correct, the **reverse blocking thyristor**) is a device which blocks the flow of current when the anode is negative with respect to the cathode, and can be triggered into conduction at any point when the anode is positive. A simple single-phase circuit which utilizes this feature is shown in Fig. 11.5(a).

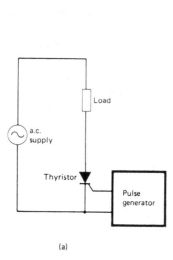

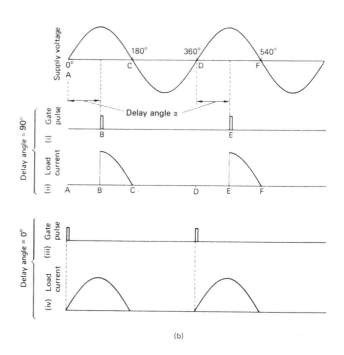

**Fig. 11.5** (a) Phase-controlled single-phase thyristor circuit and (b) voltage waveforms across a resistive load for various delay angles

The load is connected in series with the thyristor, which is triggered into conduction by the pulse generator connected between its gate and cathode, as shown in Fig. 11.5(b). The pulse generator produces a short pulse of a few microseconds duration at an angle known as the **delay angle**, $\alpha$, with respect to the commencement of the supply voltage waveform. Since the point in the waveform at which the gate pulse is applied is a function of the phase angle, the circuit is said to be **phase controlled**.

Even when the anode is positive with respect to the cathode, the thyristor does not conduct until the gate pulse is applied. Once it has been applied (at point B), the thyristor turns on and conducts; from this point onwards in the positive half-cycle, the circuit operates as though it were a half-wave rectifier circuit. If the load is resistive, the current waveform follows the supply voltage waveshape.

When the anode becomes negative with respect to the cathode (point C in Fig. 11.5), the thyristor enters its reverse blocking mode, and current no longer flows through it. It remains in this state until the next pulse is applied at point E.

If the gate pulse is applied at the same delay angle during each cycle [$\alpha = 90°$ in diagrams (i) and (ii) in Fig. 11.4(b)], the mean value of the load voltage remains constant. Should the delay angle be reduced, the thyristor is triggered into conduction at an earlier point in the cycle, and the mean output voltage is increased; if the delay angle is increased, the mean output voltage is reduced.

The maximum output voltage is obtained when $\alpha = 0°$, as shown in the lower two waveforms in Fig. 11.5(b); if $\alpha = 180°$ (or greater), the thyristor stops conducting altogether.

It can be shown that the mean voltage, $V_L$, across the load for the circuit in Fig. 11.5 is

$$V_L = 0.225 V_S(1 + \cos \alpha) \qquad [11.3]$$

where $V_S$ is the r.m.s. supply voltage and $\alpha$ is the delay angle of the gate pulse (this equation only applies in the range $0° < \alpha < 180°$).

When $\alpha = 0°$ (maximum output voltage), the output voltage is

$$V_L = 0.225 V_S(1 + \cos 0°) = 0.45 V_S$$

which also corresponds to eqn [11.1] (which is the output voltage for a half-wave rectifier circuit supplying a resistive load). In the case where $\alpha = 180°$ the output voltage is

$$V_L = 0.225 V_S(1 + \cos 180°) = 0$$

The reader will find it an interesting exercise to plot the variation in the mean output voltage across a resistive load as $\alpha$ varies over the range $0°–180°$. Using an r.m.s. supply voltage of 100 V and $\alpha = 45°$, it will be useful to draw the waveform of the output voltage, and to use the mid-ordinate rule to determine the output voltage. Its value should then be compared with the answer obtained from eqn [11.3].

**Worked example 11.2**  A phase-controlled, single-phase thyristor circuit is energized by a 240 V a.c. supply. If the load is a resistor, and the mean voltage across it is 65 V, calculate the delay angle $\alpha$ of the gate pulse.

*Solution*  From eqn [11.3]

$$V_L = 0.225 V_S(1 + \cos \alpha)$$

or

$$65 = 0.225 \times 240(1 + \cos \alpha) = 54(1 + \cos \alpha)$$

that is

$$\cos \alpha = \frac{65}{54} - 1 = 0.2037$$

or

$$\alpha = \cos^{-1} 0.2037 = 78.25°$$

**Problems**

11.1 Draw a circuit diagram of a single-phase, half-wave rectifier circuit and, with the aid of waveform diagrams, explain the operation of the circuit.

11.2 A single-phase, half-wave rectifier circuit is supplied at 240 V. If the load resistance is 100 Ω, calculate the average load current. Determine also the peak inverse voltage across the diode.

[1.08 A; 339.4 V]

11.3 With the aid of circuit and waveform diagrams, describe the operation of (a) a single-phase, full-wave rectifier circuit and (b) a full-wave bridge circuit. Give a typical application for each circuit, indicating what advantage one has over the other in that particular application.

11.4 Using either 'low resistance' or 'high resistance' as an answer, state (a) the resistance of a forward biased diode, (b) the resistance of a reverse biased diode.

11.5 Give three disadvantages of a half-wave rectifier circuit when compared with a full-wave rectifier circuit.

11.6 A single-phase rectifier circuit provides a d.c. output of 200 V, 1 A. Calculate (a) the peak current supplied by the rectifier, (b) the r.m.s. value of the supply voltage and (c) the value of the load resistance.

[(a) 3.14 A; (b) 444 V; (c) 200 Ω]

11.7 State two advantages of (a) a full-wave centre-tap reactifier circuit and (b) a bridge rectifier circuit over a single-phase, half-wave rectifier circuit.

11.8 What is the effect on the output voltage of a rectifier if (a) the diode(s) in the circuit develop a high resistance or (b) the transformer has a high impedance.

11.9 A single-phase, half-wave thyristor circuit is energized by a 240 V supply. (a) If the mean output voltage is 100 V, determine the delay angle of the gate pulses and (b) calculate the output voltage for a delay angle of 120°.

[31.59°; 81 V]

# 12 Circuit theorems

**12.1 Introduction**

Electrical and electronic circuits can be solved by the systematic application of basic laws such as Ohm's law, Kirchhoff's laws, etc. However, in many cases, it is more useful to have a range of theorems which encapsulate selected laws which provide a rapid solution to circuit problems. We will look at a number of useful theorems in this chapter.

**12.2 Thévenin's theorem**

The concept of ideal and practical voltage sources was first discussed in Chapter 2. There it was shown that a practical voltage source (such as a battery) can be thought of as comprising an ideal voltage source in series with an internal resistance (or 'output' resistance or source resistance).

Thévenin's theorem simply states that a two-terminal *active network* (i.e. one containing one or more sources combined with circuit elements such as resistors), no matter how complex, to which a load can be connected, can be thought of as a practical voltage source. A more precise statement of the theorem is given below.

> **An active network having two terminals A and B, to which an electrical load may be connected, behaves as though the network contains a single source of e.m.f., $E_T$, having an internal resistance $R_T$. The e.m.f. is the p.d. measured between A and B with the load disconnected, and $R_T$ is the resistance of the network between A and B with the load disconnected when each source of e.m.f. within the network has been replaced by its internal resistance.**

The circuit in Fig. 12.1 diagrammatically represents Thévenin's theorem. While the theorem is stated in terms of a d.c. circuit, it can also be used in a.c. circuits. In the latter case, $Z_T$ replaces $R_T$, and the word 'impedance' is used in place of the word 'resistance'. However, the reader should be careful to observe that, in a.c. calculations, we are concerned with phase angles as well as magnitude values.

We will illustrate the use of the theorem using a d.c. circuit. An advantage of this circuit theorem is that, once the value of $E_T$ and $R_T$ have been calculated, the current flowing in a very wide range of load resistors can be determined very quickly. Thévenin's theorem can be used repeatedly in certain cases, as shown in Worked Example 12.2.

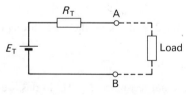

**Fig. 12.1** Thévenin's equivalent electrical circuit of an active d.c. circuit

*Worked example 12.1*    Determine Thévenin's equivalent circuit of the network between terminals A and B of Fig. 12.2(a), and calculate the current which would flow in a 20 Ω load connected to the terminals. What is the voltage between A and B when the load is connected?

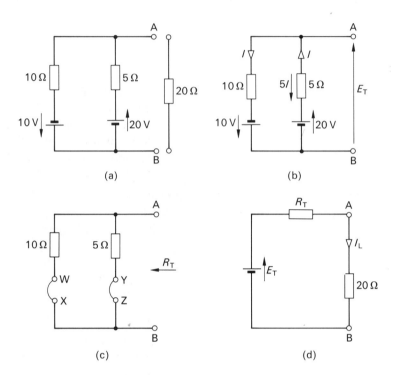

**Fig. 12.2** Circuit for Worked Example 12.1. (a) The circuit diagram, (b) the method of evaluating Thévenin's voltage, (c) calculating Thévenin's resistance and (d) solving the problem

*Solution*    The Thévenin voltage, $E_T$, is the voltage between terminals A and B *before the load is connected*; one method of calculating its value is illustrated in Fig. 12.2(b). Since there is only one closed loop in Fig. 12.2(b), we may say that

$$I = \frac{\text{total voltage acting in the loop}}{\text{total resistance of the loop}} = \frac{20 + 10}{5 + 10}$$

$$= 2\,\text{A}$$

Applying KVL between nodes A and B, and taking the path through the 20 V source, we see that

$$E_T = V_{AB} = 20 - 5I = 20 - 10 = 10\,\text{V}$$

Next we determine the Thévenin internal resistance, $R_T$. To do this, we replace each source by its internal resistance (which is zero, since they are both voltage sources), and then calculate the resistance between terminals A and B. In Fig. 12.2(c), the 10 V source is replaced by the link WX, and the 20 V source is replaced by the link YZ. The resistance between A and B is

$$R_T = \frac{5 \times 10}{5 + 10} = 3.333\,\Omega$$

That is, Thévenin's equivalent of the circuit between terminals A and B in Fig. 12.2(a) comprises a 10 V source (A positive with respect to B) in series with a 3.333 Ω resistor.

To calculate the current in the 20 Ω load, we simply connect it to Thévenin's equivalent circuit, as shown in Fig. 12.2(d). The current in the load is

$$I_L = \frac{E_T}{R_T + 20} = \frac{10}{3.333 + 20} = 0.429 \text{ A}$$

and the voltage between terminals A and B when the load is connected is

$$V_{AB} = 20I_L = 20 \times 0.492 = 8.58 \text{ V}$$

The reader will note that the voltage across the 20 Ω load is less than $E_T$, because of the voltage drop in $R_T$.

***Worked example 12.2***   Calculate the current in the 8 Ω resistor in Fig. 12.3(a).

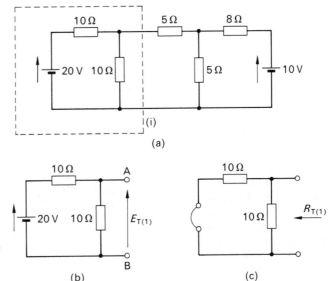

**Fig. 12.3**  Circuit for Worked Example 12.2: (a) the circuit, and the calculation of (b) $E_T$ and (c) $R_T$ for inset (i)

*Solution*   If we were to solve the circuit in Fig. 12.3(a) using loop current equations, we would need to write down three simultaneous equations, and then solve them for the current in the 8 Ω resistor. In this case we can use Thévenin's theorem to simplify the process.

Initially, we select a part of the circuit and work out its Thévenin equivalent circuit, and then reinsert it back into the circuit. By this means the circuit is simplified but unchanged.

Let us look at the part of the circuit in inset (i) in Fig. 12.3(a), which is extracted and drawn on its own in Fig. 12.3(b). Using the method of voltage division in series circuits outlined in Chapter 2, we see that the Thévenin voltage, $E_{T(1)}$, for this part of the circuit is

$$E_{T(1)} = 20 \times \frac{10}{10 + 10} = 10 \text{ V}$$

and, from Fig. 12.3(c), the Thévenin internal resistance is

$$R_{T(1)} = 10 \times 10/(10 + 10) = 5\,\Omega$$

These values are inserted in the inset (i) section of Fig. 12.4(a). Next we extract the section of the circuit in inset (ii) in Fig. 12.4(a) and calculate the Thévenin equivalent circuit values for it. It is left as an exercise for the reader to verify the values obtained, which are $E_{T(2)} = 3.333\,\text{V}$ acting in the direction in Fig. 12.4(b), and $R_{T(2)} = 3.333\,\Omega$. We simply apply Ohm's law to the circuit in Fig. 12.4(b) to calculate the current in the $8\,\Omega$ resistor, which is

$$I = \frac{10 - 3.333}{8 + 3.333} = 0.588\,\text{A}$$

which flows in the direction shown in Fig. 12.4(b).

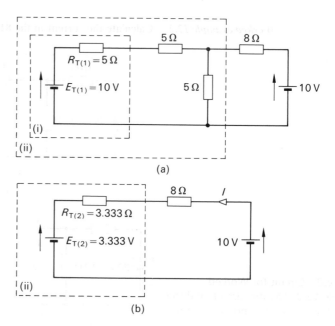

**Fig. 12.4** (a) The circuit of Fig. 12.3(a) with modified values inserted in inset (i). The circuit in inset (ii) is then connected to the Thévenin equivalent circuit. The Thévenin values for inset (ii) are calculated and used in diagram (b) to solve the problem

## 12.3 Norton's theorem

It is often more convenient to represent electronic devices, such as transistors, as a practical current source rather than as a practical voltage source. Current sources were discussed in Chapter 2, and Norton's theorem simply states that an active network, no matter how complex, having two terminals to which a load may be connected, can be thought of as a practical current source. A more precise statement of the theorem is as follows:

**An active network having two terminals A and B to which a load may be connected, behaves as though it comprises an ideal current source, $I_N$, shunted by a resistance $R_N$ (or a conductance $G_N$). Current $I_N$ is the current which flows in a short circuit applied between A and B, and $R_N$ is the resistance measured between A and B with the load disconnected and**

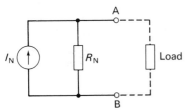

**Fig. 12.5** Norton's equivalent circuit of an active d.c. network

each source within the network replaced by its internal resistance.

The circuit in Fig. 12.5 represents Norton's theorem diagrammatically for a d.c. circuit. While the theorem is stated in terms of d.c. values, it can also be used for a.c. circuits.

As with Thévenin's theorem, once the value of $I_N$ and $R_N$ have been calculated, the current in a wide range of connected loads can be quickly determined.

***Worked example 12.3***

Determine Norton's equivalent circuit for the network between A and B in Fig. 12.2(a), and calculate the current which flows in a 20 Ω load connected between A and B.

*Solution*

Initially we will calculate $I_N$ [see Fig. 12.6(a)]. To do this, we short-circuit terminals A and B; the current $I_N$ flows in the short circuit. Since the p.d. between A and B is zero, the current in the 20 V source flows upwards in Fig. 12.6(a), and the current in the 10 V source flows downwards. Now

$$I_1 = 20/5 = 4 \text{ A}$$

and

$$I_2 = 10/10 = 1 \text{ A}$$

Applying KCL to node A, we have

$$I_N = I_1 - I_2 = 4 - 1 = 3 \text{ A}$$

flowing from A to B. The internal resistance of the Norton source is the same as for the Thévenin source (see Worked Example 12.1), and the method of obtaining it is illustrated in Fig. 12.6(b). Hence

$$R_N = 10 \times 5/(10 + 5) = 3.333 \, \Omega$$

That is, the Norton equivalent circuit consists of a 3 A ideal current source shunted by a 3.333 Ω resistor. Finally, we connect the 20 Ω resistor to the Norton equivalent circuit [see Fig. 12.6(c)], and calculate the current in the 20 Ω load using the method of current division in a parallel circuit (see Chapter 2). The value of the current is

$$I_L = I_N \frac{R_N}{R_N + 20} = 3 \times \frac{3.333}{3.333 + 20} = 0.429 \text{ A}$$

which agrees with the value obtained in Worked Example 12.1.

**Fig. 12.6** Figure for Worked Example 12.3. Calculation of (a) $I_N$ and (b) $R_N$. (c) Determination of the current in the 20 Ω resistor

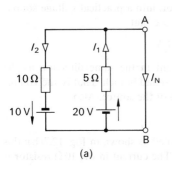

(a)

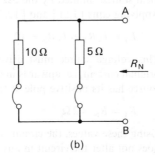

(b)

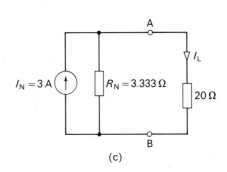

(c)

**12.4 Relationship between Thévenin's and Norton's equivalent circuits**

It was shown in Chapter 2 that each practical voltage source had an equivalent current source, and vice versa. And so it is that each Thévenin equivalent circuit has a Norton equivalent circuit, and vice versa. The relationships between the two equivalent circuits are

$$I_N = \frac{E_T}{R_T} \qquad\qquad [12.1]$$

and

$$R_N = R_T \qquad\qquad [12.2]$$

We can illustrate these equations using the results of Worked Examples 12.1 and 12.3, in which $E_T = 10\,V$ and $R_T = 3.333\,\Omega$. From eqn [12.1] we see that, for exact equivalence

$$I_N = E_T/R_T = 10/3.333 = 3\,A$$

and

$$R_N = R_T = 3.333\,\Omega$$

These values should be compared with those obtained in the worked examples.

*Worked example 12.4*    Calculate the current in the $10\,\Omega$ resistor in Fig. 12.7(a).

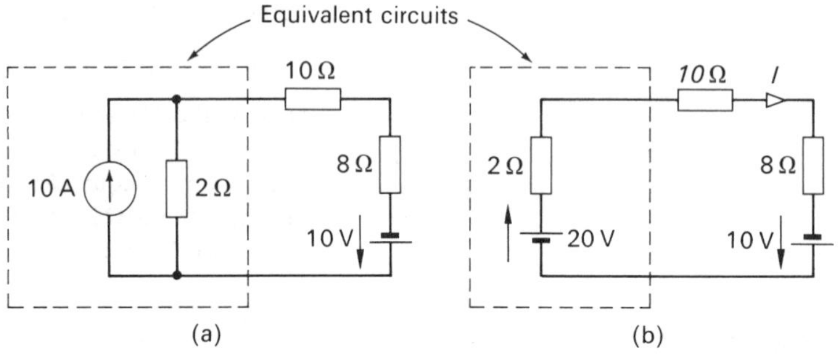

**Fig. 12.7** Figure for Worked Example 12.4: (a) the circuit diagram and (b) the circuit after the current source has been converted into a voltage source

*Solution*    Initially we will convert the practical current source, comprising the 10 A ideal source shunted by the $2\,\Omega$ resistor, into a practical voltage source. Applying eqns [12.1] and [12.2], we see that

$$E_T = I_N R_T = I_N R_N = 10 \times 2 = 20\,V$$

This voltage source must drive current in the same direction as the original current, i.e. upwards in the left-hand branch. That is, the voltage source has its positive pole at the top of the source. Also

$$R_T = R_N = 2\,\Omega$$

Using these values, the circuit is modified as shown in Fig. 12.7(b); this does not alter the circuit in any way. The current in the $10\,\Omega$ resistor is calculated as follows:

$$I = \frac{\text{net voltage}}{\text{total resistance}} = \frac{20 + 10}{2 + 10 + 8} = 1.5 \, \text{A}$$

This current flows in the direction shown in the figure. The reader will find it an interesting exercise to show that the voltage across the current source in Fig. 12.7(a) is 17 V.

**12.5  Principle of superposition**

Many practical circuits contain several voltage and current sources, and the solution of the circuit by normal methods can be complex. In many cases the **superposition principle** gives us a way out, because it says that we can calculate the current through (or the voltage across) each circuit element due to each individual source (all other sources meanwhile being eliminated in the way described below); the effective current through (or the voltage across) each element is the algebraic sum of the individual values calculated for each source.

Before looking at the general principle, we will illustrate it by means of the simple example in Fig. 12.8(a); in this case we will calculate the current in the 10 Ω resistor. Initially we remove the 5 V source and replace it with its internal resistance (which is zero since it is a battery). The current $I_1$ due to the 20 V battery is [see Fig. 12.8(b)]

$$I_1 = 20/10 = 2 \, \text{A} \quad \text{(clockwise around the circuit)}$$

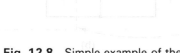

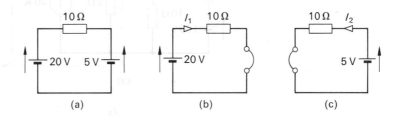

**Fig. 12.8**  Simple example of the principle of superposition

Next we return the 5 V source to the circuit and remove the 20 V source (meanwhile replacing it by its internal resistance of zero ohms). We are therefore left with the circuit in Fig. 12.8(c), and the current in the circuit is

$$I_2 = 5/10 = 0.5 \, \text{A} \quad \text{(counter-clockwise)}$$

The principle of superposition says that the current in the 10 Ω resistor is the algebraic sum of $I_1$ and $I_2$ (bearing in mind their *relative direction*) as follows:

$$I = I_1 + (-I_2) = 2 - 0.5 = 1.5 \, \text{A}$$

which flows in the direction of $I_1$. The simple application of KVL to Fig. 12.8(a) shows this to be the correct current; while this does not 'prove' the principle of superposition, it verifies its truth.

The superposition principle as it applies to the current flowing in a circuit element is stated below.

> **In a circuit containing several sources, the resultant current in any branch is the algebraic sum of the currents in that branch produced by each e.m.f. acting alone, every other source meanwhile being replaced by its respective internal resistance.**

While the exercises in this chapter are related to d.c. systems, the superposition principle is probably at its most useful in a.c. circuits, where sources of differing frequency and phase angle may be involved.

The reader should note that the principle of superposition *can only be applied to devices which have a linear relationship*, e.g. for which $V = IR$. Since power is proportional to the square (which is non-linear) of voltage or current, superposition cannot be applied directly to the calculation of power in an element.

***Worked example 12.5***  Using the principle of superposition, calculate the current, $I$, in the $10\,\Omega$ resistor in Fig. 12.9(a).

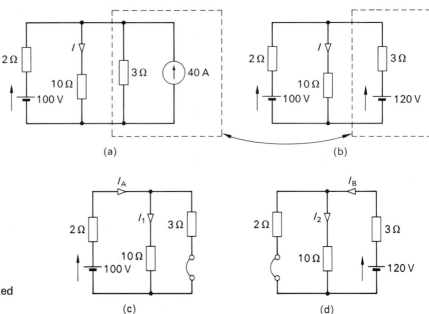

**Fig. 12.9** Figure for Worked Example 12.5

*Solution*  This is an interesting example, since it contains a practical voltage source and a practical current source. The problem is simplified if we use only one type of practical source and, in this case, we will convert the current source [shown by the broken line in Fig. 12.9(a)] into its equivalent voltage source. The equivalent voltage source (see also section 12.4) is shown by the broken line in Fig. 12.9(b). The e.m.f. associated with the new voltage source is

$$40\,\text{A} \times 3\,\Omega = 120\,\text{V}$$

and the source resistance is $3\,\Omega$.

Next, we apply the superposition principle by initially replacing the 120 V e.m.f. by its internal resistance (which is zero ohms), as shown in Fig. 12.9(c), and determine the current $I_1$ in the $10\,\Omega$ resistor in the following steps. The resistance of $10\,\Omega$ in parallel with $3\,\Omega$ is

$$10 \times 3/(10 + 3) = 2.308\,\Omega$$

and the total resistance of the circuit in Fig. 12.9(c) is

$$2 + 2.308 = 4.308\,\Omega$$

Hence the current supplied by the 100 V source is

$$I_A = 100/4.308 = 23.21\,\text{A}$$

Using the notation developed in Chapter 2 for the division of current in a parallel circuit, the proportion of $I_A$ flowing in the $10\,\Omega$ resistor is

$$I_1 = I_A \times 3/(10 + 3) = 23.21 \times 3/13 = 5.36\,\text{A}$$

Next, the 120 V source is returned to the circuit and the 100 V source is returned to the circuit and the 100 V source is removed and replaced by a short circuit, as shown in Fig. 12.9(d). The resistance of the parallel section of this circuit is

$$2 \times 10/(2 + 10) = 1.667\,\Omega$$

and the total resistance of Fig. 12.9(d) is

$$3 + 1.667 = 4.667\,\Omega$$

The current drawn from the 120 V source is

$$I_B = 120/4.667 = 25.71\,\text{A}$$

Once again, applying the method for current division in a parallel circuit gives

$$I_2 = I_B \times 2/(2 + 10) = 4.29\,\text{A}$$

Since $I_1$ and $I_2$ both flow downwards through the $10\,\Omega$ resistor, the superposition principle tells us that the current, $I$, in Fig. 12.9(a) is

$$I = I_1 + I_2 = 5.35 + 4.29 = 9.65\,\text{A}$$

## 12.6 Maximum power transfer theorem for resistive loads

When a practical voltage source is connected to a resistive load, as shown in Fig. 12.10, one of the features engineers are interested in is the condition which allows the source to supply maximum power to the load.

When the load resistance is finite, some power is consumed by the load. However, let us look at the two extreme values of load resistance, namely when $R_L = 0$ (a short circuit) and when $R_L = \infty$ (an open circuit).

When $R_L = 0$, the load current is very large indeed (since it is

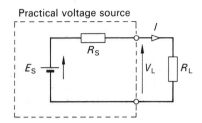

**Fig. 12.10** The maximum power transfer theorem for a resistive load

only restricted by the internal resistance, $R_S$, of the source), but the power consumed by the load is zero since $R_L$ is zero! When $R_L = \infty$, no current flows in the load and, once again, the power consumed is zero. Clearly, there is a value of resistance between the two extremes where maximum power is consumed. We will determine this condition in the following. The power consumed by $R_L$ is

$$P = I^2 R_L = \left[\frac{E_S}{R_S + R_L}\right]^2 R_L = \frac{E_S^2 R_L}{R_S^2 + 2R_S R_L + R_L^2}$$

$$= \frac{E_S^2}{(R_S^2/R_L) + 2R_S + R_L}$$

Since the value of $E_S$ is constant, maximum power is consumed by the circuit when the denominator of the above equation is a minimum. We can determine this condition by differentiating the denominator with respect to $R_L$, and equating it to zero as follows:

$$\frac{d}{dR_L}[(R_S^2/R_L) + 2R_S + R_L] = \left[-\frac{R_S}{R_L}\right]^2 + 1 = 0$$

or

$$R_L = R_S$$

That is, **maximum power is transferred into a resistive load when the resistance of the load is equal to the resistance of the source**.

The reader must be cautious when interpreting results from the maximum power transfer theorem because, when maximum power is transferred to the load, the terminal voltage is reduced by 50 per cent! If we look at the circuit in Fig. 12.10 we see that, for maximum power transfer to take place, then $R_L = R_S$; since both carry the same current, each will have one-half of $E_S$ across it.

If we plot the power consumed by a variable resistor which is connected to a practical voltage source to a base of the resistance, the resulting graph will be as shown in Fig. 12.11; $P_{max}$ is found to occur when $R_L = R_S$.

Although all our discussions have referred to a practical voltage source, the same condition also applies in the case where we use a current source. That is, maximum power transfer occurs when the resistance of the load is equal to the internal resistance of the source.

The maximum power transfer theorem is primarily applicable to electronic circuits due to the fact that the current involved is generally unacceptably large in power circuits.

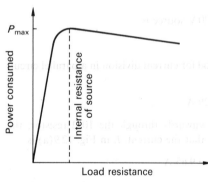

**Fig. 12.11** Graph showing the power consumed in a resistive circuit to a base of resistance

***Worked example 12.6***     The current source enclosed by the broken line in Fig. 12.12(a) is the equivalent output circuit of a transistor amplifier. What value of $R_L$ absorbs maximum power? If $I = 10$ mA, what power is absorbed by the load?

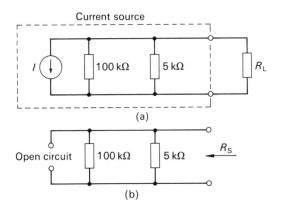

**Fig. 12.12**   Figure for Worked
Example 12.6

*Solution*   The internal or 'output' resistance of the current source is obtained by
disconnecting the load and, after replacing the current source by its
internal impedance (which is infinity), we determine the resistance
between the output terminals. This is done in Fig. 12.12(b) and we see that

$$R_S = \frac{5 \times 100}{5 + 100} = 4.76 \text{ k}\Omega$$

That is, maximum power is transferred into the load when

$$R_L = 4.76 \text{ k}\Omega$$

Since, under conditions of maximum power transfer, the load resistance
is equal to the source resistance, and it is in parallel with it, only one-half
of the source current flows in the load. That is $I_L$ is 5 mA, and the power
absorbed by the load is

$$P = I_L^2 R_L = (5 \times 10^{-3})^2 \times 4760 = 0.119 \text{ W} \quad \text{or} \quad 119 \text{ mW}$$

**Problems**

**12.1**   Determine Thévenin's equivalent circuit between terminals A and B of
Fig. 12.13.

$$[R_{TH} = 5\,\Omega, \ V_{TH} = 5\,\text{V (B positive)}]$$

**12.2**   Calculate the current flowing in a load of (a) $10\,\Omega$, (b) $2.5\,\Omega$ connected
between terminals A and B of Fig. 12.13. In each case calculate the power
consumed by the load.

[(a) 0.333 A, 1.11 W; (b) 0.666 A, 1.11 W]

**12.3**   Determine the load resistance in Fig. 12.13 which consumes maximum
power. What current flows in the load at maximum power, and what
power is consumed?

[5 $\Omega$; 0.5 A; 1.25 W]

**12.4** Determine Norton's equivalent circuit of Fig. 12.13 with respect to terminals A and B.

$$[I_N = 1\,\text{A (current flowing out of B)};\; R_N = 5\,\Omega]$$

**12.5** Determine Norton's equivalent circuit with respect to terminals A and B of Fig. 12.14.

$$[I_N = 10\,\text{A (current leaving A)};\; R_N = 1.333\,\Omega]$$

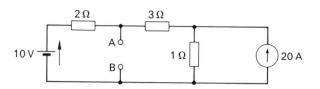

**Fig. 12.14**

**12.6** Calculate the voltage between terminals A and B of Fig. 12.14 if a load of (a) $1\,\Omega$, (b) $2\,\Omega$ is connected between the terminals. Calculate the power consumed in each case.

$$[(a)\; 5.714\,\text{V},\; 32.65\,\text{W};\; (b)\; 8\,\text{V},\; 32\,\text{W}]$$

**12.7** What load resistance must be connected between terminals A and B of Fig. 12.14 to consume maximum power? What current flows in the load and what power is consumed?

$$[1.333\,\Omega;\; 5\,\text{A};\; 33.325\,\text{W}]$$

**12.8** Determine Thévenin's equivalent circuit with respect to terminals A and B of Fig. 12.14.

$$[E_T = 13.33\,\text{V (A positive)};\; R_T = 1.333\,\Omega]$$

**12.9** Use the principle of superposition to determine the current in each resistor in Fig. 12.15.
[Current in $10\,\Omega = 1.318\,\text{A}$ (upwards); current in $8\,\Omega = 1.477\,\text{A}$ (downwards); current in $20\,\Omega = 0.159\,\text{A}$ (upwards)]

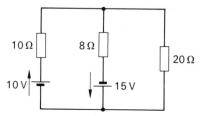

**Fig. 12.15**

# 13 Transients in electrical circuits

## 13.1 Introduction

A **transient** is a temporary event, and passes away with time. In electrical engineering, we are concerned with circuits containing inductors and capacitors which store or discharge energy, and with resistors which dissipate energy. The way in which a circuit responds to a sudden change in an applied voltage depends not only on the elements in the circuit, but also on the size and method of interconnection of the elements.

In most electrical circuits, transients die away after a period of time known as the **settling time** of the circuit. When this has occurred, the circuit enters its **steady-state** operating period.

In this chapter we concentrate on transients in d.c. circuits which occur when a voltage is either applied to or removed from R–C or R–L circuits.

## 13.2 Capacitor-charging current

It was shown in Chapter 3 that when a voltage is applied to a capacitor, a *charging current* flows into the capacitor. If the switch in the circuit in Fig. 13.1 is in position Q, the capacitor discharges any energy it stores, and the voltage across the capacitor is zero. That is, the capacitor has **zero initial conditions**. If the position of the switch blade is changed from contact Q to contact P at $t = 0$, a charging current, $i$, flows in the circuit.

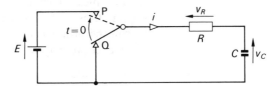

**Fig. 13.1**  Charging a capacitor

In the following we will use calculus to derive the equations for the voltage, $v_C$, across the capacitor and for the charging current, $i$. Readers who, at this stage, have not studied the calculus method, can move directly on to eqns [13.5] and [13.6].

We have shown in Chapter 3 that if the voltage between the plates of a capacitor changes by a small amount $dv_C$ in $dt$ seconds, then the corresponding change in charge, $dq$, on the capacitor is

$$dq = C \, dv_C$$

or

$$dq = i\ dt$$

or

$$i\ dt = C\ dv_C$$

That is, the charging current is

$$i = C\ dv_C/dt \qquad\qquad [13.1]$$

$$= C \times \text{rate of change of voltage across the capacitor}$$

We will now use mathematics as an engineers' 'tool' to determine the way in which the current changes during the **transient period** of operation of the circuit. The p.d. across the resistor is

$$v_R = Ri = RC\ dv_C/dt$$

but

Supply voltage, $E$ = p.d. across $C$ + p.d. across $R$

$$= v_C + RC\ dv_C/dt$$

or

$$E - v_C = RC\ dv_C/dt$$

that is

$$\frac{dt}{RC} = \frac{dv_C}{E - v_C} \qquad\qquad [13.2]$$

Since both $dv$ and $(E - v_C)$ have the dimensions of voltage, it follows that the expression on the right-hand side of the equation has the dimensions of (voltage/voltage), i.e. it is *dimensionless*. Since the two sides of the equation are equal to one another, it follows that the left-hand side of the equation must also be dimensionless. That is, the dimensions of the denominator are equal to those of the numerator, i.e. those of time. For this reason we call the product $RC$ the **time constant** of the circuit, and call it $\tau$. Hence we may write eqn [13.2] as follows:

$$\frac{dt}{\tau} = \frac{dv_C}{E - v_C} \qquad\qquad [13.3]$$

If both sides of the equation are integrated the result is

$$\frac{t}{\tau} = -\ln(E - v_C) + A \qquad\qquad [13.4]$$

where ln means the logarithm to base e (e = 2.71828), and $A$ is the *constant of integration*. Since we need to solve the equation, we must evaluate $A$, which is done by inserting a set of known values into the equation, the simplest being that $v_C = 0$ when

$t = 0$ (this is the initial condition in the circuit mentioned at the outset of this section). When this is inserted into eqn [13.4] we get

$$A = \ln E$$

Inserting this into eqn [13.4] gives

$$\frac{t}{\tau} = -\ln(E - v_C) + \ln E = \ln \frac{E}{E - v_C}$$

or

$$\frac{E}{E - v_C} = e^{-t/\tau}$$

Solving for $v_C$ yields

$$v_C = E(1 - e^{-t/\tau}) \qquad [13.5]$$

From eqn [13.1], we see that the instantaneous value of the charging current is

$$i = C \frac{\mathrm{d}v_C}{\mathrm{d}t} = CE \frac{\mathrm{d}}{\mathrm{d}t}(1 - e^{-t/\tau}) = \frac{CE}{\tau} e^{-t/\tau}$$

$$= \frac{E}{R} e^{-t/\tau} \qquad [13.6]$$

Equation [13.6] allows us to calculate the value of the charging current at any instant of time. At the moment, we will work out the current at two points in time, namely when $t = 0$ and when $t = \infty$. The first of these gives us the *initial current*, $I_0$, and the second gives us the *steady-state current* (that is, the current when all the transients have died away). To do this, we merely insert the appropriate value of time in eqn [13.6] as follows.

When $t = 0$

$$i = I_0 = \frac{E}{R} e^{-0/\tau} = \frac{E}{R} e^0 = \frac{E}{R} \qquad [13.7]$$

That is, the initial charging current is $E/R$. This allows us to write eqn [13.6] as follows:

$$i = I_0 e^{-t/\tau} \qquad [13.8]$$

When $t = \infty$

$$i = \frac{E}{R} e^{-\infty/\tau} \approx \frac{E}{R} e^{-\infty} = 0 \qquad [13.9]$$

Hence, when the capacitor is fully charged, the current has fallen to zero.

Graphs of $v_C$ and $i$ are shown in Figs 13.2(a) and (b), respectively. Since the voltage across the resistor is $v_R = IR$, the curve for the voltage across the resistor has the same general

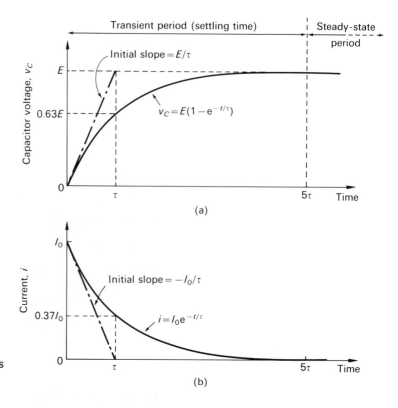

**Fig. 13.2**  Capacitor charging curves for an initially discharged capacitor

shape as the instantaneous current waveform in Fig. 13.2(b). It should be noted that the voltage across $R$ at $t = 0$ is $v_R = I_0 R = E$, and the final value of the voltage across $R$ is zero (since the final value of the charging current is zero).

In the case of circuits that have a response which is represented by the exponential equations developed here, we say that the transients have *settled out* when the instantaneous voltage (or current) has risen from zero to 99 per cent (or the current has fallen from 100 to 1 per cent) of its final value.

If we take a time of $t = 5\tau$, and use it in eqn [13.5] we get

$$v_C = E(1 - e^{-5\tau/\tau}) = E(1 - e^{-5}) = 0.993E$$

This shows that an exponentially rising transient of the shape in Fig. 13.2(a) has settled out in just under a period of $5\tau$.

Also, inserting $t = 5\tau$ into eqn [13.8] gives

$$i = I_0 e^{-t/\tau} = I_0 e^{-5\tau/\tau} = I_0 e^{-5} = 0.0067 I_0$$

which, once again, shows that an exponentially falling transient of the shape in Fig. 13.2(b) has settled out in just under a period of $5\tau$. That is

**the transients have settled out, or can be thought to have decayed, within a period of $5\tau$ after the beginning of the charging period.**

Beyond this period, we can reasonably say that the capacitor is fully charged, and the circuit has entered its steady-state operating period. This is illustrated in Fig. 13.2.

It is useful to be able to define the time constant, $\tau$, of the $RC$ circuit in terms of the initial rate of charge of the capacitor as follows:

> **The time constant, $\tau$, is the time which would be taken for the capacitor voltage to reach the final value ($E$) if the initial rate of change of $v_C$ had been maintained.**

This is shown by the initial slope line in Fig. 13.2(a). This can be shown to be correct if eqn [13.5] is differentiated with respect to time, and then inserting $t = 0$ in the resulting equation. Similarly, it takes a time equal to the time constant for the charging current of an initially discharged capacitor to reach zero [see also Fig. 13.2(b)].

### 13.2.1 Summary of equations for capacitor charge

$$i = I_0 e^{-t/\tau}$$

where $I_0 = E/R$ and $\tau = RC$ ($\tau$ in seconds if $R$ is in ohms and $C$ in farads).

$$v_R = iR = E\,e^{-t/\tau}$$

$$v_C = E(1 - e^{-t/\tau})$$

For zero initial conditions ($v_C = 0$ at $t = 0$)

$$v_C \text{ at } t = 0 \text{ is } E$$

$$v_C \text{ at } t = \infty \text{ is } 0$$

Both $i$ and $v_R$ have decayed to 0.37 of their initial value when $t = \tau$; $v_C$ has risen to 0.63 of its final value when $t = \tau$. The transients have decayed when $t = 5\tau$.

**13.3 Sketching exponential curves**

As we have seen above, there are two general types of exponential curve, and these are:

1. A rising exponential curve (see eqn [13.5] and fig. 13.2(a)) of the general form $y = Y(1 - e^{-t/\tau})$.
2. A falling exponential curve (see eqns [13.6] and [13.8] and Fig. 13.2(b)) of the general form $y = Y e^{-t/\tau}$.

Initially, we look at the *rising exponential curve*. The simplest method of sketching the curve is to produce a short table of salient points on the graph which, when joined together, give a reasonably smooth curve. This gives us two options, namely:

(a) results at intervals of $t = \tau$; and
(b) results at intervals of $t = 0.7\tau$.

**Table 13.1** $y = Y(1 - e^{-t/\tau})$ for intervals of $t = \tau$

| $t$ | $y$ |
| --- | --- |
| 0 | 0 |
| $\tau$ | $0.63Y$ |
| $2\tau$ | $0.86Y$ |
| $3\tau$ | $0.95Y$ |
| $4\tau$ | $0.98Y$ |
| $5\tau$ | $0.99Y$ |

**Table 13.2** $y = Y(1 - e^{-t/\tau})$ for intervals of $t = 0.7\tau$

| $t$ | $y$ | |
| --- | --- | --- |
| 0 | 0 | |
| $0.7\tau$ | $0.5Y$ | or $Y \times 1/2$ |
| $1.4\tau$ | $0.75Y$ | or $Y \times 3/4$ |
| $2.1\tau$ | $0.875Y$ | or $Y \times 7/8$ |
| $2.8\tau$ | $0.94Y$ | or $Y \times 15/16$ |
| $3.5\tau$ | $0.97Y$ | or $Y \times 31/32$ |
| $4.2\tau$ | $0.986Y$ | or $Y \times 63/64$ |
| $4.9\tau$ | $0.99Y$ | |

**Table 13.3** $y = Y e^{-t/\tau}$ for intervals of $t = \tau$

| $t$ | $y$ |
| --- | --- |
| 0 | 0 |
| $\tau$ | $0.37Y$ |
| $2\tau$ | $0.14Y$ |
| $3\tau$ | $0.05Y$ |
| $4\tau$ | $0.02Y$ |
| $5\tau$ | $0.07Y$ |

**Table 13.4** $y = Y e^{-t/\tau}$ for intervals of $t = 0.7\tau$

| $t$ | $y$ |
| --- | --- |
| 0 | 0 |
| $0.7\tau$ | $0.5Y = Y/2$ |
| $1.4\tau$ | $0.25Y = Y/4$ |
| $2.1\tau$ | $0.125Y = Y/8$ |
| $2.8\tau$ | $0.061Y \approx Y/16$ |
| $3.5\tau$ | $0.03Y \approx Y/32$ |
| $4.2\tau$ | $0.015Y \approx Y/64$ |
| $4.9\tau$ | $0.007Y \approx Y/128$ |

While method (b) appears to give a rather odd interval of time it has, in the opinion of the author, a number of advantages over method (a).

Using the time interval $t = \tau$ in equation

$$y = Y(1 - e^{-t/\tau})$$

we get the results in Table 13.1, and using the interval $t = 0.7\tau$ we get the results in Table 13.2. The interesting thing about an exponential curve is that *it changes in value by an equal proportion for an equal time interval.* Looking at Table 13.1, in the first interval of $\tau$, the value of $t$ changes by 0.63 or 63 per cent of its maximum value. Accordingly, the value of $y$ will change by 63 per cent of the remaining change [this is $(1 - 0.63)Y = 0.37Y$] in the next period of $\tau$. That is, when $t = 2\tau$ the value of $y$ is

$$0.63Y + (0.63 \times 0.37)Y = 0.86Y$$

and so on. The reader can easily verify the remaining results in Table 13.1 in this way.

Looking at Table 13.2 we see that $y$ has changed by 50 per cent in the first $0.7\tau$ interval. Clearly, the value of $y$ changes by 50 per cent during the next $0.7\tau$ interval, so that its value when $t = 1.4\tau$ is

$$y = 0.5Y + 0.5(1 - 0.5)Y = 0.75Y$$

and its value when $y = 2.1\tau$ is

$$y = 0.75Y + 0.5(1 - 0.75)Y = 0.875Y$$

and so on. In fact, with very little experience, it is possible to plot a fairly accurate exponential curve on a piece of paper (which need not be graph paper) quite quickly. This is the method used by the author when sketching the original curves for this book.

We now look at a *falling exponential curve* [see Fig. 13.2(b)] of the general form $y = Y e^{-t/\tau}$, where $Y$ is the *initial value* of the curve at $t = 0$. As before, we will calculate values for points which are at time intervals of (a) $t = \tau$ and (b) $t = 0.7\tau$. These are listed in Tables 13.3 and 13.4, respectively. Once again, we see that the transient has settled out when $t = 5\tau$. We also see that, when $t = \tau$, the value of the falling exponential curve has reduced in value to 37 per cent of its initial value. This is repeated for each increment of $\tau$. That is, after $2\tau$ the value of $y$ is

$$0.37 \times 0.37Y = 0.14Y$$

after $3\tau$ it is $0.37^3 Y = 0.05Y$, etc.

We also see from Table 13.4 that the curve reduces by 50 per cent for each $0.7\tau$ interval, so that after $1.4\tau$ the value of the curve is $0.5^2 Y = 0.25Y$, after $2.1\tau$ its value is $0.5^3 Y = 0.125Y$, etc. This fact makes it very easy to plot a falling exponential curve.

### 13.3.1  Rise-time and fall-time of exponential curves

When dealing with electronic and electrical circuits, engineers are concerned about the value of what is known as the rise-time and the fall-time of waveforms, which are defined as follows.

**The rise-time is the time taken for a waveform to rise from 10 to 90 per cent of the final value. The fall-time is the time taken for the waveform to fall from 90 to 10 per cent of the initial value.**

These definitions refer to any type of waveform, and here we are concerned about rising and falling exponential curves.

*The rise-time, $t_r$, of a rising exponential curve*

The equation of a rising exponential curve is $y = Y(1 - e^{-t/\tau})$, where $Y$ is the final value of the wave. When the wave has risen by 10 per cent, i.e. $y = 0.1Y$ when $t = t_1$, then

$$0.1Y = Y(1 - e^{-t_1/\tau})$$

or

$$0.1 = 1 - e^{-t_1/\tau}$$

that is

$$e^{-t_1/\tau} = 0.9$$

Taking logarithms to base e of both sides of the equation gives

$$t_1 = -\tau \log_e 0.9 = 0.1\tau$$

If the wave has risen by 90 per cent, i.e. $y = 0.9Y$ when $t = t_2$ then, by a similar mathematical process, it can be shown that

$$t_2 = 2.3\tau$$

Hence the rise-time, $t_r$, of a rising exponential curve is

$$t_r = t_2 - t_1 = 2.2\tau$$

*Fall-time, $t_f$, of a falling exponential curve*

The equation of the curve is $y = Y e^{-t/\tau}$, where $Y$ is the initial value of the curve. It can be shown, using a similar reasoning to the above, that the curve has fallen to 90 per cent of its original value in a time of $t_1 = 0.1\tau$, and has fallen to 10 per cent of its original value in $t_2 = 2.2\tau$. Hence, the fall-time of a falling exponential curve is

$$t_f = t_2 - t_1 = 2.2\tau$$

*Summary*

The rise- and fall-times of exponential curves are $2.2\tau$.

*Worked example 13.1*　A series $R$–$C$ circuit of the type in Fig. 13.1 is energized by a 20 V d.c. supply. If $R = 1\,\text{k}\Omega$ and $C = 0.5\,\mu\text{F}$ (the capacitor being initially uncharged), determine (a) the circuit time constant, (b) the initial value of the charging current, (c) the initial rate of rise of voltage across the capacitor, (d) the initial rate of fall of charging current, (e) the settling time of the transients, (f) the voltage across $C$ and across $R$ when $t = 0.25\,\text{ms}$, and the charging current at this time, (g) the time taken for the capacitor voltage to rise to 16 V, (h) the rise-time of $v_C$ and (i) sketch the curves of $v_C$, $v_R$ and $i$.

*Solution*　(a) The time constant of the circuit is

$$\tau = RC = 1000 \times 0.5 \times 10^{-6} = 5 \times 10^{-4}\,\text{s} \quad \text{or} \quad 0.5\,\text{ms}$$

(b) From eqn [13.7], the initial charging current is

$$I_0 = E/R = 20/1000 = 0.02\,\text{A}$$

(c) The initial rate of change of the capacitor voltage is

$$\frac{E}{\tau} = \frac{20}{0.5 \times 10^{-3}} = 40\,000\,\text{V/s} \quad \text{or} \quad 40\,\text{kV/s}$$

(d) Figure 13.2(b) shows that the initial *rate of reduction* of the charging current is

$$\frac{I_0}{\tau} = \frac{0.02}{0.5 \times 10^{-3}} = 40\,\text{A/s}$$

(e) The settling time for the transient to decay is

$$5\tau = 5 \times 0.5\,\text{ms} = 2.5\,\text{ms}$$

(f) When $t = 0.35\,\text{ms}$, the voltage across the capacitor is

$$v_C = E(1 - e^{-t/\tau}) = 20(1 - \exp(-[0.35 \times 10^{-3}/0.5 \times 10^{-3}]))$$
$$= 10\,\text{V}$$

*Note*: Since $0.35\,\text{ms} = 0.5\tau$ we could, from the results of Table 13.2, have said that $v_C$ is 50 per cent of $E$, or 10 V!
　The voltage across $R$ at this time is

$$v_R = E - v_C = 20 - 10 = 10\,\text{V}$$

and

$$i = v_R/R = 10/1000 = 0.01\,\text{A}$$

(g) The expression for the voltage across the capacitor is $v_C = E(1 - e^{-t/\tau})$. If $v_C = 16\,\text{V}$ when $t = t_1$, then

$$16 = 20(1 - e^{-t_1/\tau})$$

or

$$0.8 = 1 - e^{-t_1/\tau}$$

hence

$$e^{-t_1/\tau} = 1 - 0.8 = 0.2$$

Taking logarithms to base e of both sides gives

$$-\frac{t_1}{\tau} = \log_e 0.2 = -1.6094$$

or

$$t_1 = 1.6094\tau = 1.6094 \times 0.5 \text{ ms} = 0.805 \text{ ms}$$

Very approximately we can verify this solution from the values for $v_C$ in Table 13.3.

(h) The rise-time for $v_C$ is

$$t_r = 2.2\tau = 2.2 \times 0.5 \text{ ms} = 1.1 \text{ ms}$$

That is, it takes 1.1 ms for $v_C$ to rise from 2 V (10 per cent of 20 V) to 18 V (90 per cent of 20 V). Similarly, the fall-time of $i$ and $v_R$ is 1.1 ms.

(i) Using time intervals of

$$t = 0.7\tau = 0.7 \times 0.5 \text{ ms} = 0.35 \text{ ms}$$

we can (using the notation developed earlier) write down the values to be used to plot $v_C$, $v_R$ and $i$ as shown in Table 13.5. The graphs are plotted in Fig. 13.3.

**Table 13.5** Data for the graphs for Worked Example 13.1

| $t(ms)$ | $v_C$ $(V)$ | $v_R = E - v_C$ $(V)$ | $i = v_R/R$ $(mA)$ |
|---|---|---|---|
| 0 | 0 | 20 | 20 |
| 0.35 | 10 | 10 | 10 |
| 0.7 | 15 | 5 | 5 |
| 1.05 | 17.5 | 2.5 | 2.5 |
| 1.4 | 18.75 | 1.25 | 1.25 |
| 1.75 | 19.38 | 0.62 | 0.62 |
| 2.1 | 19.69 | 0.31 | 0.31 |
| 2.45 | 19.85 | 0.15 | 0.15 |

**13.4 Capacitor discharge**

Suppose that the switch blade in Fig. 13.4 has been in position P long enough for the capacitor to be fully charged. At this time $v_C = E$, $i = 0$ and $v_R = iR = 0$; since the capacitor is fully charged, the initial condition in the circuit is $v_C = E$.

When the switch blade is changed from position P to position Q, the capacitor will discharge its energy into the resistor. We will now derive the equations for the circuit; the reader who has not studied calculus should now move on to eqns [13.11]–[13.13].

Applying KVL to Fig. 13.3 we see, during the discharge period, that

$$v_R + v_C = 0$$

or

$$v_R = -v_C$$

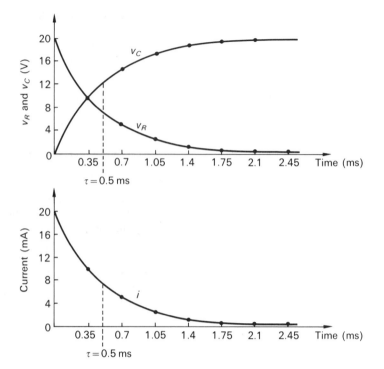

**Fig. 13.3** Graph of $v_C$, $v_R$ and $i$ to a base of time for Worked Example 13.1

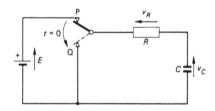

**Fig. 13.4** Capacitor discharge

But

$$v_R = Ri = RC \, dv_C/dt = \tau \, dv_C/dt$$

The reader will note from the direction of the $v_R$ arrow in Fig. 13.3 that *we have assumed the current is still flowing into the upper plate of the capacitor.*

Rearranging terms which are time functions on one side of the equation, and voltage functions on the other, we get

$$\frac{dt}{\tau} = -\frac{dv_C}{v_C}$$

Integrating both sides of the equation gives

$$\frac{t}{\tau} = -\ln v_C + A \qquad [13.10]$$

where $A$ is the constant of integration. To determine the value of $A$, we need to insert the initial conditions in the circuit into eqn [13.10], that is $v_C = E$ when $t = 0$. This gives

$$0 = -\ln E + A$$

or

$$A = \ln E$$

Reinserting this into eqn [13.10] yields

$$\frac{t}{\tau} = -\ln v_C + \ln E = \ln \frac{E}{v_C}$$

From which it follows that

$$e^{t/\tau} = E/v_C$$

or

$$v_C = E\,e^{-t/\tau} \qquad\qquad [13.11]$$

That is to say, $v_C$ *follows a falling exponential curve.* When $t = 0$, we see that

$$v_C = E\,e^{-0} = E$$

and when $t = \infty$ we get

$$v_C = E\,e^{-\infty} = 0$$

As explained earlier, this curve takes about five time constants to reach about 1 per cent of its original value, by which time we say that the transient has *settled out*.

If we apply simple circuit theory to Fig. 13.4, we see that for $t > 0$

$$v_R + v_C = 0$$

or

$$v_R = -v_C$$

that is

$$iR = -v_C = -E\,e^{-t/\tau}$$

hence

$$i = -\frac{E}{R}\,e^{-t/\tau} \qquad\qquad [13.12]$$

The negative sign associated with $i$ implies that the current flows in the opposite direction to the charging current, i.e. out of the top plate in Fig. 13.4. That is, the current is a *discharge current*. Once again the equation for the current corresponds to a falling exponential curve, and its initial value at $t = 0$ is

$$i_0 = -\frac{E}{R}\,e^0 = -\frac{E}{R}$$

and its final value when $t = \infty$ is

$$i_\infty = -\frac{E}{R}\,e^{-\infty} = 0$$

and it takes a period of about $5\tau$ for the transient to have decayed to practically zero.

Similarly it can be shown that

$$v_R = iR = -E\ e^{-t/\tau} \qquad\qquad [13.13]$$

which, again, is a falling exponential curve. Also, from the above

$$v_C = -v_R = E\ e^{-t/\tau}$$

### 13.4.1  Summary of equations for capacitor discharge

$$i = -\frac{E}{R}\ e^{-t/\tau} \quad \text{(discharge current)}$$

where

$$\tau = RC$$
$$v_R = -E\ e^{-t/\tau}$$
$$v_C = E\ e^{-t/\tau}$$

*Note*: $i$, $v_R$ and $v_C$ have decayed to 0.37 of their initial value when $t = \tau$. The transients have decayed when $t = 5\tau$.

**Worked example 13.2**  A fully charged $10\ \mu F$ capacitor is discharged through a $10\ k\Omega$ resistor. If the capacitor voltage has fallen to $45\ V$ in the first $0.12\ s$ after the discharge commences, calculate the initial voltage across the capacitor. Sketch curves of the capacitor voltage and the capacitor current to a base of time.

*Solution*  The time constant of the circuit is

$$\tau = RC = 10\,000 \times 10 \times 10^{-6} = 0.1\ s$$

Now

$$v_C = E\ e^{-t/\tau}$$

and, in this case, we know that $v_C = 45\ V$ when $t = 0.12\ s$, then

$$45 = E\ e^{-0.12/0.1} = 0.301E$$

or

$$E = 149.5\ V$$

That is, the initial voltage across the capacitor is 149.5 V.

To plot the curve for $v_C$ we only need to know two values, namely the initial voltage across the capacitor and the time constant of the circuit. These respectively are $E = 149.5\ V$ and $\tau = 0.1\ s$. To plot the discharge current we need to know, in addition, the initial value of the discharge current, which is

$$\frac{\text{Initial voltage across capacitor}}{\text{Resistance of the circuit}} = \frac{149.5}{10\,000}$$

$$= 0.01495\ A \quad \text{or} \quad 14.95\ mA$$

Taking a time interval of $0.7\tau = 0.07\ s$, we construct Table 13.6 in the manner described earlier, noting that both $v_C$ and $i$ halve for each $0.7\tau$ interval (see also Section 13.3). We set the time limit of the graph to about $5\tau = 0.5\,\text{s}$, and the graphs are plotted in Fig. 13.5.

**Table 13.6**   Table for Worked Example 13.2

| $t(s)$ | $v_C(V)$ | *Discharge current (mA)* |
|---|---|---|
| 0 | 149.5 | 14.95 |
| 0.07 | 75 | 7.5 |
| 0.14 | 37.5 | 3.75 |
| 0.21 | 18.75 | 1.875 |
| 0.28 | 9.38 | 0.938 |
| 0.35 | 4.69 | 0.469 |
| 0.42 | 2.35 | 0.235 |
| 0.49 | 1.18 | 0.118 |

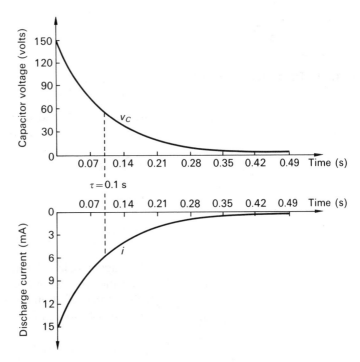

**Fig. 13.5**  Graph of $v_C$ and $i$ to a base of time for Worked Example 13.2

### 13.5  Measuring voltage and current under transient conditions

Care must be exercised when selecting instruments for the measurement of voltage and current under transient conditions. Every instrument (even an electronic instrument) has some internal resistance, and when it is connected into a circuit it may give results which differ from those expected from the theory of the circuit.

For example, a milliammeter has quite a high resistance, and will cause the current to be less than one would expect from a knowledge of the values in the circuit. A voltmeter or an oscilloscope has a resistance which may shunt some current from the circuit, and may upset the results expected from the circuit.

The general rule is to be very cautious about results obtained from a circuit in which transients are involved!

## 13.6   Growth of current in an L–R circuit

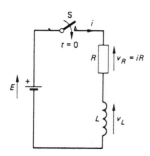

**Fig. 13.6**   Transients in an L–R circuit

When switch S in Fig. 13.6 is closed at $t = 0$, current begins to flow in the circuit and an e.m.f. is induced in inductor $L$. By Lenz's law, the induced e.m.f. opposes the flow of current, and when KVL is applied to the circuit we get

$$E = v_R + v_L = iR + L\frac{\mathrm{d}i}{\mathrm{d}t}$$

where $\mathrm{d}i/\mathrm{d}t$ is the rate of change of current in the circuit. Hence

$$E - iR = L\frac{\mathrm{d}i}{\mathrm{d}t}$$

or

$$\frac{E}{R} - i = \frac{L}{R}\frac{\mathrm{d}i}{\mathrm{d}t} = \tau\frac{\mathrm{d}i}{\mathrm{d}t} \qquad [13.14]$$

where $\tau = L/R$ and is the *time constant* of the circuit. If we let $E/R = I$, then eqn [13.14] takes the form

$$I - i = \tau\frac{\mathrm{d}i}{\mathrm{d}t}$$

that is

$$\frac{\mathrm{d}t}{\tau} = \frac{\mathrm{d}i}{I - i}$$

Once again we have 'separated the variables' so that dimensionless quantities are on opposite sides of the equations. Integrating both sides of the equation gives (readers not familiar with integration can proceed to eqn [13.16])

$$\frac{t}{\tau} = -\ln(I - i) + A \qquad [13.15]$$

where $\ln(I - i)$ is the natural logarithm (log to base e) of $(I - i)$, and $A$ is the constant of integration. We can determine the constant of integration by inserting the initial conditions in the circuit into eqn [13.15]. That is $i = 0$ when $t = 0$, as follows:

$$0 = -\ln I + A$$

or

$$A = \ln I$$

It we put this in eqn [13.15] we get

$$\frac{t}{\tau} = -\ln(I - i) + \ln I = \ln \frac{I}{I - i}$$

Taking antilogarithms gives

$$e^{t/\tau} = \frac{I}{I - i}$$

or

$$I - i = I\, e^{-t/\tau}$$

After some manipulation of the equation we get

$$i = I(1 - e^{-t/\tau}) \qquad\qquad [13.16]$$

which, as mentioned earlier, is the equation of an exponentially rising curve. The *final value of the current in the circuit* occurs when $t = \infty$, and is

$$i = I(1 - e^{\infty/\tau}) = I(1 - 0) = I = E/R$$

The voltage across the resistance is given by the equation

$$v_R = iR = \frac{E}{R}(1 - e^{-t/\tau}) \times R$$

$$= E(1 - e^{-t/\tau}) \qquad\qquad [13.17]$$

The graphs of both $i$ and $v_R$ follow a rising exponential curve, and are illustrated in Figs 13.7(a) and (b), respectively.

Also, when the switch S is closed, the loop equation for the circuit is

$$E = v_R + v_L$$

or

$$v_L = E - v_R = E - E(1 - e^{-t/\tau})$$

$$= E\, e^{-t/\tau} \qquad\qquad [13.18]$$

That is to say, the equation for $v_L$ follows a falling exponential curve, as shown in Fig. 13.7(c).

A little more analysis shows that if the initial rate of rise of current were maintained, the final current ($= E/R$) would be reached after a time of $t = \tau$. That is

Initial rate of rise of $i = I/\tau$

Also

Initial rate of rise of $v_R = E/\tau$

and

Initial rate of fall of $v_L = E/\tau$

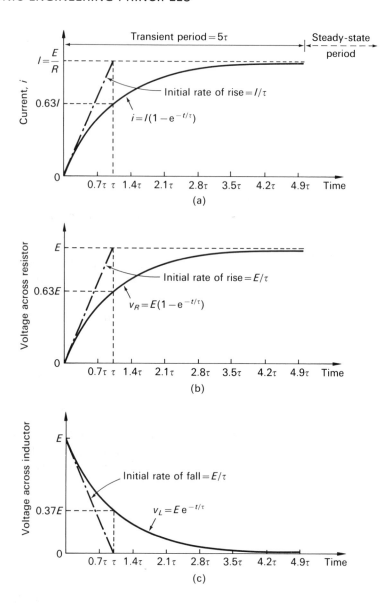

**Fig. 13.7** Growth of current in an L–R circuit

### 13.6.1 Summary of equations for growth of current in an L–R circuit

$$i = I(1 - e^{-t/\tau})$$

where

$$I = E/R \quad \text{and} \quad \tau = L/R$$

$$v_R = iR = E(1 - e^{-t/\tau})$$

$$v_L = E\,e^{-t/\tau}$$

With zero initial conditions ($i = 0$ when $t = 0$)

$$v_L = E \quad \text{at } t = 0$$

$$v_L = 0.63E \quad \text{at } t = \tau$$

$$v_L = 0 \quad \text{at } t = \infty$$

Both $i$ and $v_R$ have risen to 0.63 of their final value when $t = \tau$, at which time $v_L = 0.37E$. The transients have decayed when $t = 5\tau$.

*Worked example 13.3*  A coil of resistance $5\,\Omega$ and inductance $2\,H$ is suddenly connected to a $20\,V$ d.c. supply. Calculate (a) the time constant of the circuit, (b) the final value of the current in the coil, (c) the value of the current $0.28\,s$ after the supply is connected, (d) the current $0.4\,s$ after the supply is connected, (e) the time taken for the transients to have settled out and (f) the rise-time of the current.

*Solution*  (a) The time constant of the circuit is

$$\tau = L/R = 2/5 = 0.4\,s$$

(b) The final value of the current is

$$I = E/R = 20/5 = 4\,A$$

(c) From eqn [13.16], the current when $t = 0.28\,s$ is

$$i = I(1 - e^{-t/\tau}) = 4(1 - e^{-0.28/0.4}) = 4(1 - 0.5)$$

$$= 2\,A$$

*Note*: $0.7\tau = 0.7 \times 0.4 = 0.28\,s$. As mentioned earlier in the chapter, we know that the current has risen to 50 per cent of its final value, i.e. to $2\,A$, in this time.

(d) When $t = 0.4\,s$ the current is

$$i = I(1 - e^{-t/\tau}) = 4(1 - e^{-0.4/0.4}) = 4(1 - 0.37)$$

$$= 2.52\,A$$

In this case, $t = \tau$, so that we know that the current has risen to 63 per cent of its final value, i.e. it has risen to $0.63 \times 4 = 2.52\,A$.

(e) It was shown in section 13.2 that the transients have died away after $5\tau$, that is

$$\text{Settling time} = 5\tau = 5 \times 0.4 = 2\,s$$

(f) From section 13.3.1 we see that

$$\text{Rise-time, } t_r = 2.2\tau = 2.2 \times 0.4 = 0.88\,s$$

That is, it takes $0.88\,s$ for the current to rise from $0.4\,A$ (10 per cent of $4\,A$) to $3.6\,A$ (90 per cent of $4\,A$).

## 13.7 Decay of current in an inductive circuit

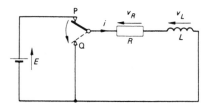

**Fig. 13.8** Decay of current in an inductive circuit

Suppose that the blade of the switch in Fig. 13.8 has been in position P long enough for the current to have reached its steady-state condition ($I_0 = E/R$), when it is suddenly switched to position Q.

We will assume for the moment that the switch is ideal, and that it takes zero time for the blade to change over from P to Q. This statement may seem unimportant but, as we shall see in section 13.8, the assumption that the switch is ideal has important practical implications.

When the switch blade is in position Q, the $L$–$R$ circuit is short-circuited, and the loop equation is

$$v_R + v_L = 0 \qquad\qquad [13.19]$$

or

$$iR + L\frac{di}{dt} = 0$$

that is

$$iR = -L\frac{di}{dt}$$

Separating the variables gives

$$\frac{R}{L}\,dt = -\frac{di}{i}$$

or

$$\frac{dt}{\tau} = -\frac{di}{i}$$

Integrating both sides gives (readers not familiar with integration should move on to eqn [13.21])

$$\frac{t}{\tau} = -\ln i + A \qquad\qquad [13.20]$$

where $\ln i$ is the natural logarithm of $i$, and $A$ is the constant of integration. The latter can be evaluated by inserting the initial conditions in the circuit ($i = I_0$ when $t = 0$) into eqn [13.20] as follows:

$$0 = -\ln I_0 + A$$

or

$$A = \ln I_0$$

Inserting this into eqn [13.20] yields

$$\frac{t}{\tau} = -\ln i + \ln I_0 = \ln \frac{I_0}{i}$$

Taking antilogarithms of both sides gives

$$e^{t/\tau} = \frac{I_0}{i}$$

or

$$i = I_0/e^{t/\tau} = I_0 \, e^{-t/\tau} \qquad\qquad [13.21]$$

From previous work, the reader will see that this is the equation of a falling exponential curve but, very importantly, it has a positive value! That is, the current continues to flow in the same direction through the inductor in Fig. 13.8.

Let us take a look at why the current continues to flow in the same direction when the $L-R$ circuit is short-circuited. Lenz's law states that when the flux in an inductor changes, an e.m.f. is induced in the circuit, and *this e.m.f. acts to maintain the flux acting in the same direction.* That is, the induced e.m.f. acts to maintain the current *in the original direction.*

Since the equation is that of a falling exponential curve, it follows that the transient will have settled out in a time of $5\tau$.

The voltage across the resistor is given by

$$v_R = Ri = R \times \frac{E}{R} e^{-t/\tau} = E \, e^{-t/\tau}$$

and, from eqn [13.19]

$$v_L = -v_R = -E \, e^{-t/\tau} \qquad\qquad [13.22]$$

Graphs showing the way in which $i$, $v_R$ and $v_L$ change are given in Fig. 13.9.

### 13.7.1 Summary of equations for the decay of current in an $L-R$ circuit

$$i = I_0 \, e^{-t/\tau}$$

where

$$I_0 = E/R \quad \text{and} \quad \tau = L/R$$
$$v_R = iR = E \, e^{-t/\tau}$$
$$v_L = -v_R = -E \, e^{-t/\tau}$$

The transients have decayed in approximately $5\tau$.

**13.8 The effect of a practical switch when the circuit is broken**

As mentioned in section 13.7, when the current in an inductive circuit is changed, an e.m.f. of value $L\,di/dt$ is induced in the coil which tends to maintain the current flowing in the original direction. However, when a practical switch is used to break the current, it takes a small (but finite) time for the contacts to open.

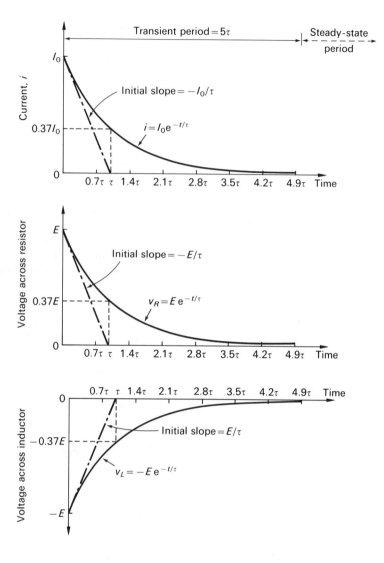

**Fig. 13.9** Graphs showing the decay of current in an *L–R* circuit

Take, for example, the switch in Fig. 13.8 which disconnects the coil from the battery and short-circuits the coil. If the coil is carrying a current of 1 A and it takes, say, 1 ms for the contacts to open then, if the inductance of the coil is 5 H the e.m.f. induced in the coil is

$$L\frac{di}{dt} = 5 \times \frac{1}{1 \times 10^{-3}} = 5000 \text{ V}$$

This e.m.f. acts to maintain the current flow in the coil, and the result is severe sparking at the contacts of the switch! Moreover, if a semiconductor device is connected in the circuit, the voltage may be sufficient to destroy the semiconductor device. In fact, the switch itself may be a semiconductor device.

**13.9 An approximate integrator circuit**

The series $R-C$ circuit in Fig. 13.10(a), in conjunction with the electronic switch S, can be used as an approximate integrator circuit. The switch can be thought of as a mechanical contact which rotates at a constant speed, so that it short-circuits the capacitor at regular time intervals of $T$. In practice the switch is a voltage-sensitive electronic device such as a diac (a bidirectional breakdown diode) or a unijunction transistor (UJT) or a neon tube.

Initially the capacitor begins to charge at a rate dictated by the supply voltage and the time constant ($RC$) of the circuit. Before the capacitor has time to fully charge, the switch S (which is assumed to have zero resistance) discharges it, and the charging process commences once again. The net result is a *sawtooth waveform* across the capacitor, as shown in Fig. 13.10(b).

The circuit can be thought of as an *approximate integrator* for the following reason. Since the supply voltage, $E$, is constant it gives, when integrated,

$$\int E \, \mathrm{d}t = Et$$

That is, the integral of $E$ is a wave which rises constantly with time. In our case we see that the circuit only produces approximate

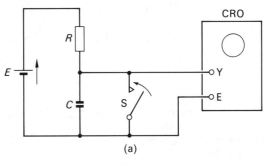

(a)

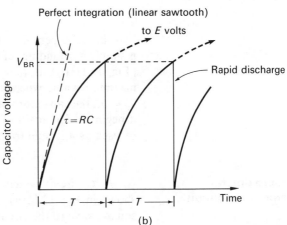

(b)

**Fig. 13.10** (a) Basic integrator circuit and (b) the 'sawtooth' wave appearing across the capacitor

integration because the voltage across the capacitor follows an exponential curve.

This type of circuit can be used as a *simple time-base generator* for an oscilloscope, its advantage being its simplicity, and its disadvantage being that the voltage across the capacitor does not rise linearly. The net result is that the displayed waveform on the oscilloscope tends to be compressed at the right-hand side of the display.

When a square wave is applied to an $R$–$C$ integrator circuit, as shown in Fig. 13.11(a), the waveform of the voltage across the capacitor will (depending on the time constant of the integrator circuit) take on one of a number of possible forms.

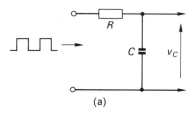

(a)

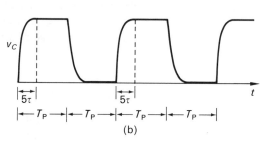

(b)

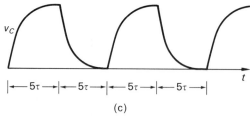

(c)

(d)

**Fig. 13.11** (a) An $R$–$C$ integrator circuit with a square wave applied to it. Typical waveforms for (b) $T_P > 5\tau$, (c) $T_P = 5\tau$, (d) $T_P < 5\tau$

If the pulse period $T_P$ is greater than $5\tau$ or $5RC$, the net result is simply a 'rounding' of the leading and trailing edge of the pulse [Fig. 13.11(b)]. If $T_P = 5\tau$, the pulse just manages to reach its maximum value when it begins to fall again [Fig. 13.11(c)]. If $T_P < 5\tau$, the capacitor voltage will not have risen very much when it begins to fall again [Fig. 13.11(d)]; in fact, such a circuit can be used as a simple triangular wave generator.

**13.10 An approximate differentiator circuit**

An $R$–$C$ circuit which achieves approximate differentiation is shown in Fig. 13.12(a). Assuming that we are applying an ideal square wave to the circuit, *the rate of change of the leading edge*

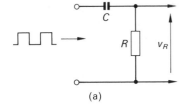

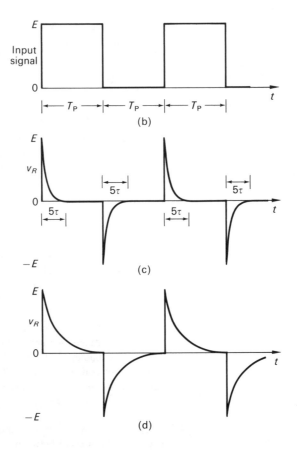

**Fig. 13.12** (a) An $R$–$C$ differentiator circuit. (b) Input waveform; $v_R$ when (c) $T_P > 5\tau$, (d) $T_P = 5\tau$

of the wave is infinitely large. The output voltage of an ideal differentiator would reflect this, and give an infinite value of output voltage. Quite clearly, a practical circuit cannot hope to cope with this value, and the maximum output voltage one can expect is the maximum value of the input voltage to the circuit, that is a voltage equal to $+E$ on the leading edge and $-E$ on the trailing edge.

If the time constant of the circuit ($\tau = RC$) is small when compared with the pulse length, $T_P$ [see Fig. 13.12(c)], the output pulse from the differentiator will have decayed very quickly (in fact after a period equal to $5\tau$), and provides a realistic approach to a differentiator. If $T_P = 5\tau$ the output pulse from the circuit will have just decayed when the trailing edge of the input wave is presented to the input [Fig. 13.12(d)]. In this case, the output is not acceptable as the differentiated version of the input.

For the $R$–$C$ circuit in Fig. 13.12(a) to function as a differentiator, we must ensure that $5\tau \ll T_P$. To be on the safe side we should make $T_P$ about $100\tau$ so that, for a simple differentiator

$$\tau = RC \approx T_P/100$$

For a square wave with $T_p = 1$ ms ($f = 500$ Hz), we need an $RC$ circuit with a time constant of $1$ ms$/100 = 10\,\mu$s (or less). If $C = 0.01\,\mu$F then a suitable value of $R$ is $R = 10\,\mu$s$/0.01\,\mu$F $= 1$ kΩ.

## Problems

**13.1**   A 10 μF capacitor is connected in series with a 100 kΩ resistor to a 100 V d.c. supply. Calculate the time constant of the circuit, and draw to scale a graph showing how the capacitor-charging current and the voltage across the capacitor vary with time. Assume the capacitor to be initially discharged.

[1.0 s]

**13.2**   A capacitor of 0.1 μF is connected in series with a 10 kΩ resistor. Calculate the time constant of the circuit. If the combination is suddenly connected to a 100 V d.c. supply, determine (a) the initial rate of rise of the p.d. across the capacitor, (b) the initial value of the charging current and (c) the final value of energy stored in the capacitor.

[0.001 s; (a) 100 kV/s; (b) 0.01 A; (c) 0.5 mJ]

**13.3**   A 10 μF capacitor is connected in series with a 10 MΩ resistor, and an electronic voltmeter with an infinitely high internal resistance is connected across the capacitor. If the capacitor is initially discharged, and the circuit is suddenly connected to a 100 V d.c. supply, determine the voltage indicated by the voltmeter 69.3 s after the supply is connected.

[50 V]

**13.4**   A 10 μF capacitor has an insulation resistance of 20 MΩ (which is, effectively, connected between the capacitor terminals). If the capacitor is charged to 200 V d.c., determine the time required after being disconnected from the supply for the p.d. between the terminals to fall to 80 V.

[2 h, 32 min and 8 s]

**13.5**   A circuit comprising a 10 μF capacitor, a resistor $R$ and an electronic voltmeter are connected in parallel to a 200 V d.c. supply. When the supply is disconnected, the voltmeter reading falls to 100 V in 56 s. When the resistor is disconnected and the test repeated, the time taken for the same reduction in voltage is 400 s. Determine the resistance of the voltmeter and of the resistor $R$.

[57.7 MΩ; 9.4 MΩ]

**13.6**   A coil has a resistance of 100 Ω and inductance 1.0 H. If the coil is connected to a 100 V d.c. supply determine (a) the steady-state current in the coil, (b) the initial rate of change of current in the coil, (c) the time constant of the circuit, (d) the final value of energy stored in the magnetic field, (e) the time taken for the current to reach 0.8 A and (f) the value of the current 5 ms after the supply has been connected.

[(a) 1 A; (b) 100 A/s; (c) 10 ms; (d) 0.5 J; (e) 0.16 s; (f) 0.393 A]

**13.7**   Draw a graph showing how the current in the coil in problem 13.6 varies with time from the instant the supply is connected.

**13.8**  A 200 V d.c. supply is suddenly connected to a coil which has a time constant of 5 ms. If the current in the coil rises to 0.2 A in 7 ms, determine (a) the steady-state current in the coil, (b) the resistance of the coil and (c) the inductance of the coil.

[(a) 0.265 A; (b) 754.7 $\Omega$; (c) 3.77 H]

**13.9**  A coil of inductance 1.0 H and resistance 80 $\Omega$ is connected to a d.c. supply for a sufficient time to allow the current to reach its steady-state value. When the coil is disconnected from the supply, it is arranged that a resistance of $R$ ohms is connected to the terminals of the coil. If the current falls to 40 per cent of its initial value in 10 ms, calculate the value of $R$.

[11.74 $\Omega$]

**13.10**  A coil of inductance 4 H and resistance 80 $\Omega$ is connected in parallel with a 200 $\Omega$ resistor of negligible inductance, the circuit being connected to a 200 V d.c. supply. Determine after 0.07 s (a) the current in the inductor and (b) the current drawn from the supply. Calculate also (c) the current drawn from the supply when the current in the coil has reached its steady-state value.

If the supply is suddenly disconnected from the coil, determine for the instant of time immediately after disconnecting the supply (d) the current in the coil and (e) the voltage across the coil. Also estimate (f) the time taken for the current to decay to 10 per cent of its value at the instant of disconnection from the supply.

[(a) 1.88 A; (b) 2.88 A; (c) 3.5 A; (d) 2.5 A; (e) 500 V; (f) 0.115 s]

**13.11**  An approximate integrator of the form in Fig. 13.11 employs $R = 1$ k$\Omega$ and $C = 1$ $\mu$F. If $E = 10$ V determine either graphically or by calculation, by how much the voltage across the capacitor deviates from that of an ideal integrator after 0.7 ms.

[2 V]

**13.12**  A circuit of the type in Fig. 13.12 is used to differentiate a 1 kHz square wave. If $C = 0.01$ $\mu$F, calculate a suitable value of $R$.

[500 $\Omega$]

# 14 Electrical machines

## 14.1 Introduction

In this chapter we look at the principle of rotating electrical machines. Broadly speaking, there are two categories of electrical machine, namely **generators** which convert mechanical energy into electrical energy, and **motors** which convert electrical energy into mechanical energy.

In general, a machine is designed for one particular purpose, i.e. it is designed either as a generator or as a motor, and it is in that type of application where it is most efficient. However, a machine can usually act either as a generator or as a motor and, in some applications, such as control engineering (see also Chapter 16), a machine can be called upon to change its role within a fraction of a second.

All electrical machines have a magnetic system which produces the main magnetic field, and a set of windings which produce the e.m.f. or the torque (depending on whether we are considering generator or motor action). Initially we will look at the classification of magnetic systems and windings, because we can find several types in both d.c. and a.c. machines.

## 14.2 Magnetic systems

Electrical machines have a fixed part known as the **frame** or **stator**, and a rotating part known as the **armature** or **rotor**. Magnetic systems of machines are described either as being **salient systems** (meaning 'jutting out') or as being **cylindrical systems** (cylindrical about the axis of rotation), as illustrated in Fig. 14.1.

Particular designs of magnetic system lend themselves to specific machine designs, and the list in Table 14.1 shows a typical range. The reader is asked to note that the a.c. synchronous motor listed in the table is the conventional industrial machine, and not the 'electric clock' motor.

**Fig. 14.1** Magnetic system of a machine: (a) salient stator and rotor, (b) cylindrical stator and rotor

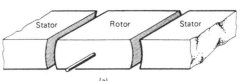

**Table 14.1** Machine construction

| Type of machine | Stator | Rotor |
| --- | --- | --- |
| D.c. machine | Salient | Cylindrical |
| A.c. induction motor | Cylindrical | Cylindrical |
| A.c. synchronous motor | Cylindrical | Salient |

**14.3 Winding classification**

Electrical machine windings are classified as either concentrated windings or as distributed windings, as follows.

**Concentrated windings**

A concentrated winding is a coil on a salient pole, as shown in Fig. 14.2, the function of the winding being to produce part of the magnetic field of the machine. The main e.m.f. or current *is not induced in a concentrated winding.*

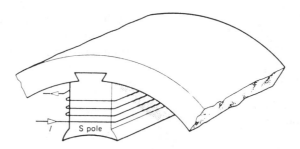

**Fig. 14.2** Concentrated winding

**Distributed windings**

These windings are distributed around the magnetic circuit and are housed in slots which are cut around part (or all) of the circumference of the stator or rotor. Distributed windings are further subdivided into cage windings, phase windings and commutator windings as follows.

*Cage winding*
This comprises conductors around the edge of a cylindrical rotor, as shown in Fig. 14.3. The conductors are short-circuited at the ends by means of stout *end-rings*. This type of winding is widely used in 'squirrel' cage induction motors.

*Phase winding*
A phase winding is a distributed winding in which groups of the coils are connected to the phases of a polyphase supply.

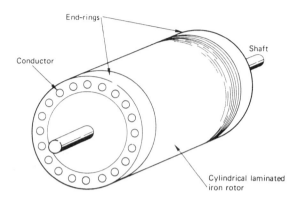

**Fig. 14.3** A rotor with a cage winding

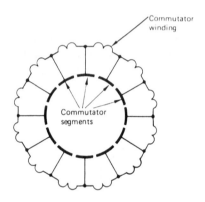

**Fig. 14.4** The principle of a commutator winding

*Commutator winding*
A commutator winding is wound continuously around the rotor of a machine without any breaks in it (see Fig. 14.4). Connections are made to the winding at regular intervals, each connection being made to a *segment* on the *commutator* of the machine. The commutator consists of copper segments assembled in the form of a cylinder, each segment not only being insulated from one another but also from the general body of the machine. More information is given in section 14.4.1.

**14.4  E.m.f. equation**

If a machine produces a flux per pole of $\Phi$, and has $p$ *pole-pairs*, then in one revolution of the armature each conductor cuts a flux of $2p\Phi$. At a speed of $n$ rev/s, the time taken to complete one revolution is $1/n$ seconds, and the average e.m.f. per revolution is

$$\text{e.m.f.} = \frac{\text{flux cut per revolution}}{\text{time for one revolution}} = \frac{2p\Phi}{1/n} = 2p\Phi n$$

If the machine has $Z$ conductors arranged in $c$ parallel paths, then the number of conductors in series is $Z/c$, so that the total e.m.f. induced in the armature is

$$E = 2p\Phi n \times \frac{Z}{c} = \frac{2p}{c}\,\Phi Z n \qquad [14.1]$$

The speed in rev/s is related to the speed, $\omega$, in rad/s by $\omega = 2\pi n$, hence eqn [14.1] can be rewritten in the form

$$E = \frac{2p}{c}\,\Phi Z\,\frac{\omega}{2\pi} = \left[\frac{pZ}{c\pi}\right]\Phi\omega \qquad [14.2]$$

In a given machine, the factors $p$, $Z$ and $c$ are fixed, so that we may rewrite eqn [14.2] in the form

$$E = K_E \Phi \omega \qquad\qquad [14.3]$$

where $K_E = pZ/c\pi$ is the **e.m.f. constant** of the machine. Since $K_E$ is a constant, we may reduce eqn [14.3] to the simple proportional form

$$e \propto \Phi \omega \propto \Phi n \propto \Phi N \qquad\qquad [14.4]$$

The relationship in eqn [14.4] holds good for both motors and generators. In the case of a motor, the induced e.m.f. opposes the applied voltage and is known as the **back e.m.f.** of the motor.

### 14.4.1  Lap and wave windings

There are two important types of commutator winding, namely:

1. *Lap winding* (frequently used in high-current machines), which provides as many parallel paths through the armature as there are poles ($c = 2p$) on the machine.
2. *Wave winding* (frequently used in high-voltage machines), which provides one pair of parallel paths ($c = 2$) for current flow through the armature.

*Worked example 14.1*    The no-load terminal voltage of a d.c. generator is 250 V when working at its rated speed and nominal flux density. If the speed is reduced by 15 per cent, and the flux per pole is increased by 10 per cent, calculate the new value of the no-load terminal voltage.

*Solution*    The new speed $N_2$ is given by

$$N_2 = N_1 \times 0.85 \text{ rev/min}$$

where $N_1$ is the rated speed. The new flux per pole is

$$\Phi_2 = \Phi_1 \times 1.1$$

where $\Phi_1$ is the nominal flux per pole. From eqn [14.4] we see that

$$E_1 \propto \Phi_1 N_1 \quad \text{and} \quad E_2 \propto \Phi_2 N_2$$

hence

$$\frac{E_2}{E_1} = \frac{\Phi_2 N_2}{\Phi_1 N_1}$$

or

$$E_2 = E_1 \frac{\Phi_2 N_2}{\Phi_1 N_1} = 250 \times \frac{1.1\Phi_1 \times 0.85 N_1}{\Phi_1 N_1}$$

$$= 233.75 \text{ V}$$

**14.5  Torque equation**

If $T$ is the torque developed by the armature of a motor (or, in the case of a generator, is applied to it by an external prime mover), $P$ is the power and $\omega$ is its rotational speed in rad/s then

$$P = T\omega$$

also, the power developed by the armature is

$$P = EI_a$$

where $E$ is given by eqn [14.2]. That is

$$P = T\omega = \left[\frac{pZ}{c\pi}\right]\Phi\omega I_a$$

or

$$T = \left[\frac{pZ}{c\pi}\right]\Phi I_a \qquad\qquad [14.5]$$

$$= K_T \Phi I_a \qquad\qquad [14.6]$$

where $K_T = pZ/c\pi$ is the **torque constant** of the machine. The relationship in eqn [14.6] holds good for either a generator or a motor. In the case of the motor it is the torque developed by the armature, and in the case of the generator it is the torque needed to drive the armature. Since $K_T$ is a constant, we can reduce eqn [14.5] to a simple proportional relationship as follows:

$$T \propto \Phi I_a \qquad\qquad [14.7]$$

*Worked example 14.2*

A d.c. motor develops a torque of 500 N m when carrying an armature current of 25 A. What torque is produced if the flux density is increased by 10 per cent and the armature current is increased to 30 A?

*Solution*

Assuming that the iron circuit of the motor does not begin to saturate with the increase in flux and in armature current, then

$$B_2 = 1.1B_1$$

where $B_1$ and $B_2$ are the initial and final value of flux density, respectively. That is

$$\Phi_2 = 1.1\Phi_1$$

From eqn [14.7] we see that

$$T_1 \propto \Phi_1 I_{a1} \quad \text{and} \quad T_2 \propto \Phi_2 I_{a2}$$

hence

$$\frac{T_2}{T_1} = \frac{\Phi_2 I_{a2}}{\Phi_1 I_{a1}} = \frac{1.1\Phi_1 \times 30}{\Phi_1 \times 25} = 1.32$$

or

$$T_2 = 1.32T_1 = 660 \text{ N m}$$

**14.6 Use of a rotating machine either as a generator or as a motor**

We see from eqns [14.3] and [14.6] that $K_E$ and $K_T$ both have the value $pZ/c\pi$, or $K_E = K_T$. This implies that a rotating machine can be used as a generator at one moment in time, and as a motor the next. Moreover, from eqn [14.3] $K_E = E/\Phi\omega$, and from eqn [14.6] $K_T = T/\Phi I_a$. Since $K_E = K_T$ it follows that

$$\frac{E}{\Phi\omega} = \frac{T}{\Phi I_a}$$

or

$$\frac{E}{\omega} = \frac{T}{I_a} \qquad [14.8]$$

This implies that if a given machine generates an e.m.f. of $4\,\text{V}/(\text{rad/s})$ when operating as a generator, then it is capable of producing a torque of $4\,\text{N m/A}$ when operating as a motor.

In practice this will not quite be the case since a machine designed to operate as a generator will not be quite so efficient when working as a motor.

**14.7 Efficiency of electrical machines**

The **efficiency**, $\eta$, of an electrical machine is given by

$$\text{Efficiency, } \eta = \frac{\text{output power}}{\text{input power}} \text{ per unit (p.u.)} \qquad [14.9]$$

In many cases the efficiency is given as a per cent value, where

Per cent efficiency $= 100 \times$ per unit efficiency

or

$$\text{Efficiency, } \eta = \frac{\text{output power}}{\text{input power}} \times 100 \text{ per cent}$$

The value of input power and output power, in practice, always differs because of the power loss in the machine, so that

Input power = output power + power loss

The power loss in a machine can be grouped as follows:

1. **copper loss** ($I^2R$ loss), which occurs in the armature and field circuits;
2. **iron loss**, which is due to hysteresis and eddy current loss in the magnetic circuit;
3. **mechanical power loss**, which is due to friction in the bearings, and to windage loss (the power loss due to circulation of air in the machine).

*Worked example 14.3* Determine the efficiency of an electrical motor which provides a mechanical output of 25 kW, and consumes 28.5 kW of electrical power.

If the copper loss amounts to 2 kW, calculate the total value of other power losses in the motor.

*Solution*  The efficiency of the motor is

$$\text{Efficiency} = \frac{\text{output power}}{\text{input power}} = \frac{25\,\text{kW}}{28.5\,\text{kW}}$$

$$= 0.88 \text{ p.u.} \quad \text{or} \quad 88 \text{ per cent}$$

The total losses in the motor are

$$\text{Total loss} = \text{input power} - \text{output power}$$

$$= 28.5 - 25 = 3.5 \text{ kW}$$

hence

$$\text{Remaining losses} = \text{total power loss} - \text{copper loss}$$

$$= 3.5 - 2 = 1.5 \text{ kW}$$

## 14.8  D.c. machine construction

The general form of construction for both d.c. motors and d.c. generators is shown in Fig. 14.5. The fixed part of the machine (the **frame**) consists of a cylindrical iron **yoke**, which is usually fabricated from rolled steel. A number of **pole cores** (two of them in Fig. 14.5) are bolted on to the yoke, and carry the main **field windings**. Each pole has a **pole tip** attached to it, its function being twofold: firstly, it secures the field winding in position and, secondly, it increases the cross-sectional area of the magnetic path. The latter ensures a much more uniform magnetic flux distribution around the armature.

Many d.c. machines have a **compole** (alternatively known as a **commutating pole** or **interpole**) between each pair of main poles, as shown in Fig. 14.5. The primary function of the compoles is

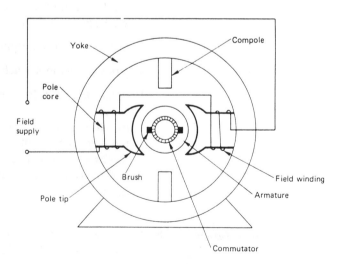

**Fig. 14.5**  D.c. machine construction

to improve the current commutation, i.e. to reduce the sparking which sometimes occurs between the brushes and the commutator when the machine carries load.

The armature (see Fig. 14.6) consists of many circular iron laminations (about 0.5 mm thick) which are clamped together to form a cylinder. The laminations are insulated from one another by a coat of varnish, and have slots in their circumference to house the armature winding. Since the armature is made of iron, which is a conductor, and it rotates in the field system of the motor, an e.m.f. is induced in it. This gives rise to **eddy currents** in the armature, which produce a power loss and heating in the machine. In turn, this has the effect of reducing the efficiency of the motor.

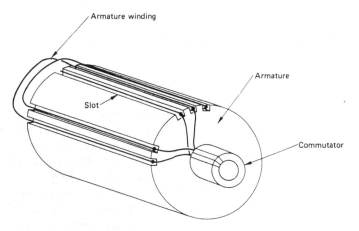

**Fig. 14.6** Armature construction

The armature winding is connected to the **commutator** of the machine; current is either supplied to or collected from the commutator by means of **carbon brushes** which are in contact with the commutator, each brush being held in place by a metal **brush box**. The brush box is designed to allow for a radial movement of the brushes to compensate for wear of the brushes; pressure between the brushes and the commutator is maintained by means of spring pressure.

The number of brushes on the machine depends not only on the rating of the machine, but also on the design of the commutator winding (remember, some designs of commutator winding provide more parallel paths through the armature winding). In a small machine, such as a hand-held drilling machine, one pair of brushes may be sufficient to deal with the current. A large machine of several kilowatts may have several sets of brushes, with four or six brushes on the same brush arm.

**14.9  Simple d.c. generator and the commutator action**

We will use the simple machine in Fig. 14.7 to explain the operation of a d.c. generator and the action of the commutator.

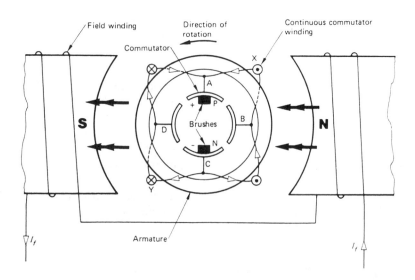

**Fig. 14.7** Simple d.c. generator

The generator has its field winding on its salient pole stator, and the armature (the rotor of the machine) supports a continuous armature winding. The armature itself is simply a cylinder of iron, the armature winding being coiled around it until, finally, the 'end' and the 'start' of the winding are joined together. This is a very early design of winding, and proves to be an excellent vehicle for understanding the operation of the generator.

The armature is driven round by means of a **prime mover** such as a steam turbine, and the field current $I_f$ is provided by some external means such as a battery. In this way we establish the means of electricity generation, namely the movement of conductors in a magnetic field.

Since the armature is a cylinder of iron which has a very low magnetic reluctance, it acts as a magnetic short circuit, and *no magnetic flux penetrates through to the conductors inside the armature*. That is, all the e.m.f. is induced in the outer conductors, and none is induced in the inner conductors. The inner conductors merely serve to conduct the current from the 'end' of one of the outer conductors to the 'start' of the next. Modern armature windings eliminate the need for the 'inner' conductors.

With the armature rotation in Fig. 14.7, the application of Fleming's right-hand rule gives the direction of induced current shown in the figure. That is, the induced e.m.f. on conductors on the left-hand side of the armature causes current to flow from C via D to A. On the right-hand side, current flows from C via B to A. So far as the external circuit is concerned, A is positive with respect to C. We therefore place brush P at the position where the highest potential occurs, and brush N where the negative potential occurs.

As the armature rotates, the conductors successively pass under the N- and S-poles, so that the direction of the induced e.m.f. in each conductor reverses or *alternates*. Nevertheless, brush P is *always* connected to the most positive potential on the armature winding, and brush N is always connected to the most negative potential. That is, even though an *alternating voltage* is induced in the armature winding, a *direct voltage* or *unidirectional voltage* always exists between brushes P and N. The purpose of the commutator is therefore twofold as follows:

1. It enables a connection to be made between the external circuit and a point of fixed electrical potential on the winding.
2. It allows the current in the winding to reverse while maintaining a unidirectional voltage between terminals P and N.

In particular, the reader should note from Fig. 14.7 that *the brushes are connected to a point on the armature winding which is between the poles.*

In a modern machine the reader will find that the brushes are physically positioned so that they are in line with the pole centre. The reason is that modern armature windings have a 'skew' on them, and a conductor which is between the poles on the armature is connected to a point on the commutator which is physically in line with the pole centre.

It follows that the commutator enables *rectifying action* to occur; that is, it converts an alternating current in the armature to a direct current in the brushes. In a motor the opposite action occurs, and is known as an *inverting action* or *inversion*. In this case, the current supplied from a d.c. source is converted by the commutator into an alternating current in the armature.

## 14.10 The equivalent circuit of the armature of a d.c. machine

To simplify a circuit for calculations, engineers represent many complex elements by means of an **equivalent circuit** which, for most purposes, is electrically equivalent to the original circuit.

There are three essential elements 'inside' the armature of a d.c. machine; these are:

1. the induced e.m.f., $E$ (which is the 'back' e.m.f. in a motor);
2. the armature resistance, $R_a$; and
3. the voltage drop $V_b$ in the brushes.

Each of these is shown in Fig. 14.8 both for a generator and a motor. The largest of these values is the e.m.f., $E$. The voltage drop arrow across the armature resistance, $R_a$, will always oppose the flow of current through the armature, as will the voltage drop in the brushes. The equations for the terminal voltage $V_T$ for the generator and motor are, respectively

*Generator terminal voltage*

$$V_T = E - I_a R_a - V_b \qquad [14.10]$$

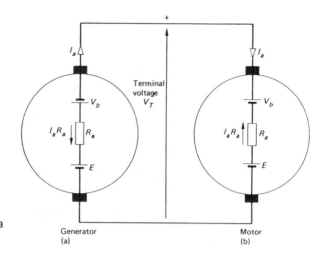

**Fig. 14.8** Equivalent circuit of the armature of (a) a generator and (b) a motor

*Motor terminal voltage*

$$V_T = E + I_a R_a + V_b \tag{14.11}$$

Generally speaking, the voltage drop $V_b$ is either combined with $E$, or it is ignored (which is usually permissible since $E \gg V_b$).

### 14.11 D.c. machine connections

The 'name' given to a d.c. machine depends on the way in which the field windings are connected. The main types are:

1. separately excited;
2. shunt wound;
3. series wound; and
4. compound wound.

Any of these types can be used with motors or generators but, in the following, we refer to motors. The following modifications should be made when referring to generators:

(a) the power source is rotational power provided by the prime mover;
(b) the generator supplies power to an electrical load (that is, the generator can be thought of as providing current to the 'supply source' or battery in the figures).

In a **separately excited** d.c. motor (Fig. 14.9), the field winding and the armature are supplied from independent sources. The field winding has many turns of fine wire, and has a high resistance; this ensures that the field current has a small value (small, that is, when compared with the armature current).

**Shunt-wound** machines (Fig. 14.10) have a common supply source for both the field winding and the armature. Once again, the field winding has a high resistance and carries a 'low' value

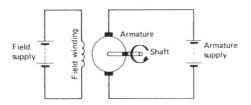

**Fig. 14.9** Separately excited d.c. machine

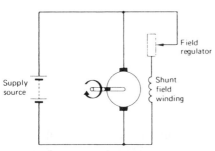

**Fig. 14.10** Shunt-wound d.c. machine

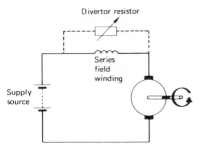

**Fig. 14.11** Series-wound d.c. machine

### 14.12 Characteristics of a separately excited d.c. generator

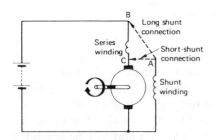

**Fig. 14.12** A compound-wound d.c. machine

of current. The field current can be varied by means of a **field regulator** resistor in series with the shunt field winding.

In a **series-wound** d.c. motor (see Fig. 14.11), the field winding is in series with the armature. Since the armature current has a high value, the series winding carries this current, and it must have a low resistance. That is, the winding consists of a very few turns of large cross-sectional area. When it is necessary to control the current in the series winding, a low-resistance **divertor resistor** is shunted across the winding (*note*: the divertor resistor should *never be reduced to zero value*, otherwise no current would flow in the winding with consequent damage to the motor).

A **compound-wound** d.c. machine (see Fig. 14.12) has two field windings, namely a shunt winding and a series winding. Each main pole of the machine carries part of the shunt winding and part of the series winding, the net magnetic flux being dependent not only on the current in each winding but also on the relative 'direction' of the current (see below). The machine is said to have a **long-shunt** connection when point A on the shunt winding is connected to point B on the series winding, and is said to have a **short-shunt** connection when point A is connected to point C.

Furthermore, depending on the connections of the shunt- and series-wound windings, the magnetic flux by the series winding may either assist (**cumulative compound wound**) or oppose (**differentially compound wound**) the shunt winding flux.

It was shown in eqn [14.4] that the generated e.m.f. is $E \propto \Phi\omega$. If the field winding is supplied at constant current, i.e. the flux remains constant, the equation reduces to $E \propto \omega$. That is, the generated e.m.f. is zero at zero speed, and it increases linearly with speed, as shown in Fig. 14.13.

Moreover, if the excitation current is increased to give a greater magnetic flux (that is $\Phi_2 > \Phi_1$) then, at any given value of speed, the e.m.f. is increased (see the graphs for flux $\Phi_1$ and flux $\Phi_2$). It follows that *the generated e.m.f. is proportional to the armature speed*.

If, in eqn [14.4], the speed remains constant, the equation reduces to $E \propto \Phi$, which implies that *the generated e.m.f. is proportional to the field flux*. If it were possible to plot the generated e.m.f. to a base of magnetic flux we would, in fact, get the straight line $E-\Phi$ graph in Fig. 14.14. However we cannot, in practice,

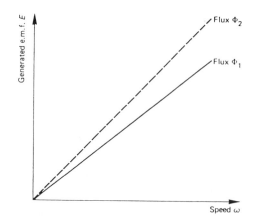

**Fig. 14.13** E.m.f.–speed characteristic for a separately excited d.c. generator for two values of field flux $\Phi_1$ and $\Phi_2$

increase the magnetic flux in a uniform manner, and all we can do is to increase the field current uniformly. Since the field current is related to the magnetic flux by the $B–H$ curve of the iron circuit of the machine, the $E–I_f$ curve of the machine follows the $B–H$ curve in the way shown in the figure.

The initial value of induced e.m.f. (corresponding to $I_f = 0$) depends on the residual flux in the iron circuit. For a generator which does not have a load connected to its terminals, the $E–I_f$ curve in Fig. 14.14 is known as the **open-circuit characteristic** of the machine.

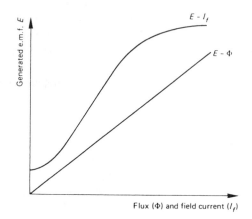

**Fig. 14.14** $E–\Phi$ and $E–I_f$ curves for a separately excited d.c. generator at a constant armature speed

If the armature speed is increased (at constant excitation), there is a general upward lift of the characteristic in Fig. 14.14.

The **load characteristic** is a graph of the *terminal voltage* plotted to a base of armature current (or load current), which is shown as the $V_T–I_a$ graph in Fig. 14.15 (which is also known as the *external characteristic* of the generator). A graph of the generated e.m.f. to a base of $I_a$ (see the $E–I_a$ plot) is known as the *internal characteristic*.

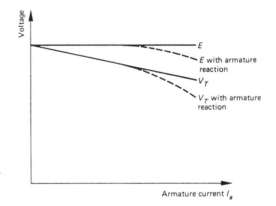

**Fig. 14.15** Load characteristic of a separately excited d.c. generator running at constant speed and excitation current

Equation [14.4] shows that if the speed and excitation current remain constant, then $E$ remains constant when the load current changes. Unfortunately, the current carried by the armature conductors also produces a magnetic flux (known as **armature reaction**), and this has the effect of distorting the magnetic flux in the generator. As the armature current increases, so does the distorting effect of the armature reaction. The net result is a reduction of the total flux produced by the field system as the armature current increases. Since the generated e.m.f. depends on the total flux, its value falls as the armature current increases. The general effect on both the external and internal characteristics of the machine is shown by the broken line in Fig. 14.15.

It was shown earlier that the terminal voltage of a generator is given by $V_T = E - I_a R_a$. That is, as the armature current increases, the terminal voltage initially reduces in a linear manner, as shown by the $V_T$–$I_a$ graph as the full line in Fig. 14.15. However, at higher values of current, armature reaction reduces the generated voltage, $E$, so that there is a fairly rapid reduction in $V_T$ as $I_a$ increases.

*Worked example 14.4*    The no-load terminal voltage of a separately excited d.c. generator is 200 V when the flux per pole is 15 mWb and the armature speed is 1000 rev/min.

(a) If the armature speed is increased to 1200 rev/min and the flux per pole is reduced to 12 mWb, calculate the new value of the no-load terminal voltage.

(b) If, with the value of flux in part (a) and an armature current of 25 A, armature reaction reduces the flux per pole by 5 per cent, calculate the terminal voltage if the armature resistance is 0.2 Ω.

*Solution*    (a) The induced e.m.f. equation is $E \propto \Phi N$. If $E_1$, $\Phi_1$ and $N_1$ are the original values of e.m.f., flux and speed, and $E_2$, $\Phi_2$ and $N_2$ are the respective final values, then

$$\frac{E_2}{E_1} = \frac{\Phi_2 N_2}{\Phi_1 N_1}$$

or

$$E_2 = E_1 \Phi_2 N_2 / \Phi_1 N_1$$

$$= 200 \times 12 \times 10^{-3} \times 1200/(15 \times 10^{-3} \times 1000)$$

$$= 192\,\text{V}$$

(b) If armature reaction reduces the flux in part (a) by 5 per cent, then the new flux is $\Phi_3 = 0.95\Phi_2$. At constant speed, the relationship between e.m.f. and flux is $E \propto \Phi$, then

$$\frac{E_3}{E_2} = \frac{\Phi_3}{\Phi_2} = \frac{0.95\Phi_2}{\Phi_2} = 0.95$$

or

$$E_3 = 0.95 \times 192 = 182.4\,\text{V}$$

This means that armature reaction has reduced the generated voltage by nearly 10 V. From eqn [14.10], the terminal voltage is

$$V_T = E - I_a R_a = 182.4 - (25 \times 0.2) = 177.4\,\text{V}$$

## 14.13 Characteristic of a shunt-wound d.c. generator

Once a shunt-wound d.c. generator [see Fig. 14.16(a)] has been operated, its field has some residual magnetism, so that whenever it is next used the residual magnetism causes a field current to begin to flow. Consequently, the generated voltage begins to build up from a **residual** value along a characteristic which follows the magnetization curve of the magnetic circuit [see Fig. 14.16(b)]. When the generator is unloaded, the general shape of thhe $E$–$I_f$ curve is known as the **no-load characteristic**.

Steady operating conditions are achieved at the intersection of the no-load characteristic and a straight line of slope $R_T$, where $R_T$ is the *total resistance of the field circuit*. The point where the two intersect is known as the *operating point*. When the field current is $I_{f1}$, the generated e.m.f. is $E_1$.

Reducing the resistance of the field regulator reduces the total resistance of the field circuit, allowing the generated voltage to increase slightly; increasing the resistance of the field regulator has the effect of reducing the generated voltage. If the resistance of the regulator is increased to a value where the slope of the $R_T$ line is tangential to the open-circuit characteristic, the generator will fail to excite. At this point, the value of the field circuit resistance is known as the **critical resistance**, $R_C$, as shown in Fig. 14.16(b).

When we close switch S, we can plot the **load characteristic** of the generator. The load characteristic in Fig. 14.16(c) is drawn

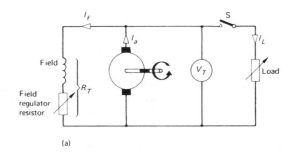

(a)

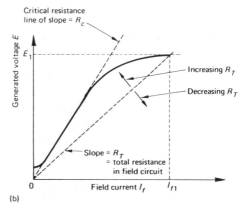

(b)

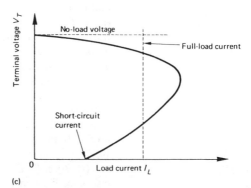

(c)

**Fig. 14.16** Shunt-wound generator: (a) test circuit, (b) the no-load characteristic and (c) the load characteristic

with the armature driven at constant speed with the excitation adjusted to produce the normal no-load voltage (the field resistance remaining unchanged for the duration of the load characteristic test). Initially, as the load current increases, the terminal voltage begins to reduce due to:

1. $I_a R_a$ drop in the armature resistance; and
2. armature reaction.

The *full-load current* of the generator is generally quoted as a value where stable operating conditions are achieved, that is when the slope of the load characteristic is positive.

As the load current is increased further, a point will be reached where the combined effects of the $I_a R_a$ drop and armature reaction cause the terminal voltage to reduce very rapidly. Beyond this point a condition will be reached where, even if the load resistance is reduced, *the load current also reduces*. Finally, if the terminals are short-circuited, the load current falls to a value which is less than the full-load current of the machine. This condition should be avoided at any cost, as the action of short-circuiting the generator can damage it.

**Worked example 14.5**  The open-circuit characteristic of a d.c. shunt generator running at a speed of 800 rev/min is as follows:

| Field current (A) | 2 | 3 | 4 | 5 | 6 | 7 | 8 |
|---|---|---|---|---|---|---|---|
| E.m.f. (V) | 215 | 310 | 390 | 460 | 510 | 550 | 580 |

(a) Determine the no-load armature voltage at a speed of 800 rev/min if the total resistance of the field circuit is $80\,\Omega$. (b) If the speed is reduced by 10 per cent, estimate the no-load voltage with a field circuit resistance of (i) $80\,\Omega$, (ii) $75\,\Omega$.

*Solution*  The open-circuit characteristic for a speed of 800 rev/min is plotted in Fig. 14.17. The characteristic for the speed of $0.9 \times 800 = 720$ rev/min is obtained by multiplying the e.m.f. values in the above table by 0.9, and plotting the e.m.f. values to a base of field current.

(a) *Field resistance of* $80\,\Omega$: Two points are needed to plot the field resistance line on the characteristic. One of these is the origin of the graph, and the other is the value of voltage needed to produce a given value of field current. Using a field current of 7 A, the voltage needed across the field circuit is $(7 \times 80) = 560$ V. A line drawn from the origin of the graph through the point 7 A, 560 V gives the appropriate field resistance line. The intersection of this line with the open-circuit characteristic gives the open-circuit voltage to which the generator builds up at 800 rev/min. In this case, the voltage is found to be 540 V (see Fig. 14.17).

(b) (i) *Total field resistance* $= 80\,\Omega$, *speed* $= 720$ *rev/min*: The field resistance line remains as in part (a), and the point where it cuts the open-circuit characteristic for 720 rev/min gives a generated voltage of 430 V (see Fig. 14.17).

*Note*: While the speed has been reduced by 10 per cent, the generated e.m.f. has been reduced by 20.4 per cent!

(ii) *Total field resistance* $= 75\,\Omega$, *speed* $= 720$ *rev/min*: The field resistance line for $75\,\Omega$ is plotted in a similar manner to that described earlier, and the intersection of this line (see Fig. 14.17) with the open-circuit characteristic gives a generated e.m.f. of 470 V.

**14.14  Series-connected d.c. generator**

The generated voltage–excitation current curve follows the general shape of the magnetization curve of the iron circuit of the machine but, in this case, the excitation current is the armature current of the machine (see Fig. 14.18). The curve for the terminal

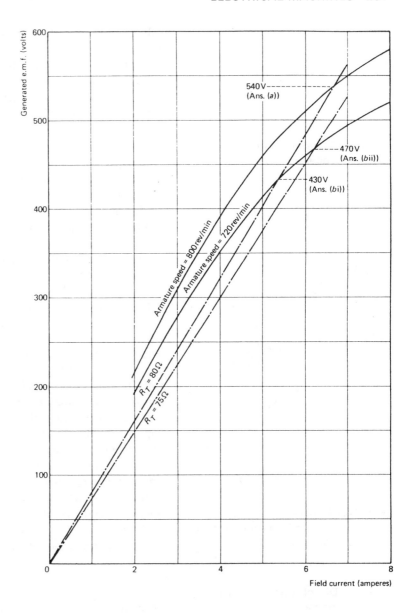

**Fig. 14.17** Graph for Worked Example 14.5

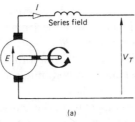

(a)

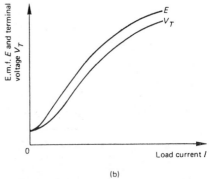

(b)

**Fig. 14.18** (a) Series-wound generator and (b) its load characteristic

voltage $V_T$ lies slightly below the generated e.m.f. curve, the difference between the two being due to the voltage drop in the armature circuit. As with other machines, armature reaction also reduces the voltage.

Since the generated voltage varies with load current, the series generator is unsuitable for use as a general-purpose generator.

## 14.15 Compound-wound d.c. generator

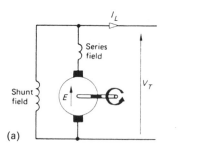

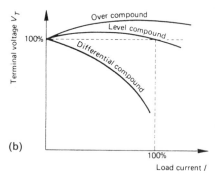

**Fig. 14.19** (a) One form of compound-wound generator and (b) its load characteristic

A compound-wound generator has both series and shunt windings, giving a wide range of operating characteristics (see Fig. 14.19).

Machines in which the magnetic flux produced by the series winding aids that produced by the shunt winding are classed as **cumulative-compound generators**; included in this group are **level-compound** (or **flat-compound**) machines and **over-compound** machines. In these machines the effect of the series winding flux is to induce a voltage in the armature which compensates not only for the $I_a R_a$ voltage drop but also for armature reaction effects. In a level-compound machine, the additional induced voltage just compensates for the $I_a R_a$ drop and for armature reaction under full-load conditions, so that the full-load voltage is equal to the no-load voltage. In the case of an over-compound machine, the extra induced e.m.f. results in a full-load voltage which is greater than the no-load terminal voltage.

Machines in which the flux produced by the series winding opposes the flux produced by the shunt winding are known as **differentially compound** machines. In this case, an increase in load current results in a rapid reduction of the net flux (and therefore a reduction in terminal voltage). These machines have an inherent limitation to the output current, and are useful in applications where the short-circuit current must be limited in value; one application for this type of machine is a generator for a transportable electric welding machine.

## 14.16 Direct current motor starter

At the instant a d.c. motor is connected to the power supply, the armature is stationary and the induced e.m.f. in the armature (the 'back e.m.f.') is zero. Since the armature has a very low resistance, the resulting surge of armature current at the instant of connection is very large indeed. In the majority of cases, the current is large enough to damage the commutator at the point where it is in contact with the brushes.

To prevent damage from this cause, every d.c. motor (with the exception of small hand-held motors) has a 'starter' associated with it, the purpose of the starter being to insert some resistance in series with the motor during the starting period. This has the effect of reducing the current on start-up, and allowing the armature to begin to rotate under controlled conditions. Once the armature begins to rotate, a back e.m.f. builds up in the armature and the resistance in the starter can be reduced. Finally, under full-speed conditions, the resistance in the starter can be reduced to zero, and the armature connected directly to the supply.

A basic shunt motor starter is shown in Fig. 14.20(a). When the starting arm is in the OFF position, the machine is disconnected from the power supply. Note that the the shunt field winding is connected to the armature via current-limiting resistors $R_1$, $R_2$ and $R_3$; this connection ensures that any magnetic energy in the field circuit when the motor is switched off is discharged. When the motor is first switched on, the armature is connected to the supply via $R_1$, $R_2$ and $R_3$ and, as the motor speed increases, the starting arm is moved from position A to position B, then to C and, finally, to D.

A practical version of the starter is shown in Fig. 14.20(b).

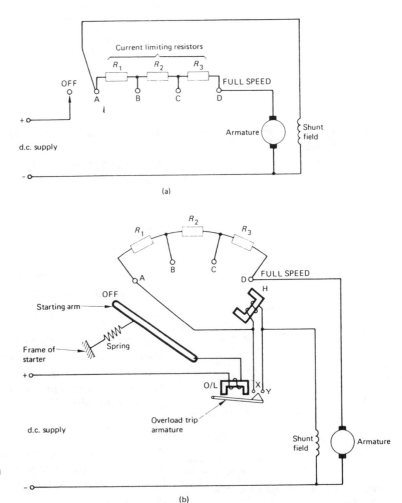

**Fig. 14.20** (a) Basic circuit of a shunt motor starter and (b) a practical starter

Prior to starting, the starting arm is held in the OFF position by a spring, one end of the spring being secured to the frame of the starter. During starting, the shunt field current passes through the coil of electromagnet H, whose function it is to provide sufficient magnetic pull to hold the starting handle against the force of the spring when the starter is in the FULL-SPEED position.

In the event of the supply voltage dropping in value much lower than its nominal value, the current in electromagnet H will not be sufficient to hold the starting handle in the FULL-SPEED position, when the pull of the spring causes the starting handle to return to the OFF position. Coil H therefore provides a simple form of **undervoltage protection**.

**Overcurrent** or **overload protection** is provided by the O/L coil in association with electromagnet H. The O/L coil carries the armature current and, for normal values of armature current, the magnetic pull provided by the coil is not sufficient to attract its armature. When an overload occurs, the magnetic pull of the O/L coil attracts its armature, and causes contacts X and Y to be short-circuited. This short-circuits coil H, allowing the starting handle to return to the OFF position under the influence of the spring.

## 14.17 Characteristics of a shunt motor

### Torque–armature current characteristic

The total torque or gross torque, $T_g$, produced by a d.c. motor is given by

$$T_g = K_T \Phi I_a \qquad [14.12]$$

where $K_T$ is the torque constant of the motor, $\Phi$ the magnetic flux per pole and $I_a$ the armature current. If the field current is maintained at a constant value and if, for the moment, we can ignore the effects of armature reaction, the gross torque is related to the armature current by

$$T_g \propto I_a \qquad [14.13]$$

That is, the $T_g$–$I_a$ graph is a straight line passing through the origin (see Fig. 14.21). However, due to the effects of armature reaction, the effective value of the flux $\Phi$ reduces as the armature current increases. Thus, if we regard $\Phi$ as a variable, eqn [14.12] can be reduced to

$$T_g \propto \Phi I_a \qquad [14.14]$$

Since armature reaction causes $\Phi$ to reduce with increasing armature current, it follows that the gross torque also drops a little from a straight line graph as the armature current increases (see Fig. 14.21).

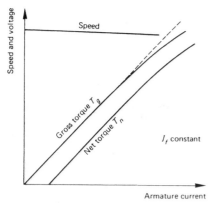

**Fig. 14.21** Characteristics of a shunt motor

When the armature rotates, power is consumed not only by the friction of the bearings but also by the ventilating system of the motor (which acts to keep the motor cool). The former is known as **friction loss** and the latter as **windage loss**. The motor provides the torque necessary to overcome these losses, so that the net torque $T_n$ at the output shaft is

$$T_n = T_g - (\text{friction torque} + \text{windage torque})$$

Consequently, the curve relating net torque to armature current is parallel to, but slightly below, the corresponding gross torque curve (see Fig. 14.21).

### Speed–armature current characteristic

The relationship between the back e.m.f., $E$, of the motor, the flux per pole, $\Phi$, and the armature speed, $\omega$, is

$$E = K_E \Phi \omega$$

where $K_E$ is the e.m.f. constant of the machine. That is

$$E \propto \Phi \omega$$

therefore

$$\omega \propto \frac{E}{\Phi} \qquad [14.15]$$

The supply voltage to the motor is given by

$$V_T = E + I_a R_a$$

that is

$$E = V_T - I_a R_a \qquad [14.16]$$

Substituting eqn [14.16] into eqn [14.15] gives

$$\omega \propto \frac{V_T - I_a R_a}{\Phi} \qquad [14.17]$$

If $\Phi$ does not change with armature current, then eqn [14.17] reduces to

$$\omega \propto V_T - I_a R_a \qquad [14.18]$$

Equation [14.18] is one of a straight line with a negative slope, i.e. the speed of the armature reduces in a linear manner with increase of armature current, as shown in Fig. 14.21. However, the effect of armature reaction is to reduce the flux per pole at higher values of armature current. It should be noted in eqn [14.17] that $\omega \propto 1/\Phi$, so that any reduction in $\Phi$ brings about an increase in armature speed. Consequently, the net effect of armature reaction is a levelling out of the speed–$I_a$ curve. In fact, the speed of a shunt motor varies by less than about 5 per cent from no-load to well over the rated full-load current. This type of motor is therefore classified as a constant-speed machine.

**Worked example 14.6**     A 450 V d.c. shunt motor has an armature circuit resistance of 0.15 Ω; at no-load the armature current is 8 A and the armature speed is 250 rev/min. If the full-load current is 150 A, calculate the full-load speed if (a) the flux per pole remains constant, (b) the flux per pole is reduced by 4 per cent by armature reaction and (c) the flux per pole is increased by 10 per cent above the no-load value by means of a field regulator.

*Solution*     The conditions in the problem are $V_T = 450$ V, $R_a = 0.15$ Ω, $I_{a0} = 8$ A, $N_0 = 250$ rev/min, $I_{aFL} = 150$ A.

Under no-load conditions, the armature-induced voltage $E_0$ is

$$E_0 = V_T - I_{a0}R_a = 450 - (8 \times 0.15) = 448.8 \text{ V}$$

and at full load the induced armature voltage is

$$E_{FL} = V_T - I_{aFL}R_a = 450 - (150 \times 0.15) = 427.5 \text{ V}$$

(a) The flux per pole is $\Phi_1$, and is constant. Now

$$E_0 \propto \Phi_1 N_0 \qquad\qquad [14.19]$$

and

$$E_{FL} \propto \Phi_1 N_{FL} \qquad\qquad [14.20]$$

where $N_{FL}$ is the full-load speed of the armature, hence

$$\frac{E_{FL}}{E_0} = \frac{\Phi_1 N_{FL}}{\Phi_1 N_0} \qquad\qquad [14.21]$$

or

$$N_{FL} = N_0 \times E_{FL}/E_0 = 250 \times 427.5/448.8$$
$$= 238.1 \text{ rev/min}$$

(b) Under no-load conditions the flux per pole is $\Phi_1$, and under loaded conditions the flux per pole is $\Phi_L = 0.96\Phi_1$. Equation [14.21] is modified in this case to

$$\frac{E_{FL}}{E_0} = \frac{\Phi_L N_{FL}}{\Phi_1 N_0} = \frac{0.96\Phi_1 N_{FL}}{\Phi_1 N_0}$$

hence

$$N_{FL} = N_0 E_{FL}/0.96E_0 = 250 \times 427.5/(0.96 \times 448.8)$$
$$= 248 \text{ rev/min}$$

Note that in the absence of armature reaction [solution (a)], the speed drops by 4.76 per cent at full load, but with armature reaction the speed drops [solution (b)] by only 0.8 per cent.

(c) In this case $\Phi = 1.1\Phi_1$. Equation [14.21] becomes

$$\frac{E_{FL}}{E_0} = \frac{\Phi N_{FL}}{\Phi_1 N_0} = \frac{1.1\Phi_1 N_{FL}}{\Phi_1 N_0} = \frac{1.1 N_{FL}}{N_0}$$

that is

$$N_{FL} = N_0 E_{FL}/1.1 E_0 = 250 \times 427.5/(1.1 \times 448.8)$$

$$= 216.5 \text{ rev/min}$$

That is, increasing the current in the field coil has reduced the speed of the machine.

## 14.18  Characteristics of a series motor

### Torque–armature current characteristic

The gross torque, $T_g$, produced by a series motor is

$$T_g = K_T \Phi I_a$$

where $K_T$ is the torque constant of the motor, $\Phi$ the flux per pole and $I_a$ the armature current. In this case, the armature current also flows through the series field winding and, at normal values of armature current, the magnetic flux produced by the field system can be taken to be proportional to the field current. That is

$$T_g \propto \Phi I_a \propto I_a^2 \qquad [14\ 22]$$

Hence, doubling the armature current quadruples the gross torque (see Fig. 14.22).

However, at high values of armature current, the non-linear effects of the $B$–$H$ curve of the iron circuit must be taken into account. The net result, at high values of armature current, is that the machine suffers from the onset of magnetic saturation, and there is a departure from the square law predicted by eqn [14.22], and the torque–armature current graph becomes more linear at high values of $I_a$.

The curve relating the net shaft torque or useful torque, $T_n$, to armature current is generally parallel to the $T_g$–$I_a$ curve, but lies slightly below it (see Fig. 14.22). The difference in torque between the two curves is accounted for by the friction and windage loss of the motor.

As with the shunt machine, armature reaction has the effect of reducing the flux per pole as the armature current increases. This, in turn, reduces the torque at higher values of armature current.

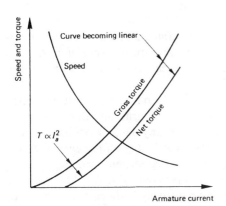

**Fig. 14.22** Series motor characteristics

### Speed–armature current characteristic

The relationship between the 'back' e.m.f. in the armature, the flux per pole and the armature speed $\omega$ is

$$E = K_E \Phi \omega$$

where $K_E$ is the e.m.f. constant of the machine, $\Phi$ the flux per pole and $\omega$ the armature speed. Hence

$$\omega \propto \frac{E}{\Phi} \qquad [14.23]$$

Moreover, for the series machine

$$E = V_T - I_a R_a$$

where $V_T$ is the supply voltage, $I_a$ the armature current and $R_a$ the armature resistance. It follows that

$$\omega \propto \frac{V_T - I_a R_a}{\Phi} \qquad [14.24]$$

At low values of armature current, the value of the product $I_a R_a$ is much less than the value of $V_T$, so that the numerator of eqn [14.24] can, in many cases, simplify to $V_T$. Additionally, since the supply voltage $V_T$ is usually constant then, to a first approximation, we may say that

$$\omega \propto \frac{1}{\Phi} \qquad [14.25]$$

Also, at normal values of $I_a$, the flux is proportional to $I_a$, and eqn [14.25] can be represented in the form

$$\omega \propto \frac{1}{I_a} \qquad [14.26]$$

That is, when the series motor is running under no-load conditions (when $I_a$ is low), the speed of the armature is very high. As load is applied to the motor (and $I_a$ increases), the speed of the motor falls in value (see Fig. 14.22).

Generally speaking, a mechanical load is permanently connected to the series motor to ensure that the armature does not 'race' under conditions of light load. Many large machines have a centrifugal overspeed trip fitted which cuts off the current to the motor in the event of an overspeed condition occurring.

In the case of small series motors (such as hand-held machines), the no-load power loss is comparatively high, ensuring that they consume a relatively high current at no-load.

Series motors are frequently used for traction (milk floats, battery locomotives, cars, etc.) and crane drives which require a starting torque up to five times the full-load torque. This torque is developed by the large value of starting current.

*Worked example 14.7*  A 250 V series motor which has an armature resistance of 0.1 Ω and a field winding resistance of 0.025 Ω, develops its full-load torque at a speed of 450 rev/min when the motor current is 40 A.

Calculate, for half full-load torque (a) the current drawn by the motor and (b) the armature speed. Neglect the effects of armature reaction and magnetic saturation.

*Solution*  The conditions in the problem are $V_T = 250$ V, $R_T$ = total resistance of the armature circuit $= 0.1 + 0.025 = 0.125\,\Omega$, $N_{FL} = 450$ rev/min, $I_{aFL} = 40$ A.

Since the effects of armature reaction and magnetic saturation can be neglected, then $\Phi \propto I_a$. Hence

Torque, $T \propto \Phi I_a \propto I_a^2$

(a) If $I_{aFL}$ and $T_{FL}$ are the respective full-load values of the armature current and torque, and $I_{a2}$ and $T_2$ are the respective values for half full load, then

$$\frac{T_2}{T_{FL}} = \frac{I_{a2}^2}{I_{aFL}^2}$$

or

$$I_{a2} = I_{aFL}\sqrt{(T_2/T_{FL})} = I_{aFL}\sqrt{(\tfrac{1}{2}T_{FL}/T_{FL})} = I_{aFL}\sqrt{\tfrac{1}{2}}$$

$$= 0.707 I_{aFL} = 0.707 \times 40 = 28.3\ \text{A}$$

(b) For a d.c. machine, the induced e.m.f. is

$$E \propto \Phi N$$

If $E_{FL}$, $\Phi_{FL}$ and $N_{FL}$ are the respective values of e.m.f., flux and speed under full-load conditions, and $E_2$, $\Phi_2$ and $N_2$ are the respective half full-load values, then

$$\frac{E_2}{E_{FL}} = \frac{\Phi_2 N_2}{\Phi_{FL} N_{FL}}$$

hence

$$N_2 = \frac{\Phi_{FL} E_2}{\Phi_2 E_{FL}} \times N_{FL}$$

and since $\Phi \propto I_a$, then

$$N_2 = \frac{I_{aFL} E_2}{I_{a2} E_{FL}} \times N_{FL}$$

Now, the value of $E_2$ is given by

$$E_2 = V_T - I_{a2} R_T = 250 - (28.3 \times 0.125) = 246.5\ \text{V}$$

and the value of $E_{FL}$ is

$$E_{FL} = V_T - I_{aFL} R_T = 250 - (40 \times 0.125) = 245\ \text{V}$$

therefore

$$N_2 = \frac{40 \times 246.5}{28.3 \times 245} \times 450 = 640\ \text{rev/min}$$

Since the motor is running under half full-load conditions, the armature current in case (b) is less than in case (a), so that the motor speed has increased in order to generate the required value of back e.m.f.

**14.19  Universal motors**

Small series motors (with ratings up to about 400 W or so) are suitable for operation from either a d.c. or a single-phase a.c. supply; such machines are described as **universal motors**. When supplied from an a.c. source, the field flux and the armature current reverse simultaneously, and a unidirectional torque is developed by the armature. The field systems of these machines are laminated in order to minimize the iron loss when operating from an a.c. supply.

**14.20  Introduction to three-phase a.c. motors**

Perhaps the most popular motor for industrial applications is the **three-phase cage rotor induction motor**. The reason for its popularity is that it is cheap, robust, efficient and reliable. Basically, it is a constant-speed machine and, in its simplest form, only needs a switch to control it.

Up to the time that power semiconductor circuits were developed, the main advantage of the d.c. motor over the induction motor was that the speed and direction of rotation of the d.c. motor were much more easily controlled. Today, we can control the speed and direction of the induction motor relatively easily.

In the next few sections we look at the principle of operation of the induction motor and its characteristics.

**14.21  Production of a rotating magnetic field**

The induction motor depends for its operation on the production of a *rotating magnetic field* by the three-phase stator winding. The general principle is illustrated in Fig. 14.23. The stator is wound with three coils [Fig. 14.23(a)]; in the case considered here the windings are star-connected, and it is assumed for the moment that they are concentrated windings (in practice, each winding is a phase winding which is distributed around part of the stator, and the windings may either be star- or delta-connected).

When current $I_R$ flows in the 'red' winding, it produces an m.m.f. $F_R$, $I_Y$ produces $F_Y$ in the 'yellow' winding, and $I_B$ produces $F_B$ in the 'blue' winding. Moreover, we will assume that a *positive value* of current in any coil results in an m.m.f. which acts radially outwards from the centre of the motor, and a negative value of current gives an m.m.f. which acts radially inwards towards the centre of the motor.

The current waveforms in the windings are shown in Fig. 14.23(b). The individual m.m.f.s at instant W in the current waveforms are depicted in Fig. 14.23(c), numerical values of the m.m.f.s being listed in Table 14.2. Note that since $I_Y$ has a negative value at point W on the current waveform, then $F_Y$ also has a negative value, i.e. $F_Y$ acts radially inwards towards the centre of the rotor. At the same time $I_B$ is positive, so that $F_B$ acts radially outwards from the centre of the motor. The total m.m.f. at this

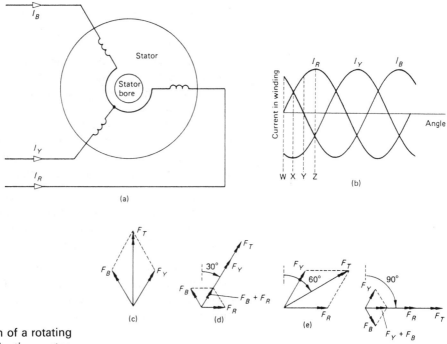

**Fig. 14.23**  Production of a rotating magnetic field in an induction motor

instant of time is given by the *phasor sum* of $F_R$ (which is zero at this time), $F_Y$ and $F_B$ in Fig. 14.23(c), in which the directions of the m.m.f.s are related to the physical position of the three coils.

It should be pointed out that the m.m.f. arrow $F_T$ in Fig. 14.23(c) points vertically upwards, implying that the motor develops one pole-pair (i.e. one N-pole and one S-pole).

Applying a similar argument to the currents at instant X in Fig. 14.23(b), we find that the value of the total m.m.f. is unchanged at $F_T$, but that the pole-pair has rotated through 30° in a clockwise direction.

If we consider the m.m.f.s produced at instant Y in Fig. 14.23(b), we find that the total m.m.f. is unchanged at $F_T$, but has rotated

**Table 14.2**

| Point on Fig. 14.23 | m.m.f. | | | Comment |
|---|---|---|---|---|
| | $F_R$ | $F_Y$ | $F_B$ | |
| W | 0 | −0.866 max | +0.866 max | See diagram (c) |
| X | 0.5 max | −1.0 max | +0.5 max | See diagram (d) |
| Y | 0.866 max | −0.866 max | 0 | See diagram (e) |
| Z | 1.0 max | −0.5 max | −0.5 max | See diagram (f) |

60° in a clockwise direction from the original position. A similar consideration at point Z in Fig. 14.23(b) shows that a further clockwise rotation of 30° occurred.

In this case we are dealing with a two-pole machine (a one pole-pair machine), and the reader will observe that each revolution of the rotating field corresponds to one cycle of the supply frequency. Clearly, the higher the frequency of the supply system, the faster the speed of rotation of the magnetic field.

The above discussion shows that when a balanced three-phase supply is connected to the stator windings of a three-phase induction with the winding arrangement in Fig. 14.23(a) then:

1. a single pair of magnetic poles is produced;
2. the pole-pair rotates around the axis of the rotor; and
3. the speed of rotation of the magnetic field depends on the frequency of the supply system.

Applying a similar argument to that above, it can be shown that **the direction of rotation of the magnetic field can be reversed by interchanging the connections between *any pair* of lines and stator windings**, as shown in Fig. 14.24.

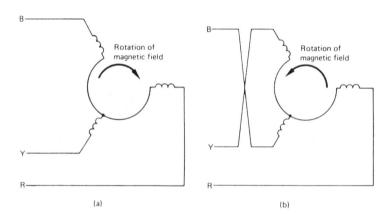

**Fig. 14.24** Reversal of the direction of rotation of the rotating field

(a)                          (b)

**14.22  Production of torque in an induction motor**

The basic principle of the production of mechanical force by a rotor conductor is illustrated in Fig. 14.25. Diagram (a) shows the magnetic field produced by the stator windings; in this case the field moves from right to left across the conductor. The conductor itself is part of a complete circuit (see section 14.23), so that the e.m.f. induced in the conductor causes a current to flow in it.

To determine the direction of the induced e.m.f., we simply apply Fleming's right-hand rule (remember, we are dealing with an *induced e.m.f.* at this stage). Fleming's rule *assumes that the conductor moves relative to the magnetic field* and, since the

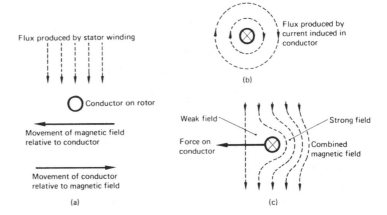

**Fig. 14.25** Production of force by a rotor conductor

magnetic field moves from right to left then, for the purpose of Fleming's rule, *the relative motion of the conductor with respect to the magnetic field is from left to right* [see Fig. 14.25(a)]. We therefore conclude that the direction of the current induced in the conductor is *into the page* [see Fig. 14.25(b)]. Once this current is established, it produces its own magnetic flux, as shown in Fig. 14.25(b).

The combined magnetic fields of the stator winding and the conductor are shown in Fig. 14.25(c). The reader should note that the magnetic field produced by the current in the conductor opposes the stator flux on the left-hand side of the conductor, and assists it on the right-hand side. Consequently the conductor experiences a force, *F*, which is in a direction from the region of strong field to the region of weak field, i.e. **the conductor moves in the direction of rotation of the magnetic field produced by the stator winding** [see also Fig. 14.25(a)].

If the direction of rotation of the stator field is reversed [see Fig. 14.24(b)], then the force experienced by the rotor conductor also reverses. This is the method used to reverse the direction of rotation of an induction motor.

## 14.23 The cage rotor induction motor

The most popular motor used by industry is the **cage rotor induction motor** (also known as the *squirrel cage induction motor*), so named because of the method of assembly of the rotor conductors [see Fig. 14.26(a)].

The rotor conductors are short-circuited at their ends by stout end-rings. In many small- and medium-size motors, the end-rings have small fins on them to act as fan blades which circulate air inside the motor to improve the ventilation. Larger machines have separate fan blades secured to the shaft for the same purpose.

The rotor conductors are embedded in soft iron in the form of iron laminations [see Fig. 14.26(b)]. The function of the iron is

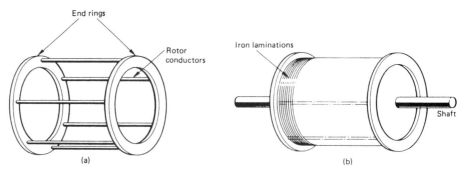

**Fig. 14.26** Rotor construction of a cage rotor induction motor

to reduce the reluctance of the iron circuit, the laminated construction reducing the power loss due to eddy currents which are induced in the motor.

As described earlier, the torque developed by the rotor conductors causes the rotor to rotate in the same direction as that of the rotating field. However, as the rotor accelerates it reduces the relative velocity between it and the rotating field; as this reduces, so the magnitude of the current in the rotor also reduces. It was explained earlier that the torque acting on the rotor is dependent on the magnitude of the rotor current so that, as the rotor accelerates, the torque developed by the rotor reduces. The faster the rotor rotates, the smaller the torque produced by the rotor.

Consider for the moment what happens if the rotor rotates at the same speed as that of the rotating field. In this event, the rotor conductors no longer move relative to the field, and have no e.m.f. induced in them, and the rotor current (and torque) falls to zero! In this case the rotor begins to slow down until it develops sufficient torque to provide the requirements of the load together with the friction and windage loss of the motor.

That is to say, under normal circumstances, **the rotor of the induction motor runs at a speed which is slightly less than that of the rotating field**.

The speed of rotation of the magnetic field is known as the **synchronous speed**, $\omega_s$ rad/s (or $n_s$ rev/s or $N_s$ rev/min). If the speed of rotation of the rotor is $\omega$ rad/s, then the speed difference $(\omega_s - \omega)$ rad/s is known as the *slip* of the rotor. A more important parameter is the **fractional slip**, $s$, which is defined as

$$\text{Fractional slip, } s = \frac{\text{slip}}{\text{synchronous speed}}$$

$$= \frac{\omega_s - \omega}{\omega_s} \qquad [14.27]$$

Since the fractional slip is the ratio of two speeds, it is dimensionless, and is sometimes given as a 'per unit' value.

Alternatively, its value is multiplied by 100, and is given in 'per cent'. The slip can also be calculated using speeds of rev/min or rev/s as follows:

$$\text{Fractional slip, } s = \frac{N_s - N}{N_s} = \frac{n_s - n}{n_s}$$

**Worked example 14.8**    If the speed of the rotating field of an induction motor is 3000 rev/min and the speed of the rotor is 298.5 rad/s, calculate the value of the fractional slip of the motor.

*Solution*    In this case $N_s = 3000$ rev/min and $\omega = 298.5$ rad/s. The speed of the rotor is converted to rev/min as follows:

$$N = \frac{\omega}{2\pi} \times 60 = \frac{298.5}{2\pi} \times 60 = 2850 \text{ rev/min}$$

From eqn [14.25]

$$\text{Fractional slip}, s = (N_s - N)/N_s$$

$$= (3000 - 2850)/3000$$

$$= 0.05 \text{ per unit or 5 per cent}$$

**14.24  Relationship between speed, number of poles and supply frequency**

It was shown in section 14.21 that, in the case of a two-pole machine (i.e. a one pole-pair machine), the rotating field completes one revolution each time the supply frequency completes one cycle. If the motor is redesigned so that it produces $p$ pole-pairs, then the magnetic field rotates through $1/p$ of a revolution for each cycle of the supply frequency. That is to say, the flux rotates through $f/p$ revolutions per second. Since this is equal to the synchronous speed $n_s$ in rev/s, then

$$n_s = \frac{f}{p} \text{ rev/s}$$

or

$$f = n_s p \text{ hertz} \qquad [14.28]$$

and since $\omega_s = 2\pi n_s$, then

$$f = \omega_s p/2\pi \qquad [14.29]$$

**Worked example 14.9**    Determine the synchronous speed of a 10-pole, 50 Hz induction motor (a) in rev/s, (b) in rev/min.

*Solution*    In this case $p = 10/2 = 5$ pole-pairs, and $f = 50$ Hz, then from eqn [14.28]
(a) $n_s = f/p = 50/5 = 10$ rev/s
(b) $N_s = 60 n_s = 60 \times 10 = 600$ rev/min

*Worked example 14.10*   Calculate the rotor speed of a four-pole, three-phase induction motor which is supplied at 50 Hz, the fractional slip of the motor being 4.5 per cent.

*Solution*   Here we have $p = 4/2 = 2$ pole-pairs, $f = 50$ Hz and $s = 4.5$ per cent or 0.045 p.u. From eqn [14.28]

$$n_s = f/p = 50/2 = 25 \text{ rev/s}$$

and

$$s = (n_s - n)/n_s$$

hence

$$sn_s = n_s - n$$

or

$$n = n_s(1 - s) = 25(1 - 0.045)$$

$$= 23.875 \text{ rev/s} \quad \text{or} \quad 1432.5 \text{ rev/min}$$

## 14.25  Torque–speed characteristic of an induction motor

The shape of the torque–speed characteristic of an induction motor depends to a great extent on the resistance of the rotor winding of the motor. The reader should refer to more advanced texts for the theory relating to the shape of the characteristic, but it is not out of place to discuss the general shape of the curve here.

The majority of three-phase cage rotor induction motors have a low rotor resistance, and have a torque–speed curve of type A in Fig. 14.27. The *starting torque*, $T_s$, is typically 90 per cent of the full-load torque of the motor. As the motor speeds up, the torque developed increases until, at speed $\omega_m$, it develops its maximum torque, $T_m$, which is typically 2.5 times the full-load torque of the motor.

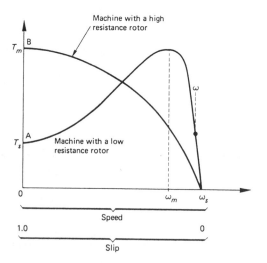

**Fig. 14.27**  Torque–speed characteristic of a cage rotor induction motor

As the speed of the motor increases further, the torque reduces in value until it falls to zero at the synchronous speed of the motor. However, as mentioned in section 14.23, the rotor runs at speed $\omega$, which is slightly less than the synchronous speed of the machine.

Under normal operating conditions, the motor operates on the near vertical part of the torque–speed curve (curve A) in the region of $\omega$. Since this part of the characteristic is a near-constant speed region (the fractional slip in this region is very small), we can regard the induction motor as being a constant-speed motor. Should the load torque increase, the operating point moves further up the curve to give a higher torque and a slightly lower speed, and vice versa for a lower torque demand.

Some applications require the use of an induction motor having a higher value of rotor resistance. A typical torque–speed curve for this type of motor is shown in curve B in Fig. 14.27. In this case the starting torque of the motor is equal to the maximum torque of the motor and, as the motor speeds up, the developed torque reduces continuously. This type of characteristic is very useful when a high starting torque is needed.

**14.26  Introduction to stepper motors**

Unlike a.c. and d.c. motors, stepper motors are designed for **direct digital control** (DDC), the signals for the speed and direction of rotation being produced by a digital computer or control system. Applications of stepper motors include both linear and rotating machines, such as machine tool control, coordinate positioning systems, $X-Y$ plotters, etc.

The digital control system applies a series of pulses to the motor windings, each pulse causing the rotor to move through a unit step displacement. Typical angular movements per step are 1.8°, 2.5°, 3.75°, 7.5°, 15° or 30°. The two general configurations used for stator windings are the **unipolar** and **bipolar** drives, simplified versions being illustrated in Fig. 14.28. Since only one-half of the winding of the unipolar machine is used, it produces less torque at low stepping rates, but shows better high-speed performance.

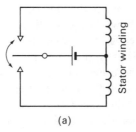

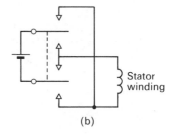

**Fig. 14.28** Stepper motor with (a) a unipolar drive and (b) a bipolar drive

(a)                    (b)

**14.27 Stepper motor construction**

There are several types of stepper motor, including the permanent magnet type, the variable reluctance type and the hybrid type.

In the **permanent magnet type**, the rotor has many permanent magnetic poles on it; the stator has electromagnets on it, and the magnetic pull between the stator and rotor poles is organized to produce the stepping action.

For simplicity, let us assume that both the stator and the rotor can be cut along the axis of the motor and rolled out as shown in Fig. 14.29(a).

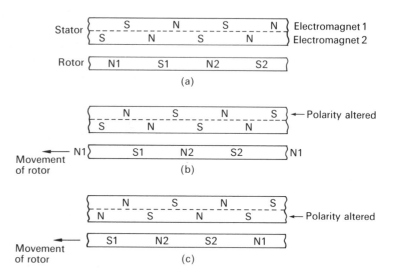

**Fig. 14.29** Principle of operation of a permanent magnet stepper motor

The stator has two sets of coils on it, one set being housed around the circumference of one-half of the stator, and the other set being housed around the other half. The two sets are staggered as shown in Fig. 14.29, and each is independently controlled by a computer. The rotor has a number of permanent magnets on it and, for the sake of simplicity, we will consider a four-pole motor. Initially, the stator poles are energized to give the polarities shown in Fig. 14.29(a).

Each pole on the rotor will align itself to give the greatest 'pull', so that each N-pole on the rotor aligns itself between a pair of stator S-poles, and each rotor S-pole aligns itself between a pair of stator N-poles.

When the polarity of one set of stator poles is reversed, as shown in Fig. 14.29(b), the permanent magnet poles on the rotor realign themselves to give the greatest magnetic 'pull' once more. That is, the rotor moves one step to the 'left'. If the polarity of the other set of stator poles is then reversed, as shown in Fig. 14.29(c), the rotor steps to the 'left' once again.

The *speed of the rotor* is controlled by the rate at which the polarity of the stator poles is reversed, i.e. the rate at which pulses are applied to the stator windings. The *direction of rotation of the rotor* is controlled by the sequence in which the polarity of the stator poles is reversed. For example if, in Fig. 14.29, the 'lower' set reversed afterwards, then the direction of rotation of the rotor would be reversed (the reader should verify that this is the case).

Due to the nature of a permanent magnet machine it is the case that, even when it is unpowered, a torque needs to be applied to it in order to cause the rotor to be displaced from its rest position because of the residual magnetism in the motor. This is known as the **detent torque**. Not all stepper motors have this characteristic.

In a **variable reluctance stepper motor**, the rotor is made of soft iron, and has a number of poles on it which is unequal to the number of stator poles (which are electromagnets). When the stator windings are excited, the rotor poles align themselves with the path of least magnetic reluctance. When the stator winding excitation is changed, the position of least reluctance changes, and the rotor moves through one angular step to align with the next position.

As before, the frequency at which the stator pulse is applied controls the speed of rotation, and the sequence in which the stator pulse is applied controls the direction of rotation.

Since the rotor teeth are of soft iron, they retain very little magnetism, and there is practically no detent torque.

The **hybrid stepper motor** is a hybrid between the permanent magnet type and the variable reluctance motor. The stator has several salient poles, each carrying a coil, and each main pole is divided into many smaller poles by means of longitudinal slots cut in the face of the pole. The rotor consists of an axial cylindrical permanent magnet, with slotted pole pieces on its outer face, the pole pieces producing a permanent magnet effect on the rotor generally similar to the stator arrangement in, say, Fig. 14.29(a). This arrangement enables the hybrid stepper motor to have a much smaller angular step than is possible with other types.

The basis of a microprocessor-based stepper motor drive is shown in Fig. 14.30. Diagram (a) illustrates the microprocessor section, in which the **random access memory** (RAM) contains the program controlling the motor. The start/stop signal is usually applied to the microprocessor via an **interrupt connection** on the **central processing unit** (CPU); as the name implies, whenever a signal appears on this line the program is interrupted, and the microprocessor immediately responds.

The CPU sends and receives data along its **data bus**, and we use five of the data bus lines in this application. Four are used to control the four 'phases' of the stepper motor, and the fifth is used to detect the state of a 'direction or rotation' signal from the system. The reader will note that the least significant data

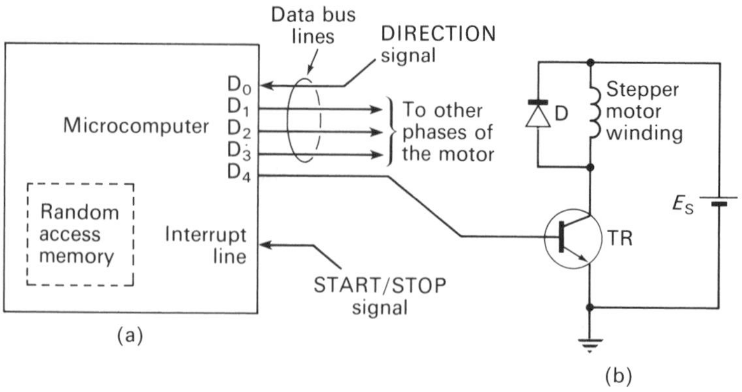

**Fig. 14.30** (a) The basis of a microprocessor-based stepper motor control system and (b) a simplified control circuit for one phase

bus line is numbered $D_0$; this is conventional in computer technology. The signals sent out on the data bus lines are logic '1' and logic '0' signals (usually corresponding to 5 V and zero, respectively), so that a logic '1' on the direction line may correspond to, say, clockwise rotation and logic '0' to anticlockwise rotation.

A simplified control circuit for one phase of the motor is shown in Fig. 14.30(b). Transistor TR acts as an electronic switch, and when a logic '1' signal is applied to the data bus line, the transistor turns 'on' and the stepper motor winding is energized. When a logic '0' is applied to the data bus, the transistor turns 'off', and the motor winding is de-energized.

When we discussed transients in Chapter 13, we showed that when the current in an inductive circuit is rapidly cut off, there was a large self-induced e.m.f. in the inductance. Diode D in Fig. 14.30(b) prevents the induced voltage from damaging the transistor; this diode is known as a *flywheel diode*.

**Problems**

**14.1**  Explain why an e.m.f. is induced in an armature of a d.c. machine irrespective of whether it acts as a generator or as a motor.

**14.2**  Describe the construction of a d.c. machine and describe the type of windings used.

**14.3**  Write a short essay describing (a) armature reaction and (b) commutation in d.c. machines. Suggest how the effects of armature reaction in d.c. machines can be reduced.

**14.4**  A d.c. shunt generator has the following no-load characteristic at a speed of 1200 rev/min:

| E.m.f. (V) | 95 | 150 | 202 | 225 | 268 | 300 | 338 | 375 |
|---|---|---|---|---|---|---|---|---|
| $I_f$(A) | | 1.0 | 1.5 | 2.0 | 2.25 | 2.75 | 3.25 | 4.0 | 5.0 |

Plot the no-load characteristic and determine (a) the no-load terminal voltage when the total resistance in the shunt field circuit is 90 Ω, and (b) the critical resistance of the shunt field circuit.

[315 V; 100 Ω]

**14.5**   A shunt generator has the following open-circuit characteristic at a constant speed of 600 rev/min:

E.m.f. (V)   220   350   415   450   490   515
$I_f$(A)        0.5   1.0   1.5   2.0   3.0   4.0

The resistance of the shunt field winding is 105 Ω. Plot the open-circuit characteristic and determine, for self-excitation without a field regulator, the no-load terminal voltage. Also obtain the resistance of a field regulator to give a terminal voltage of (i) 500 V, (ii) 460 V and (iii) 400 V.

[515 V; (i) 37.9 Ω; (ii) 99.4 Ω; (iii) 180.7 Ω]

**14.6**   The open-circuit characteristic of a shunt-wound d.c. generator running at 800 rev/min is as follows:

E.m.f. (V)   90   170   215   270   287.5   295
$I_f$(A)        1.0   2.0   3.0   5.0   7.0   9.0

Determine the following for a speed of 800 rev/min: (a) the open-circuit voltage if the resistance of the field circuit is 37.5 Ω; (b) the total resistance of the field circuit to give an e.m.f. of 250 V; (c) the critical resistance of the field circuit. If the resistance of the field circuit is 62.5 Ω estimate (d) the armature speed at which the generator just fails to excite.

[(a) 290 V; (b) 62.5 Ω; (c) 90 Ω; (d) 555 rev/min]

**14.7**   A d.c. series motor takes a current of 40 A when developing a full-load torque at 1000 rev/min. Neglecting losses and magnetic saturation, calculate the current drawn by the motor and its armature speed when developing (i) 50 per cent full-load torque and (ii) 25 per cent full-load torque.

[(i) 28.28 A, 1414 rev/min; (ii) 20 A, 2000 rev/min]

**14.8**   The speed–torque characteristic of a d.c. shunt motor is as follows:

N (rev/min)   1280   1250   1180   1130   1080   1000
T (N m)        200    400    800   1000   1200   1400

The motor drives a load having the following characteristic:

N (rev/min)   240   420   740   1200   1800
T (N m)        400   600   800   1000   1200

Estimate the speed of the combination and the torque developed by the motor at this speed. (*Note*: plot both graphs on the same piece of graph paper and see where they intersect.)

[1130 rev/min; 980 N m]

**14.9**   Show how a rotating magnetic field is produced by a three-phase stator winding in an induction motor. Hence explain how a torque is developed by the rotor.

**14.10**   Describe the construction of a three-phase cage-type induction motor.

**14.11** Calculate the synchronous speed of a four-pole, 60 Hz, three-phase induction motor in (a) rev/s, (b) rev/min, (c) rad/s.

[(a) 30 rev/s; (b) 1800 rev/min; (c) 188.5 rad/s]

**14.12** The synchronous speed of an induction motor is 3000 rev/min. If the supply frequency is 50 Hz, determine the number of stator poles.

[2]

**14.13** A special-purpose alternator generates a frequency of 400 Hz. If the rotor speed is 4800 rev/min, calculate the number of poles on the stator.

[10]

**14.14** The fractional slip of a six-pole, three-phase, 50 Hz induction motor is 5 per cent. Determine the speed of the rotor in rev/min.

[950 rev/min]

**14.15** A 12-pole, 50 Hz induction motor runs at 470 rev/min. Calculate (a) the synchronous speed of the rotating field and (b) the fractional slip of the motor.

[(a) 500 rev/min; (b) 0.06]

**14.16** A 16-pole synchronous motor runs at a speed of 450 rev/min. Determine (a) the supply frequency and (b) the speed at which it would run if operated at a frequency of 55 Hz.

[(a) 60 Hz; (b) 412.5 rev/min]

**14.17** A three-phase, four-pole, 50 Hz, induction motor operates at a fractional slip of 4 per cent. The input power to the motor is 50 kW. If the total losses in the machine amount to 3 kW, calculate the efficiency of the motor and the rotor speed.

[94 per cent; 1440 rev/min]

**14.18** Explain the operation of one form of stepper motor, and show how its speed and direction of rotation can be controlled.

# 15 Information transmission

**15.1  Introduction**

Every field of management and engineering is concerned not only about the acquisition of data, but also its transmission and decoding. In this chapter we look at many of the aspects of data transmission, including d.c., a.c., analogue, digital, asynchronous and synchronous transmission.

In particular we are concerned with systems involved in the transfer of information, including telephony, radio and television; all these can transmit analogue or digital information. The system may be **unidirectional**, that is it can transmit in one direction only (such as domestic radio and television), or it may be **bidirectional** (such as the public telephone system).

A simple bidirectional telecommunication system is illustrated in Fig. 15.1. If we consider the public telephone system, the 'source and/or receiver' is a person acting either as the source of information or as the receiver of information. The information from the 'source' is a series of sound pressure waves, and these are converted into the signal to be transmitted.

**Fig. 15.1**  The basis of a simple telecommunications system

The conversion is brought about by a **transducer**, *which converts energy of one kind into energy of another kind*. In the telephone system, the transducer at the transmitting end is a microphone which causes the variation in air pressure to alter the electrical resistance of the microphone. In turn, this causes the current through it to vary.

The signal is then transmitted through the system; it may be transmitted simply as a variation of direct current in the line, or it may be converted into alternating current, or may even be a digital signal. The telecommunication system itself could simply be a pair of wires, or it may be a radio link, or even a satellite communication system.

At the receiving end, the signal is converted from a series of electrical waves or pulses into sound pressure waves by another transducer, namely the telephone earpiece.

In the case of a telephone system we should, to be strictly accurate, show two transducers at either end of the system, one being a 'sending' or 'transmitting' transducer, and the other a 'receiving' transducer.

In a unidirectional system, such as the domestic radio system, the transducer at the sending end is a microphone, and at the receiving end is an earphone or a loudspeaker.

In the early days of telephone communications, it was necessary to have a separate cable for each communications link. Where there are more callers than telephone lines, some subscribers have to wait until a line is free. This has largely been overcome to a great extent by a method known as **multiplexing**, the basic principle being illustrated in Fig. 15.2 (see also section 15.12 for more details). The multiplexer in this case is represented by a rotating switch which connects one subscriber to another for a short period of time, after which the rotating blades of MUX 1 and MUX 2 connect another pair of subscribers, and so on.

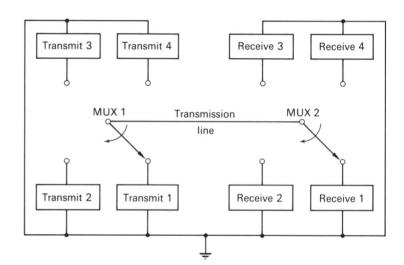

**Fig. 15.2** The principle of multiplexing

Modern multiplexers are, of course, semiconductor circuits. As shown in Fig. 15.2, it is possible for a multiplexed line to handle many signals in an apparently simultaneous manner without subscribers being aware of the fact. The signals may either be **analogue signals** in the form of bursts of different frequencies, or they may be **digital signals** in the form of groups of binary 1s and 0s.

One of the limitations of telephone systems is the fact that there is a power loss in the systems, and the signal is **attenuated** or reduced in magnitude, and this limits the distance over which signals can be transmitted. A method of overcoming this limitation is to insert amplifiers or **repeaters** at intervals in the line.

Wires are gradually being replaced by **optical fibres** made of glass, which are typically 0.1 mm in diameter. The signals which are transmitted along the line are generated by lasers and, once the optical information is in the fibre, it travels by internal reflection. In fact, optical fibres are so good that a 2 km length of fibre absorbs less light than a normal sheet of glass!

Even so, signal attenuation occurs in optical fibres, and repeaters are needed every 30 km or so; even this is a greater distance between repeaters needed by 'normal' telephone systems.

A very large proportion of trunk calls in the UK are transmitted by **microwaves**, which are very high frequency radio waves, and can be focused into a very narrow beam by means of a dish aerial. The received signal is amplified or boosted before being retransmitted to the next receiver.

The majority of international telephone calls are handled by **satellite stations**. Typically, a telephone call is passed through a local network and then to a radio telescope station, where the signal is transmitted by microwaves up to the satellite. It is then retransmitted to another Earth station to link with another subscriber. Many satellites are in a circular orbit around the Earth where, to an Earth observer, they appear to be stationary. The part of the Earth they can cover with radio waves is known as their **footprint** and, within this area, any Earth station can communicate with any other at any time.

## 15.2 D.c. and a.c. transmission signals

A d.c. signal is a unidirectional signal, and may either be a direct current or a direct voltage; that is, it has a fixed polarity [see Fig. 15.3(a)]. This type of signal has one polarity, and is also known as a **unipolar signal**.

An a.c. signal (which may either be a voltage or a current) is one which alternates about zero value, and has *zero average value* [see Fig. 15.3(b)].

Many signals have both d.c. and a.c. components in their make-up. For example, the triangular wave in Fig. 15.3(c), although nominally a d.c. signal, can be thought of as being the sum of the d.c. signal in Fig. 15.3(a) (the *average value* of the wave) and the a.c. signal in Fig. 15.3(b).

Similarly, the binary 'word' in Fig. 15.3(d) has both d.c. and a.c. components in its make-up; in fact it may be described as a **bipolar signal**, since it has both positive and negative parts in its make-up. The binary signal is a typical **ASCII code** (American Standard Code for Information Interchange), and is made up of *seven bits* (binary digits) which are either logic '0' or logic '1'. In communications systems, logic '0' is represented by a positive voltage and logic '1' by a negative voltage, each bit having an equal time slot.

The waveform in Fig. 15.3(d) can be seen to have a d.c. or mean value because the binary word contains four 1s and three 0s,

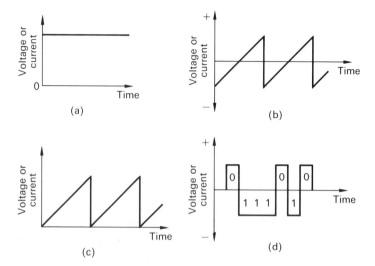

**Fig. 15.3** (a) A d.c. signal, (b) an a.c. signal, (c) a signal having both d.c. and a.c. components and (d) a binary-coded signal having both d.c. and a.c. components

resulting in a net negative voltage. The a.c. component makes up the 'shape' of the wave [as is also the case in Fig. 15.3(c)].

Engineers also say that the signal in Fig. 15.3(d) is made up of signal elements known as marks and spaces. A **mark** is either a negative voltage or the presence of a tone, and a **space** is either a positive voltage or the absence of a tone.

We can think of an electrical transmission line as a series of cascaded $R-C$ networks as shown in Fig. 15.4. Each short section of line, say a 1 m length, has a certain resistance $R$ and a certain capacitance $C$, the transmission line comprising many thousands of these sections in series with one another. As was shown in Chapter 13, when a pulse train is transmitted along the line (see Fig. 15.5), each $R-C$ network has the effect of 'blunting' the waveshape, so that the input wave [which is a 'square' wave in Fig. 15.5(a)] has a shape approaching a sine wave a little way along the line [Fig. 15.5(b)]; at the same time, the peak-to-peak amplitude of the wave is reduced. Further along the line the wave becomes more distorted and attenuated [Fig. 15.5(c)]. The greater the length of the line, the greater the distortion and attenuation.

**Fig. 15.4** Simplified equivalent circuit of a transmission line

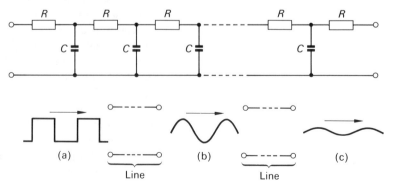

**Fig. 15.5** Transmission of a train of pulses along a transmission line

To ensure that signals can be transmitted over the public telephone system, it is necessary to introduce transformers and amplifiers. For this reason, d.c. signals cannot be transmitted. Moreover, with d.c. signals, the presence of electrical noise and interference masks the received signal, and it is difficult to restore the signal to its original condition.

A.c. signal transmission has many advantages over d.c. transmission, including:

1. the magnitude of the signal can be increased or boosted either by amplification or transformer action; and
2. a direct 'conductor' path is not necessary for the transmission of a.c. signals.

It is usually the case that most a.c. signals are complex, that is they contain not only a **fundamental frequency** but also many **harmonic frequencies**. The fundamental frequency carries the 'base' information, and the higher-frequency signals add quality or tone to the received waveshape.

The **bandwidth** of a transmission system is the range of frequencies it can handle without significant attenuation. The bandwidth required for good-quality speech is from about 30 Hz to about 10 kHz, and for music it is from about 20 Hz to 20 kHz. Generally speaking, the pitch of a signal is established by the fundamental frequency, and the volume by the amplitude. The public telephone system is designed to handle a bandwidth from 300 Hz to 3.4 kHz so that voice transmission, although adequate, is not perfect (the lower and upper frequencies are severely attenuated).

**15.3 Relationship between frequency and wavelength**

It takes a finite time for a signal to travel along a transmission line, as it does for a radio wave to travel through the atmosphere. Consequently, when the 'start' of a wave has travelled a little way along a line, the sending end voltage (or current) is part way through the cycle.

A little later, the sending end will have completed its cycle when the 'start' of the wave is some distance down the line. The distance between the 'start' of the wave and the sending end when the wave is complete is the wavelength, $\lambda$, of the wave.

Clearly, the higher the frequency of the wave, the shorter will be the wavelength. The mathematical relationship between the wavelength, the frequency, the periodic time, and the velocity of propagation of the wave is

$$v = f\lambda = \lambda/T$$

where $v$ is the velocity of the wave, $f$ the frequency and $T$ the periodic time.

A radio wave travels through free space (or a vacuum) with the velocity of light, namely $3 \times 10^8$ m/s. That is, a 40 MHz radio signal has a wavelength of

$$\lambda = v/f = 3 \times 10^8/40 \times 10^6 = 7.5 \text{ m}$$

In the case of a transmission line, the electrical signal travels with a velocity which is less than the speed of light. For example, if the velocity is 0.6 that of light, then

$$v = 0.6 \times 3 \times 10^8 = 1.8 \times 10^8 \text{ m/s}$$

and the wavelength of a 40 MHz wave in the cable is

$$\lambda = v/f = 1.8 \times 10^8/40 \times 10^6 = 4.5 \text{ m}$$

## 15.4 Analogue and digital signals

An **analogue signal** is one which varies smoothly with time; examples include a sinusoidal wave and a speech wave. A **digital signal** is one which has either of two voltage (or current) levels; a typical digital waveform is shown in Fig. 15.3(d). The signal changes between two 'logic' levels, namely logic '1' and logic '0'; the two logic levels have well-defined voltage levels which, depending on the system, may either be positive or negative.

Many devices are analogue in nature, and a variation in input signal, e.g. a sound pressure wave, produces a variation in output voltage or current from it. If the characteristic of the device was perfectly linear, then a variation in input signal produces an output waveshape which is identical to the input wave. However, practical devices do not have a perfectly linear characteristic, and so distortion is always present.

Additionally, the signal will be attenuated as it is transmitted, and it will pick up some electrical noise in the form of spurious induced voltages. The **signal-to-noise ratio** of the system is defined by the equation

$$\text{Signal-to-noise ratio} = \frac{\text{wanted signal power}}{\text{unwanted noise power}}$$

It should be noted that quality of the signal cannot be improved simply by amplifying it, because the noise component of the signal is also amplified! In the case of an analogue signal, there is a limit over which the signal can be transmitted before it is impossible to restore the original signal. Digital signals do not suffer to such an extent, because a 'noisy' signal can be 'cleaned up' to a great extent by electronic logic circuits.

## 15.5 Modulation and demodulation

**Modulation** is a process by which a characteristic of one signal (called the **carrier signal**) is altered in sympathy with a characteristic of a second signal (called the **modulating signal**), the composite signal being known as the **modulated signal**. The reverse process is known as **demodulation**, in which the modulating signal is reconstructed from the modulated signal.

One example of the use of modulation is in the transmission of speech. As mentioned earlier, speech frequencies cover a band up to about 10 kHz, which is much too low for global transmission. Consequently, we must modulate a radio frequency signal (the carrier) with the speech frequency (the modulating signal) in order to broadcast it; the speech signal is finally demodulated in the radio receiver.

To consider what is involved in the process of modulation, let us look at a general expression for a sinusoidal carrier signal. We will assume that the carrier wave is a sinusoidal voltage, and the instantaneous voltage of the wave is

$$v = V_c \sin(\omega_c t + \theta)$$

where $V_c$ is the maximum value of the carrier signal, $\omega_c$ its angular frequency and $\theta$ a phase angle (which may be zero). To superimpose information on the carrier wave we can alter either:

1. the amplitude of the carrier (amplitude modulation or a.m.);
2. the frequency of the carrier (frequency modulation or f.m.);
3. the phase angle of the carrier (phase modulation).

In **amplitude modulation**, the amplitude of the modulating signal alters the amplitude of the carriers by an amount $\delta V_c$, so that the instantaneous voltage of the carrier becomes

$$v = (V_c + \delta V_c)\sin \omega_c t \qquad [15.1]$$

where $\delta V_c$ is the change in the amplitude of the carrier caused by the modulating signal. The reader will note that the frequency of the carrier is unchanged, and we have assumed that $\theta = 0$.

When **frequency modulation** is used, only the frequency of the carrier is changed by a change in amplitude of the modulating signal, and the equation for the modulated signal is

$$v = V_c \sin(\omega_c + \delta\omega_c)t \qquad [15.2]$$

where $\delta\omega_c$ is the change in the frequency of the carrier caused by the change in amplitude of the modulating signal, the amplitude of the carrier signal remaining unchanged.

Finally, in **phase modulation** a change in the amplitude of the modulating signal produces a change in the phase angle of the carrier, as shown by the following equation:

$$v = V_c \sin(\omega_c t + \delta\theta) \qquad [15.3]$$

where $\delta\theta$ is the change in phase angle of the carrier produced by a change in amplitude of the modulating signal; neither the magnitude nor the frequency of the carrier is changed in phase modulation. The demodulator must detect this phase angle change relative to the phase angle of the unmodulated carrier.

In the following sections only the simplest analysis will be performed on eqns [15.1]–[15.3].

### 15.6 Amplitude modulation

In amplitude modulation, the carrier signal [Fig. 15.6(a)] has a fixed frequency, its amplitude being varied in the modulating process by the modulating signal [Fig. 15.6(b)] to give the amplitude-modulated wave in Fig. 15.6(c).

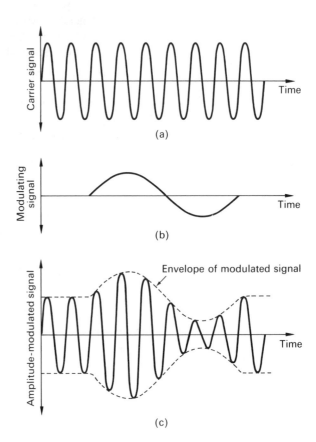

**Fig. 15.6** Amplitude modulation: (a) the carrier signal, (b) the modulating signal and (c) the modulated signal

In the demodulation process in the receiver, the modulated wave is rectified (to eliminate the negative half-cycles) and smoothed, and the resulting *envelope* of the rectified wave gives the same information as the original modulating signal.

If the carrier signal is expressed mathematically as $V_c \sin \omega_c t$, where $V_c$ is the maximum value of the wave and $\omega_c$ its angular frequency, the modulating signal or audio signal is given by $V_A \sin \omega_A t$, where $V_A$ and $\omega_c$ are the corresponding maximum value and angular frequency, then the amplitude of the modulated signal at any instant is

$$v = (V_c + V_A \sin \omega_A t)\sin \omega_c t$$

When compared with eqn [15.1] we see that

$$\delta V_c = V_A \sin \omega_A t$$

We can modify the equation as follows:

$$v = V_c\left[1 + \frac{V_A}{V_c}\sin \omega_A t\right]\sin \omega_c t$$

$$= V_c(1 + m_A \sin \omega_A t)\sin \omega_c t \qquad [15.4]$$

where $m_A$ is the **modulation factor** or **depth of modulation**.

The depth of modulation should never be greater than unity, that is $V_A$ must never be greater than $V_c$, otherwise *overmodulation and distortion* of the signal occur.

If we use the trigonometrical identity for the product of two sine waves, the above expression breaks down into

$$v = V_c \sin \omega_c t + \frac{m_A}{2}V_c \cos(\omega_c - \omega_A)t - \frac{m_A}{2}V_c \cos(\omega_c + \omega_A)t$$

The above equation tells us that, when a sinusoidal carrier signal is amplitude modulated by a *single sine wave*, we get

1. a **carrier signal** $V_c \sin \omega_c t$ of amplitude $V_c$ and frequency $\omega_c$;
2. a **lower side frequency component** of amplitude $m_A V_C/2$ and frequency $(\omega_c - \omega_A)$; and
3. an **upper side frequency component** of amplitude $m_A V_C/2$ and frequency $(\omega_c + \omega_A)$.

The **frequency spectrum** of the wave, which shows the frequencies produced by the modulation process, is illustrated in Fig. 15.7. The **minimum bandwidth** required to transmit the amplitude-modulated sine wave is the difference between the upper and lower side frequencies. That is

$$\text{Bandwidth} = (\omega_c + \omega_A) - (\omega_c - \omega_A) = 2\omega_A \text{ rad/s}$$

or

$$\text{Bandwidth} = (f_c + f_A) - (f_c - f_A) = 2f_A \text{ hertz}$$

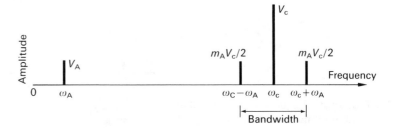

**Fig. 15.7** Frequency spectrum of a carrier which has been amplitude modulated by a single frequency

*Worked example 15.1*  A 5 V, 1 MHz carrier wave is amplitude modulated by a 2 V, 0.1 MHz sine wave. Determine (a) the depth of modulation, (b) the upper and lower side frequencies, (c) the bandwidth required to transmit the modulated signal, (d) the maximum and minimum voltage of the modulated signal and (e) the magnitude of the upper and lower side frequencies.

*Solution*  (a) The depth of modulation is

$$m_A = V_A/V_c = 2/5 = 0.4$$

(b) The upper side frequency is

$$f_c + f_A = 1 + 0.1 = 1.1 \text{ MHz}$$

and the lower side frequency is

$$f_c - f_A = 1 - 0.1 = 0.9 \text{ MHz}$$

(c) The bandwidth required is

$$(f_c + f_A) - (f_c - f_A) = 1.1 - 0.9 = 0.2 \text{ MHz}$$

(d) The maximum value of the modulated wave is calculated as follows:

$$V_c + V_A = 5 + 2 = 7 \text{ V}$$

and the minimum value of the modulated wave is

$$V_c - V_A = 5 - 2 = 3 \text{ V}$$

(e) The magnitude of each of the side frequencies is given by

$$m_A V_c/2 = 0.4 \times 5/2 = 1 \text{ V}$$

### 15.6.1  Sidebands

A speech signal is a complex signal and contains a band of frequencies (as does any other data signal), and when the carrier wave is modulated by a speech signal, each frequency in the speech wave produces a lower and an upper side frequency. That is, the speech signal produces a **lower sideband** and an **upper sideband**. The usual way of representing this situation is shown in Fig. 15.8, and the *minimum bandwidth* required to transmit the signal is

$$(f_c + f_H) - (f_c - f_H) = 2f_H \text{ hertz}$$

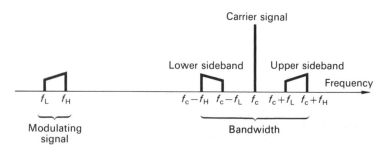

**Fig. 15.8**  Frequency spectrum of an amplitude-modulated complex signal

Clearly, if a high-quality audio signal is to be transmitted, the bandwidth of the modulating signal is large.

*Worked example 15.2* A 150 kHz carrier wave is amplitude modulated by a signal containing frequencies in the range 100 Hz to 4 kHz. Determine the maximum and minimum frequencies in the modulated wave, and the minimum bandwidth required.

*Solution* The maximum frequency in the modulated wave is

$$f_c + f_H = 150 + 4 = 154 \, \text{kHz}$$

and the minimum frequency is

$$f_c - f_H = 150 - 4 = 146 \, \text{kHz}$$

Hence the minimum bandwidth is given by

$$154 - 146 = 8 \, \text{kHz}$$

## 15.7 Frequency modulation

We will rewrite eqn [15.2], which expresses the basis of frequency modulation, in terms of the frequency in Hz as follows:

$$v = V_c \sin 2\pi (f_c + \delta f_c)t \qquad [15.5]$$

In this case the modulating signal causes the carrier frequency to change by an amount $\delta f_c$ hertz, and is illustrated in the waveforms in Fig. 15.9. When the modulating signal amplitude is zero, the modulated signal has the same frequency as the carrier wave [see Fig. 15.9(c)]. When the modulating signal voltage is positive, the frequency of the modulated signal is greater than the unmodulated carrier signal, and when the modulating signal voltage is negative, the frequency of the modulated signal is less than that of the unmodulated carrier.

The maximum frequency of the modulated signal clearly occurs when the modulating signal is at its maximum positive value; the minimum frequency of the modulated signal occurs when the modulating signal is at its maximum negative value.

An important advantage of f.m. over a.m. transmissions is a reduction of the effects of electrical noise. Noise is generated in many electrical systems (e.g. ignition systems, electrical motors, fluorescent lights), and is 'added' to the modulated signal by induction. Since an f.m. receiver expects to receive a constant-amplitude signal, it contains amplifiers which can 'clip' the received signal to this amplitude. In this way the amount of noise produced by signal amplitude variation is significantly reduced.

There are a number of terms in common usage with f.m., and these are defined below.

The **frequency swing** is the change in frequency produced by a given modulation signal amplitude, and corresponds to $\delta f_c$ in eqn

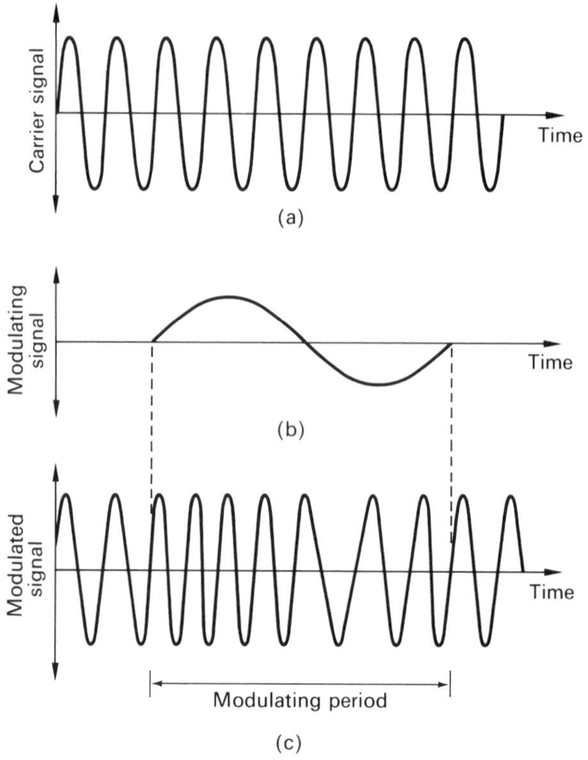

**Fig. 15.9** Frequency modulation: (a) the carrier signal, (b) the modulating signal and (c) the modulated signal

[15.5]. The **frequency deviation** is the maximum value of the frequency swing for which the system has been designed, i.e. it corresponds to $\delta f_{c(max)}$. The **modulation index**, $m_f$, is given by the ratio $\delta f / f_A$, where $f_A$ is the frequency of the modulating (audio) signal.

From a mathematical viewpoint, it can be shown that an f.m. signal produces an infinite number of side frequencies, i.e. the required bandwidth is infinite! Fortunately, for all practical purposes, the minimum bandwidth is

$$2 \times (\text{frequency deviation} + \text{modulating frequency})$$

$$= 2 \times (\delta f_{c(max)} + f_A) = 2f_A(m_f + 1)$$

*Worked example 15.3*   A 1 MHz carrier signal is frequency modulated by a 15 kHz sinusoidal signal, which produces a frequency swing of 5 kHz. Calculate (a) the modulation index, (b) the maximum and minimum frequency of the modulated wave. (c) Estimate the minimum bandwidth of the modulated wave.

*Solution*   (a) The modulation index is given by

$$m_f = \text{frequency swing/modulating frequency}$$

$$= 5/15 = 0.333$$

(b) The maximum modulated frequency is

Carrier frequency + frequency swing

$$= 1 \times 10^6 + 5 \times 10^3 \text{ Hz} = 1.005 \text{ MHz}$$

and the minimum frequency is

Carrier frequency − frequency swing

$$= 1 \times 10^6 - 5 \times 10^3 \text{ Hz} = 0.995 \text{ MHz}$$

(c) The estimated value of the bandwidth is given by

$$2f_A(m_f + 1) = 2 \times 15\,000(0.333 + 1) \approx 40 \text{ kHz}$$

## 15.8  Frequency-shift keying (f.s.k.)

Another, and perhaps more simplified, form of f.m. is **frequency-shift keying**, which is widely used to transmit logic (binary) signals over the telephone system. As explained earlier, d.c. signals cannot be transmitted over long distances on the telephone system; one method of overcoming this disadvantage is to connect the logic system to the telephone line via a **modem** (*mo*dulator/*dem*odulator). This enables a binary '0' to increase the frequency of the carrier wave to a higher frequency (typically 2.1 kHz), and a logic '1' causes the carrier frequency to be reduced (typically 1.3 kHz).

In this way a **binary word** comprising a series of 0s and 1s is converted by the modem into an f.s.k. string of high and low frequencies. A modem at the receiving end converts the frequencies back into appropriate logic levels.

## 15.9  Phase modulation

As shown in section 15.5, the phase modulation signal has the effect of modifying the phase angle of the carrier signal without altering either the amplitude or frequency of the carrier. It was shown that a mathematical expression representing **phase modulation** is

$$v = V_c \sin(\omega_c t + \delta\theta) \qquad\qquad [15.3 \text{ repeated}]$$

where the value of the phase angle $\delta\theta$ depends on the magnitude of the modulating signal. For a single sine wave modulating signal of frequency $\omega_A$, the phase angle is

$$\delta\theta = m_p \sin \omega_A t$$

where $m_p$ is the phase **modulation index**, and the maximum value of the phase angle between the carrier and the modulated wave is the **phase deviation**.

Although, mathematically, the bandwidth of a phase-modulated wave is very large indeed it can, for all practical purposes, be thought to consist of a carrier plus an upper and a lower sideband.

**15.10 Pulse modulation**

While many signals in life are analogue signals, e.g. speech, modern data transmission signals tend to be digital. We therefore need to look at systems which can convert an analogue signal into a digital signal; we look here at two of them.

If the amplitude of the analogue signal in Fig. 15.10(a) is **sampled** by means of an electronic circuit, then the measured value at the instant of sampling can be converted into a digital equivalent of the analogue voltage. If the sampled voltage is converted into a series of pulses of constant width and variable amplitude [see Fig. 15.10(b)], it forms the simple basis of **pulse amplitude modulation** (p.a.m.).

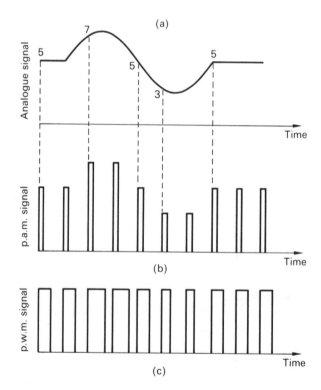

**Fig. 15.10** (a) Analogue signal, (b) corresponding p.a.m. signal and (c) the corresponding p.w.m. signal

The accuracy with which an analogue signal is converted into a series of pulses depends, of course, on the number of times that the analogue signal is sampled. Taking the sine wave in Fig. 15.10(a) together with the number of times it is sampled, the received signal in Fig. 15.10(b) could be confused into thinking that it was receiving a square wave!

In fact, the *minimum sampling frequency* that can be used is known as the **Nyquist frequency**, which is twice the highest frequency in the complex signal being sampled. Thus, if the highest frequency in a modulating wave is known to be 10 kHz, then the minimum sampling frequency that may be used is 20 kHz; that is the waveform must be sampled at least every 50 μs. If the sampling frequency is less than the Nyquist frequency, then **aliasing** occurs. This means that the receiving system thinks that a much lower frequency is being transmitted and, accordingly, reproduces a lower-frequency signal.

One of the limits of p.a.m. transmission is that the system 'sees' the p.a.m. signal as a series of d.c. pulses, which cannot conveniently be transmitted over the telephone system. However, this limitation is overcome by the use of *pulse-code modulation* which is described in section 15.11.

Another form of pulse modulation is **pulse-width modulation** (or p.w.m.) in which the transmitted pulses have a constant height, the *width of the pulses* being related to the amplitude of the signal being sampled [see Fig. 15.10(c)].

A range of pulse modulation techniques have been used including **pulse-position modulation** (p.p.m.), in which constant-width pulses are produced, the 'position' of each pulse relative to the unmodulated position being controlled by the modulating signal.

## 15.11 Pulse-code modulation (p.c.m.)

**Table 15.1** Simple 4-bit pure binary code

| Signal voltage | Binary code | | | |
|---|---|---|---|---|
| | $2^3$ | $2^2$ | $2^1$ | $2^0$ |
| 0 | 0 | 0 | 0 | 0 |
| 1 | 0 | 0 | 0 | 1 |
| 2 | 0 | 0 | 1 | 0 |
| 3 | 0 | 0 | 1 | 1 |
| 4 | 0 | 1 | 0 | 0 |
| 5 | 0 | 1 | 0 | 1 |
| 6 | 0 | 1 | 1 | 0 |
| 7 | 0 | 1 | 1 | 1 |
| 8 | 1 | 0 | 0 | 0 |
| 9 | 1 | 0 | 0 | 1 |

In a **pulse-code modulation** system, the signal being transmitted is initially sampled in a p.a.m. system (see section 15.10) to convert it to a series of constant-width, variable-height pulses. Each of these pulses is passed to an **encoder**, which converts the amplitude of the pulse into a binary 'word'; that is, the signal is quantized.

Using a simple 4-bit **pure binary code**, a voltage waveform could be quantized in steps of 1 V as shown in Table 15.1. Each bit in the code has a **weight** or decimal value, the **least significant bit** (l.s.b.) or $2^0$ bit having a 'weight' of unity, and the **most significant bit** (m.s.b.) or $2^3$ bit having a decimal weight of 8. This allows every decimal value in the range 0–9 to be represented by a unique binary code. Thus we may say

$$\text{Decimal } 5 = \text{decimal } (4 + 1)$$
$$= (0 \times 2^3) + (1 \times 2^2) + (0 \times 2^1) + (1 \times 2^0)$$
$$= \text{binary } 0101$$

also

$$\text{Decimal } 6 = \text{decimal } (4 + 2) = \text{binary } 0110$$

$$\text{Decimal } 7 = \text{decimal } (4 + 2 + 1) = \text{binary } 0111$$

Clearly the p.a.m. signal in Fig. 15.10(b) can be transmitted in the form of a series of binary 0011, 0101 and 0111 signals, as shown in Fig. 15.11, in which each block contains the binary version of the magnitude sample size.

In addition to sending the data, the system must also transmit a **synchronizing signal**; this signal is sent at point S in Fig. 15.11.

**Fig. 15.11** Simplified p.c.m. transmission

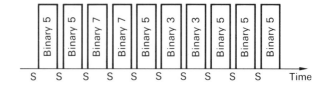

The synchronizing signal tells the receiving end that a p.c.m. data word is complete. Special codes[†] are used to allow the receiving end to detect (and, in some cases, to correct) errors in the received data.

## 15.12 Multiplexing

As mentioned earlier in the chapter, **multiplexing** is a method by which several signals can be transmitted along a single wire without loss of the identity of any of the signals. The two most popular methods are time-division multiplexing and frequency-division multiplexing.

The basis of **time-division multiplexing** (t.d.m.) was illustrated in Fig. 15.2 which, in its simplest form, consists of a pair of synchronously rotating contacts. In addition to the communications channels (channels 1–4 in Fig. 15.2), synchronizing pulses need to be incorporated to maintain synchronism between the sending and receiving ends. In the case of the telephone system in the UK, each group of eight bits occupies a time slot of 3.9 $\mu$s.

In **frequency-division multiplexing** (f.d.m.), each user is assigned a different frequency band. The transmitted signal contains several carrier waves, each of different frequency, and each being separately modulated by a user. The receiver contains tuned circuits to separate the different carrier signals, thereby maintaining integrity of the received signal.

The transmission method may use any suitable medium including wire, waveguides, fibre optics, radio, television, etc.

## 15.13 Digital communications

Data communication by digital means is a feature of everyday life, and ranges from communication between computers and peripherals to telephone conversations. We will deal here with the aspects which are of general interest to electrical and electronic engineers.

Data is transmitted in binary format, and the most usual format (though by no means the only one) is the ASCII code. In this code, each character (alphabetical or numerical) is transmitted using *eight bits of data* (known as a **byte**); seven of the bits are

---

[†] Information on error detecting and correcting codes is available in *Data Communication* by D. C. Green (Longman).

used to define the character, and the eighth bit is a **parity bit** which can be used for error-detecting purposes. The method by which data is transmitted depends on the conflicting requirements of speed and cost and, broadly speaking, use either parallel transmission or serial transmission (see sections 15.14 and 15.15, respectively).

Digital systems are organized in terms of the **word length** of the computer system, a computer word being an ordered set of characters which occupies one storage location in the memory of the computer. Depending on the computer, the word length could be 8, 16, 32, 64 bits, etc.

Since the telephone system is geared to transmitting a.c. signals rather than d.c. signals, modulation techniques figure largely in the transmission of data. Moreover, since amplitude modulation is more sensitive to noise than some other modulation techniques, it is not widely used for data transmission.

The *frequency-shift keying* method described in section 15.8 is widely used to give a *frequency-modulated version of binary data*, in which logic '1' and logic '0' produce quite distinct and different frequencies. *Phase modulation* (see section 15.9) is also widely used in data transmission, in which a change from logic '1' to logic '0' (or vice versa) produces a phase change of 180°. As with f.m., phase modulation is relatively insensitive to noise.

## 15.14 Parallel data transmission

An 8-bit parallel data link between a computer and a *peripheral* is illustrated in Fig. 15.12. There are as many lines in the **data bus** or **highway** as there are bits in the data word which is transferred between the two.

The peripheral may be a device which simply receives data (such as a printer or computer monitor), or it may supply data

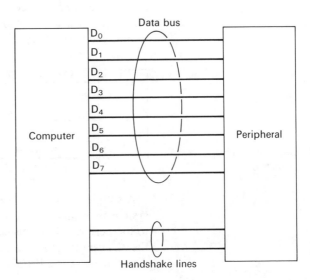

**Fig. 15.12** Parallel data transmission between a computer and a peripheral

(such as a keyboard), or it may both receive and supply data (such as a disk storage system). Consequently, the data bus is **bidirectional**, that is data may travel in either direction. The reader will note that the data bus lines are numbered from $D_0$ to $D_7$; it is the convention in computer technology to refer to the line handling the least significant bit as the $D_0$ line.

The digital system needs at least two other lines to handle data transfer between a computer and a peripheral, and these are known as **handshake lines**. One of them is the **data available line**, which is used to signal the fact that data is available for transmission; the other is the **data accepted line**, which is used to signal the fact that data has been accepted. In this way, both the computer and the peripheral are aware of the state of data transfer. The handshake lines are part of the **control bus** of the computer system.

Due to the large number of wires needed to transmit data along the parallel bus system (particularly when the word length is large), parallel transmission of data is only an economic proposition over a relatively short distance of a few metres. However, since a complete computer word can be transmitted simultaneously, very high speed data communication is possible.

### 15.15  Serial data transmission

When data is to be transmitted over any but the shortest of distances, and when the speed of transmission is not particularly important, data is transmitted in a bit-by-bit fashion in a *serial mode*.

Consider the system in Fig. 15.13, in which the computer A can either transmit data to, or receive data from, peripheral B. To transmit data in its simplest form from A to B, we need a 'transmit data' line (TXD) and a return line (earth line); to receive data from B we need a 'receive data' line (RXD) and a return line. This is known as a **full-duplex system**, which can carry out a simultaneous two-way transmission.

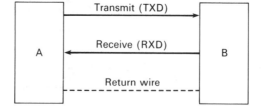

**Fig. 15.13** Basis of a full duplex serial data transmission

A simpler and less expensive method, known as a **half-duplex system**, uses only one wire (and a return wire) linking the computer to the peripheral. In this case, the line must be switched between the 'transmit' and 'return' modes, and data can flow in one direction at a time. Problems arising from serial data transmission are twofold, namely:

1. How is the data stream divided into individual bits?
2. How are the bits divided into separate computer words?

The way in which this is done depends on the way in which the data is transmitted, namely *asynchronous transmission* (or non-synchronous transmission) or *synchronous transmission*. These are dealt with in sections 15.16 and 15.17, respectively.

**15.16 Asynchronous or non-synchronous data transmission**

The basis of accurate and reliable data transmission is accurate timing, which is controlled by electronic 'clocks' within the computer and the peripherals. In asynchronous data transmission, the clock responsible for dividing the bits within a data stream is not synchronized. In fact, the clock at the receiving end must wait for the data to arrive before it is started up.

When data is not available, the transmission line is held at the logic '1' level (see Fig. 15.14). It is only when the logic level on the line falls to '0' that the receiving end becomes aware that data is available; this is known as the **start bit**. Provided that the clock frequencies at both ends of the line are within about 4 per cent of one another, then no difficulty arises in synchronizing the data. Assuming that the data is transmitted in ASCII format, the receiving end knows that the next seven bits form an ASCII 'word'; in Fig. 15.14 the Xs could either be a '0' or a '1', depending on the transmitted data.

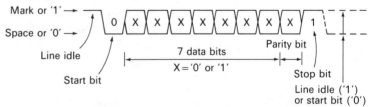

**Fig. 15.14** Asynchronous serial data transmission signal

After the data has been sent, a **parity bit** is transmitted. The logical value ('0' or '1') of this bit depends on the ASCII word already received. If the receiving ends finds that the parity bit does not match up with the ASCII word it has received, an error is *flagged* or *signalled*, and the character is rejected. What happens next depends on the organization of the computer system.

One (or optionally two) **stop bits** at the logic '1' level are next received. The stop bit carries no useful information, and acts merely as a 'space' between consecutive characters (which may arrive at any time).

If the bit pulse period is $T_b$ seconds, the minimum time period occupied by one character is

Start bit + seven data bits + parity bit + stop bit = $10T_b$

Asynchronous data transmission is relatively simple in operation, and is used with relatively low-speed peripherals such as a printer and simple terminals.

## 15.17 Synchronous serial data transmission

Synchronous serial data transmission is generally used where information is to be passed between computers in a network. In this case the information is transmitted continuously without any gaps between either consecutive characters or blocks of data; that is to say, start and stop bits are not used.

The receiving end equipment must therefore be able to detect the difference between bits (bit synchronization) and between meaningful groups of bits or characters (character synchronization).

**Bit synchronization** is achieved by using a special encoding technique in which the synchronization signal is decoded from the data signal. In this way, the receiving equipment knows from the data itself not only where it is within a computer word but also what the logic value of the bit is; moreover, it can reproduce the clock signal of the transmitting computer.

Now, it may appear that a string of 1s and 0s, without start and stop bits, is fairly meaningless to the computer. However consider the following 'sentence':

Providingthatweknowthelanguagewecan
easilysplitthisstatementintowords.

Since we all understand the English language, we can 'mentally' reinsert spaces between words in the sentence. So it is with computers, and they can split blocks of characters into separate groups of characters. This is known as **character synchronization**.

Synchronous transmission operates at a higher speed than does asynchronous transmission because there are no start or stop bits involved.

## 15.18 Protocols

A **protocol** is a set of rules to be observed when information is transferred between two parties which, in our case, is between a pair of computers or peripherals. Each protocol deals in its own way with data transfer, synchronization of data, error detection, etc. The main purpose of having international protocols is to ensure that, when equipment is purchased from different manufacturers, the purchaser can be assured that the different items of equipment will work together. We will look at one protocol for asynchronous serial data transmission and one for synchronous data transmission.

### 15.18.1 RS-232 serial protocol

The RS-232 is one of the best known standards used to interface between serial devices. Although a 25-pin connector is used to connect a device to the RS-232 interface, only three of the connections are needed in many cases. One line is needed to transmit data, one for returning data, and a signal return line (earth) as shown in Fig. 15.13.

The RS-232 interface is widely used for connecting computers and peripherals which handle data at a low to medium data rate. The voltages involved are $+3\,V$ to $+15\,V$ for logic '0', and between $-3$ and $-15\,V$ for logic '1'; in many cases, the voltage levels are in the range $\pm 6\,V$.

### 15.18.2 High-level data-link control (HDLC)

This is operated as a full-duplex protocol (although it can be used in half-duplex mode), and the link between the transmitter and the receiver(s) is over two independent channels.

When blocks of data are transmitted synchronously over one channel, acknowledgement of the receipt of the data is sent over the other channel. Transmission takes place continuously unless an error is detected by the receiving station; when this occurs, transmission stops and the system takes 'time out' before sending the data again.

Each station on the HDLC network has its own electronic address, and one station is the 'master' station or **primary station**, and is responsible for control of the network. The other stations are known as **secondary stations**, and only respond to information from the primary station.

## 15.19 Data networks

A data network interconnects computers and terminals, and this is a field of work which is changing the face of computer technology. In fact, the study of networking technology is known as network topology.

A **local area network** (LAN) is one which connects computer equipment either in the same building or within buildings on the same site (say within about 1 km of one another). Since the bandwidth of the telephone network is inadequate for the high bit rates involved in this case, interconnections are either by coaxial cable or by fibre optic link. There are a number of basic topologies suitable for use in a LAN, and these are described in sections 15.19.1–15.19.4.

A **wide area network** (WAN) is used to connect items of computer equipment which are a considerable distance apart (from, say, 1 km to a world-wide network). The interconnections may be over a standard telephone line (analogue or digital), or radio.

Many systems are **internetworked**, so that a network may comprise several LANs, or a WAN and several LANs.

### 15.19.1 Bus network and tree network

In a bus system (see Fig. 15.15), all the **nodes** (computers or peripherals) are connected to a common **data highway**. This may be a single bus linking all the nodes [Fig. 15.15(a)], or it may be a main bus system with branches linking separate computers or peripherals to it [Fig. 15.15(b)]. The system in Fig. 15.15(b) is known as a **tree network** or **cluster network**. One limitation of this system is the maximum physical length of the bus.

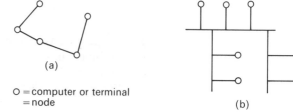

(a)

O = computer or terminal
= node

(b)

**Fig. 15.15** (a) Simple bus system and (b) a more general form of bus system or tree network

A major problem of this type of network is the way in which it deals with the case when several computers or peripherals want to put data on at the same time. This is dealt with by special bus protocols, which are beyond the scope of this book.

### 15.19.2 Ring network

Here the nodes are connected in the form of a ring, as shown by the full line in Fig. 15.16. Data is transmitted around the ring in one direction only, so that a message is passed from one node to another. The message circulates around the ring until one node recognizes its own address which is in the message.

In the early days of computing, the ring broke down when a fault developed in the ring. This has been overcome, and is no longer a problem.

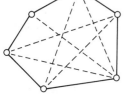

**Fig. 15.16** A ring network

### 15.19.3 Mesh network

The configuration of a mesh network is generally similar to the ring network in Fig. 15.16, but each node is connected to every other node by the broken lines shown in the figure. The network provides many links between the nodes, so that the effect of the failure of any of the links or nodes is minimal. However, the cost of such a system is high, with the consequence that some of the links to the less important nodes may be omitted.

### 15.19.4 Star network

This is a relatively simple network (see Fig. 15.17) in which all the nodes are connected to a **master station** or hub. The star network has two principal disadvantages. Firstly, since all messages pass through the central node, it must have a very large memory capacity and must be capable of working at a very high data rate in order to deal with all the nodes which need to use the system simultaneously. Secondly, if the master station develops a fault, network operation may fail; some networks can, in fact, offer a reduced service under this condition.

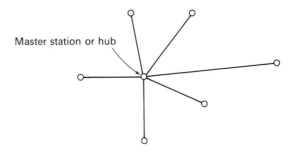

Master station or hub

**Fig. 15.17** A star network

### Problems

**15.1** In connection with telecommunication systems, explain the terms analogue, digital, unidirectional, bidirectional and transducer.

**15.2** Outline the need for multiplexing data, and describe a time-division multiplexing system (t.d.m.) and frequency-division multiplexing system (f.d.m.).

**15.3** A triangular wave rises uniformly from zero volts to 24 V. What is the average voltage of the wave?

[8 V]

**15.4** A binary system uses voltages of '0' = +10 V and '1' = −10 V. If a binary 'word' of 0111001 is produced by the system, what is the average value of voltage on the line?

[−0.7 V]

**15.5** In terms of a sinusoidal signal, explain what is meant by (a) frequency, (b) periodic time, (c) amplitude, (d) wavelength and (e) velocity of propagation?

A 15 MHz sinusoidal signal is transmitted along a cable at 80 per cent of the speed of light. Calculate the wavelength of the signal.

[16 m]

**15.6** Describe the purpose of modulation and demodulation in data transmission systems. Explain why side frequencies and sidebands occur in all forms of modulation; what is the purpose of the side frequencies and sidebands?

**15.7** An amplitude-modulated wave has a 150 kHz carrier. If the modulating signal contains frequencies ranging from 100 Hz to 5 kHz, give the frequencies involved in the amplitude-modulated frequency spectrum, and the bandwidth of the modulated wave.

[145–149.9 kHz; 150 kHz; 150.1–155 kHz; bandwidth = 10 kHz]

**15.8** Describe frequency modulation and phase modulation. What is the essential difference between the two?

**15.9** Explain the purpose of pulse modulation, and show how pulse-code modulation is used to transmit data.

**15.10** Explain the operation of time-division multiplexing and frequency-division multiplexing. Suggest suitable applications for the two types of system.

**15.11** Digital data may be transmitted either serially or in parallel. Describe the two types of system. In connection with serial transmission, outline how data is transmitted both asynchronously and synchronously.

**15.12** Explain why the use of a protocol is necessary with a data transmission system and, in particular, describe the RS-232 and the HDLC protocol.

**15.13** Data networks are vital to the transmission of computer data. Describe the terms 'local area network' and 'wide area network'. In particular, outline the technologies involved in LANs.

# 16 Control principles

**16.1   Introduction**

The background theory to control engineering was developed along independent lines in the Second World War. One line was that of **servomechanisms** (derived from the Latin *servus* meaning slave), and the other was that of **regulators** or **controllers**. Since the two were developed independently, the former in great secrecy, both have their own language and terminology. In this chapter we will attempt to use terms common to both types of system.

In the early days of control engineering, all systems were **open-loop systems** of the type in Fig. 16.1. In this type of system, an **input signal** or **reference signal** or **set-point** value, which is generally a low-power signal derived from the setting point of a control knob, is applied to an **amplifier** or **controller**, which amplifies or controls the flow of power from an independent power source. The output from the controller is applied to an **output element** which, in turn, controls a **load**. The reason that the system is described as an open-loop system is that no attempt is made by the system to 'see' if the output actually corresponds to the signal applied at the input.

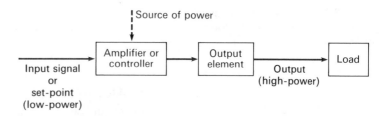

**Fig. 16.1**   Basis of an open-loop control system

The diagram in Fig. 16.1 is known as a **block diagram**, because each important element (or group of elements) is contained in a separate block. As will be seen later in the chapter, we can write down either a number or an expression inside each block to represent the relationship between the input to the block and the output from it. Such a number or expression is known as the **transfer function** of the block, where

$$\text{Transfer function} = \frac{\text{output from the block}}{\text{input to the block}}$$

A simple example of an open-loop system is a steam turbine whose speed is controlled by a simple inlet valve, the position of the handle of the steam valve corresponding to the input signal to the system (which requires little power to turn it). The steam valve itself corresponds to the amplifier or controller; in effect, the steam valve 'amplifies' the small signal applied originally to the handle to allow a large amount of power to flow to the turbine. The resulting flow of steam drives the turbine round which, in turn, drives the load which may be, say, an alternator or compressor.

While an open-loop system is simple in nature, its performance is subject to many variables. For example, the boiler steam pressure can vary, or the load applied to the turbine can alter, both of which have a significant effect on the speed of the turbine. Consequently, the output from an open-loop system can vary unpredictably.

To overcome the limitations of the open-loop system, it is necessary to monitor continuously the output from the system, and take the necessary corresponding action. That is, output information must be *fed back*, so that the system 'knows' what is happening at the output. This led to the development of **closed-loop control systems** or **negative feedback systems**, whose general block diagram is shown in Fig. 16.2.

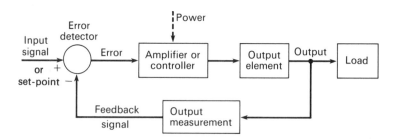

**Fig. 16.2** General block diagram of a negative feedback closed-loop control system

In this type of system, a signal proportional to the output is fed back and compared with the input signal, i.e. subtracted from it, in a section known as the **error detector**. The resulting error signal is given by

$$\text{Error, } E = \text{input signal} - \text{signal fed back} \qquad [16.1]$$

It is the error signal which is used to activate the amplifier or controller.

This type of feedback is known as **negative feedback** because the *signal feedback is subtracted from the input signal*. Certain types of system use **positive feedback**, in which the signal feedback is *added to* the input signal; oscillators are included in this category of system.

From the above we see that the essential requirements of a negative feedback control system are:

1. An *input signal*, or *set-point* which is the *reference* against which the output or **measured variable** is compared.
2. A means of measuring the *output* or *controlled variable* of the system.
3. A method of comparing the input and feedback signals (the *error detector*). The difference between the two signals is known as the *error* or **deviation**. Any deviation which remains when the system reaches its *steady-state operating condition*, i.e. after all transients have decayed, is known as the **steady-state error** or **offset**.
4. A *controller* or *amplifier*.
5. An *output element* which provides the required output.

The path in Fig. 16.2 from the error signal to the output is known as the **forward path**, and the return path from the output to the error detector is known as the **feedback path**.

The reader should note that the 'flow', indicated by the arrows on the block diagram, refers only to **information flow** or **signal flow**, and not to power flow. It is true that power flow is involved, but we are not concerned with that aspect here.

Comparing the block diagrams in Figs 16.1 and 16.2, we see that the open-loop block diagram corresponds to the forward loop of the closed-loop system. To form a closed-loop system, all we need to add is an output-measuring device and an error detector. In the case of the steam turbine, this was introduced by James Watt, who invented the centrifugal governor.

**16.2 Discussion about open-loop and closed-loop systems**

An *open-loop system* is one in which the desired output or **desired value** from a system, say liquid flow, is set by means of an input signal; it is assumed that the output will remain at the desired level, irrespective of variations either within the system or in the load or demand placed on the system.

Quite clearly, no system can maintain a constant output without some form of overall control, the simplest being manual control. That is, a human operator monitors the output from the system (thereby forming a feedback system), and provides the link with the input. We will now compare the essential features of **manual control** with **automatic control** of a feedback system.

1. The cost of manual control can be high, and the work is laborious, repetitive and soul-destroying.
2. Human reaction time (typically 0.3 s or greater) introduces a 'dead time' into the system, and is totally unsuited for a rapid response system.
3. Humans cannot operate over very long periods of time, and stressful conditions lead to a reduction in efficiency.
4. Unless the task is very simple, it is impossible to standardize the behaviour of humans.

**16.3 A non-engineering closed-loop system**

We are largely concerned in this book with engineering systems, but there are other closed-loop systems with which we are all concerned, such as financial and biological systems. In the following we will take a brief look at a typical **financial cycle** or **economic cycle**.

What is needed in the financial world is steady growth of income, expenditure and employment. However, what happens in practice is an unpredicted sequence of boom and slump; in control systems we refer to this as **instability**, and is described in engineering terms in section 16.10.

Consider an employee (which could be you) who has a certain amount of money to spend (see Fig. 16.3). After making certain decisions about his/her immediate needs, he/she spends the money in a shop; after a certain time the shop needs to restock, and places an order with a wholesaler. Later in time, the wholesaler places an order with the factory which supplies the goods; the employee, who works in the factory, continues in employment and receives his pay at the end of the week (or month). In an ideal world, this balanced operating state continues for a long time.

**Fig. 16.3** Flow in a simple monetary system

In an uncertain world, both national and international events may cause, for example, the Chancellor of the Exchequer to raise interest rates, making the employee think that saving some of his/her money looks like a 'good thing'. This means that he/she has less money to spend on goods in the shop and the latter in turn, has less money to order replacement goods. This means that the wholesaler needs to order less goods from the factory, who may have to think about 'streamlining' his operation. This could herald the onset of recession!

On the other hand, since the employee has saved some of his/her money, the bank has more money to lend to the factory to produce more goods. But because the bank rate is high, the factory management may not think it a good time to borrow money. The net result may be that the technology within the factory falls behind others, which may accelerate the recession.

One method which has been adopted to overcome the declining availability of money is to reorganize the factory. This action may, in turn, make the factory more profitable, causing the recession to turn into a boom, when more people than before can be employed!

In the financial world, as in engineering systems, we need to understand the reasons for system instability before we can correct it.

## 16.4 Regulating systems and servosystems

Strictly speaking, the division of control systems into regulators and servosystems is artificial, and is due to historic rather than scientific reasons.

We can think of *regulator systems* as those which regulate or control the output of a plant, the input signal or set value being held constant. The term 'regulator' also includes many aspects of process control systems, and we may group the following in this type:

1. voltage control regulators;
2. speed control regulators;
3. fluid flow control;
4. temperature control;
5. roll stabilization of ships;
6. liquid level control.

A *servosystem* is designed to follow a changing input or reference signal, that is it follows a changing trajectory profile. Such systems include the following:

1. missile guidance systems;
2. automatic pilots for aircraft;
3. machine-tool position-control systems;
4. tracking radar systems;
5. positioning systems for radio and optical telescopes;
6. remote position-control systems;
7. inertial guidance systems;
8. gunnery control.

## 16.5 Transfer function of a closed-loop system

Figure 16.4 shows a block diagram of a typical closed-loop system. It may, for example, represent a motor speed control system, in which the reference value, $R$, could be a voltage derived from the wiper of a potentiometer. The error signal, $E$, is the difference between $R$ and the measured value of the output (obtained, say, from a tachogenerator). The error signal is amplified by a power amplifier which controls the motor current, the motor shaft speed being the 'output' from the system.

If the value of the transfer function, $H$, in the feedback path in Fig. 16.4 is other than unity, it is known as a **non-unity feedback system**. In the special case where $H = 1$, it is known as a **unity feedback system**.

**Fig. 16.4** (a) Block diagram of a single-loop control system and (b) the equivalent block diagram

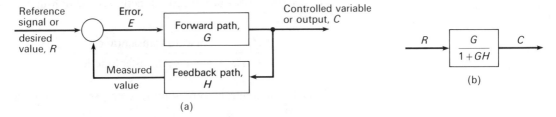

(a)

(b)

The transfer function of the forward path of the system is

$$G = \frac{\text{controller output}}{\text{controller input}} = \frac{C}{E}$$

or

Output, $C = GE$ [16.2]

also the transfer function of the feedback path is

$$H = \frac{\text{measured value}}{\text{output}} = \frac{\text{measured value}}{C}$$

or

Measured value $= CH$ [16.3]

The transfer function round the complete loop from the error signal, through the forward path to the output, and then back through the feedback path to the measured value of the output, is known as the **loop transfer function** or as the **open-loop transfer function**, and is

Loop transfer function = forward path transfer function

$\times$ feedback path transfer function

$$= \frac{C}{E} \times \frac{\text{measured value}}{C}$$

$$= \frac{\text{measured value}}{E} = GH \qquad [16.4]$$

In particular, we are interested in the **closed-loop transfer function** relating the reference value, $R$, to the controlled variable, $C$, which is determined as follows. The equation for the controlled variable is

$C = GE = G(R - \text{measured value of output})$

$= G(R - CH) = GR - CGH$

or

$C(1 + GH) = GR$

Hence the closed-loop transfer function of the non-unity feedback system is

$$\frac{C}{R} = \frac{G}{1 + GH} = \frac{\text{forward transfer function}}{1 + \text{loop transfer function}} \qquad [16.5]$$

The block diagram representing eqn [16.5] is shown in Fig. 16.4(b).

**16.6 A simple speed control system**

A basic form of speed control regulator is shown in Fig. 16.5(a). In this system, the set-point value is the voltage from the wiper of the potentiometer, and this is compared with the measured value of the output, derived from a tachogenerator. That is, the transfer function, $H$, of the tachogenerator is in the feedback path block [see Fig. 16.5(b)]. Its value can be obtained from the nameplate data, and is 0.01 V per revolution per minute.

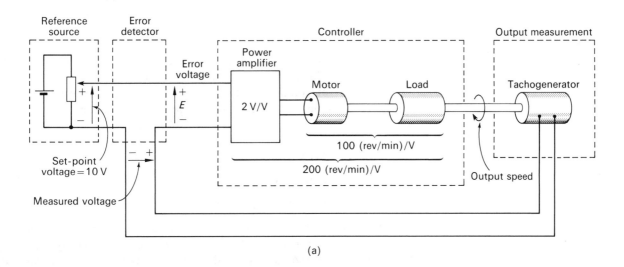

(a)

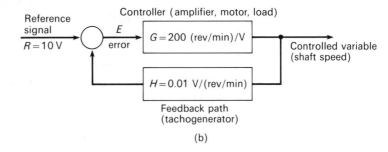

**Fig. 16.5** (a) A closed-loop speed control system and (b) its block diagram

The error detector simply consists of interconnected wires, the connections being such that the error voltage is the difference between the set-point voltage and the tachogenerator voltage.

The controller consists of a power amplifier of gain 2 V/V, and a motor and mechanical load which gives a speed of 100 (rev/min)/V applied to the motor. The overall steady-state gain of the controller in the forward path is

$$G = \text{amplifier gain} \times \text{motor and load 'gain'}$$

$$= 2 \text{ V/V} \times 1000 \text{ (rev/min)/V} = 200 \text{ (rev/min)/V}$$

The output from the system is the shaft speed of the motor and, from eqn [16.5], the closed-loop transfer function is

$$\frac{\text{Steady output speed}}{\text{Reference signal}} = \frac{C}{R} = \frac{G}{1 + GH} = \frac{200}{1 + (200 \times 0.01)}$$

$$= \frac{200}{1 + 2} = 66.67 \text{ (rev/min)/V}$$

Since the reference signal (set-point) is 10 V, then the steady output speed is

$$66.67 \text{ (rev/min)/V} \times 10 \text{ V} = 666.7 \text{ rev/min}$$

At this stage we are not aware of what this speed means in terms of the 'best' performance of the system. Let us calculate the value of the error voltage under steady-state operating conditions. At a speed of 666.7 rev/min, the tachogenerator output is

$$666.7 \times 0.01 = 6.667 \text{ V}$$

hence the error voltage is

$$E = \text{set-point voltage} - \text{tachogenerator voltage}$$

$$= 10 - 6.667 = 3.333 \text{ V}$$

That is, so long as the speed remains at 666.7 rev/min, the error voltage also remains at 3.333 V. The reason for the steady-state error is that the amplifier and motor need this voltage in order to overcome the friction, windage and other losses of the system.

If all things were perfect (which, unfortunately, they never are), the error voltage would be zero! Let us consider the ideal case that the error voltage is zero; in this case the voltage from the tachogenerator would be 10 V (so that it would just match the set-point voltage). For this to happen, the shaft speed of the tachogenerator can be calculated from the transfer function of the tachogenerator, which is

$$H = \frac{\text{tachogenerator output voltage}}{\text{tachogenerator shaft speed}} = 0.01$$

$$= \frac{10 \text{ V}}{\text{tachogenerator speed}}$$

that is

$$\text{Tachogenerator shaft speed} = 10/0.01 = 1000 \text{ rev/min}$$

At this speed, the error voltage would be zero! In Fig. 16.4, we are therefore dealing with a system having a **steady-state speed error** of

$$1000 - 666.7 = 333.3 \text{ rev/min}$$

The reason for this error has, of course, been explained above.

## 16.7 Improving the steady-state performance of a system

From eqn [16.2] we see that the controller output, $C$, is given by

$$C = GE$$

where $E$ is the error. That is

$$E = C/G$$

Quite clearly, if we increase the numerical value of $G$ (the forward path transfer function), we can reduce the error. Let us consider the effect of increasing the gain of the amplifier in section 16.6 from 2 to 10 V/V (in practice this is about the only thing we have much control over, because the other items in the plant such as the motor, load and tachogenerator are 'fixed' items). With the new value of amplifier gain, the forward transfer function has the value

$$G' = 10 \times 100 = 1000 \ \text{V/V}$$

and the closed-loop transfer function has the value

$$\frac{C}{R} = \frac{G'}{1 + G'H} = \frac{1000}{1 + (1000 \times 0.01)} = 90.91$$

If $R = 10$ V, then the new value of the shaft speed is

$$C = 90.91R = 909.1 \ \text{rev/min}$$

giving a new speed error of $(1000 - 909.1) = 90.9$ rev/min (corresponding to an electrical error voltage of 0.909 V). This represents a significant reduction in the error when compared with the previous case in section 16.6.

Based on this argument, we can continue to 'improve' the accuracy of the system simply by increasing the amplifier gain! However, there may be a penalty to pay for increasing the gain, because it may lead to instability in the system, as outlined in section 16.10.

## 16.8 Transient response of a first-order system

As explained in Chapter 13, a *first-order system* is one which has a single energy storage element in it, such as a capacitor, or an inductor. When a step change or sudden change in disturbing signal is applied, it gives rise to a transient response which has a single time constant, or exponential response (details of which were given in Chapter 13). In electrical and electronic circuits, the time constant of a first-order system tends to be fairly short (usually much less than one second), but electromechanical systems typically (due to, say, inertia) have a time constant of a few seconds, and chemical plants could have a time constant of hours or even days!

A typical change in output signal from a first-order system is generally similar to the rise in current in an $L$–$R$ circuit, or the

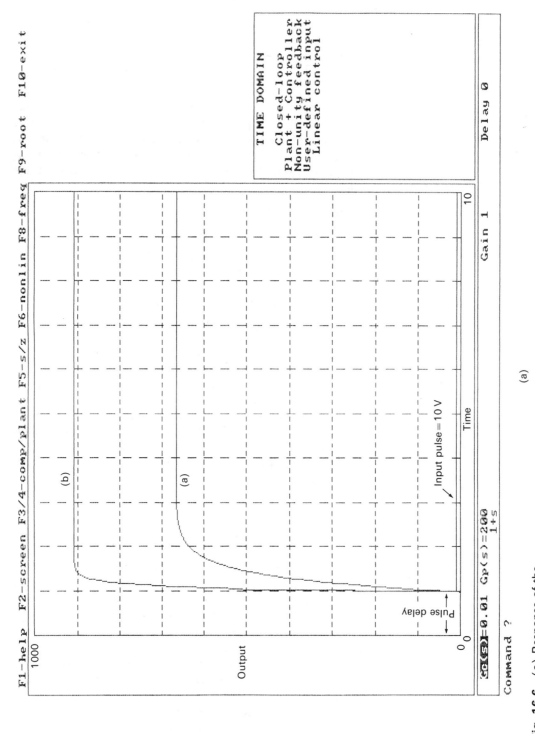

**Fig. 16.6** (a) Response of the system in Fig. 16.5 for a 10 V change in reference signal for two values of amplifier gain

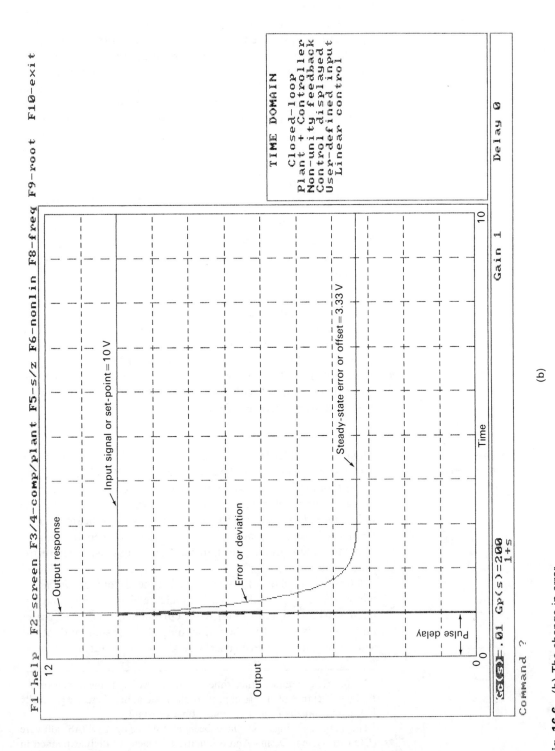

**Fig. 16.6** (b) The change in error for a forward gain of 200

fall in charging current in an $R-C$ circuit (see Chapter 13). Since we are concerned with control systems in this chapter, we will consider a simple speed control system once more of the type in Fig. 16.5, in which the *dominant time constant* may be due to the inertia of the motor and load.[†] If the mechanical time constant is, say, one minute then, for a sudden change or step change in input signal, the response may be as shown in graph (a) in Fig. 16.6(a).[‡] In this case, so that we can see the start of the output response, the change in reference signal is applied to the system one minute after the start of the graph. The graph shows not only the response for a power amplifier gain of 2 V/V [curve (a)] – see section 16.6 – but also the response for an amplifier gain of 10 V/V [curve (b)]. Two points can be noted from Fig. 16.6:

1. An increase in forward gain produces a higher steady-state shaft speed, i.e. a smaller error.
2. An increased forward gain causes the motor shaft to reach its final speed much more quickly.

An analysis for the improvement in response time (see item 2 above) is beyond the scope of this book.

At the foot of Fig. 16.6(a) we can just see the input signal to the system which, since it is only 10 V, is very much scaled down. However, in Fig. 16.6(b), this part of the graph is scaled up to show not only the input signal, but also the way in which the error signal changes with time; for an amplifier gain of 2 V/V, the steady-state error is 3.33 V.

## 16.9 Transient response of a second-order system

Most practical control systems are more complex than the first-order system described in section 16.8. They may be second- or higher-order systems having, respectively, two or more energy storage elements in them. Suppose, for example, the inductance of the motor in Fig. 16.5 produces a significant amount of energy storage, causing the system to have two energy storage elements; one is due to mechanical inertia, and the other to the inductance.

Consider the case of a second-order speed control system, in which the input signal is suddenly changed to 10 V. When the change in input signal is applied, the current in the motor increases and energy is stored in the inductance of the system and, at the same time, the load inertia accelerates, resulting in further increase

[†] There are, of course, other time constants involved due, for example, to the inductance of electrical parts of the system, but these are generally less significant than the mechanical time constant.
[‡] The graphs in Fig. 16.6 have been plotted using CODAS software (COntrol system Design And Simulation), which is briefly explained in Chapter 18.

in energy storage. As explained earlier, as the motor speed increases, the error diminishes in value. However, due to energy storage in the system, it may not be possible to prevent the motor continuing to speed up even when the error is zero! The net result is that the motor speed overshoots the desired value of speed, giving rise to a negative error.

The negative error produces a reverse torque, slowing the motor down until, finally, acceleration stops when the motor speed is well in excess of the value requested by the reference signal.

A CODAS plot of the output of such a system is shown in Fig. 16.7, where the peak speed is 1247 rev/min (the first-order response for a system with an amplifier gain of 10 V/V gain is shown for comparison). The negative error existing at this time causes the motor to decelerate until its speed falls below the required value, when the error becomes positive again. These oscillations continue until, finally, the shaft speed settles down to the steady-state speed of 909.1 rev/min.

Higher-order systems, i.e. third-, fourth-, have a type of response not too dissimilar to that in Fig. 16.7.

**16.10 System response**

The form of output response given by a system when it is excited by a step change or sudden change in input signal depends not only on the order of the system, i.e. first-, second-, third-order, but also on the system gain and energy loss within the system. For a step input, there are five particular types of output response we should be aware of (see Figs 16.8 and 16.9):

1. an overdamped system [curve (a) in Fig. 16.8];
2. a critically damped system [curve (b) in Fig. 16.8];
3. an underdamped system or damped system [curve (c) in Fig. 16.8];
4. a conditionally stable system or undamped system [curve (d) in Fig. 16.8];
5. an unstable system (Fig. 16.9).

**Overdamped system**

In this case, the energy loss within the system is large, making it particularly sluggish. In all the examples chosen, the system is third-order (the 'order' is given by the total number of 's' terms in the lower line of the edit box at the bottom of the figure). Once again, the input step change is delayed after the start of the graph so that we can see it in the diagram. The output is so sluggish in this case that it has not reached its steady-state value 19 s after the application of the input step.

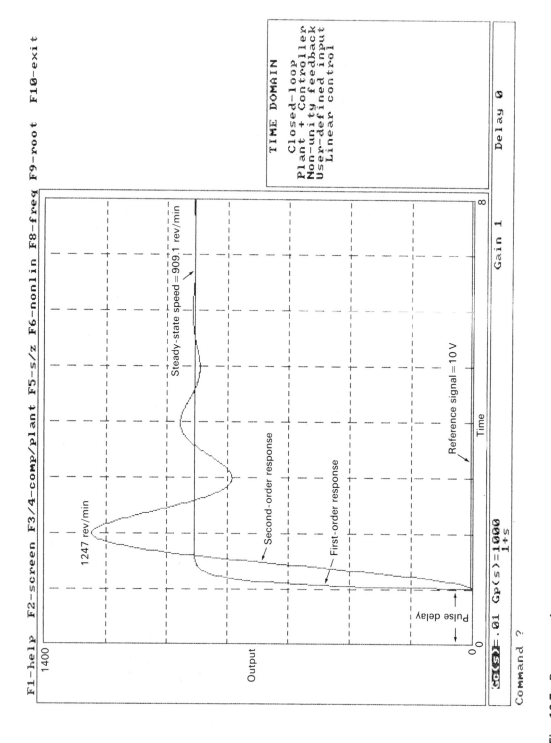

**Fig. 16.7** Response of an underdamped second-order control system. The response of a first-order system is shown for comparison

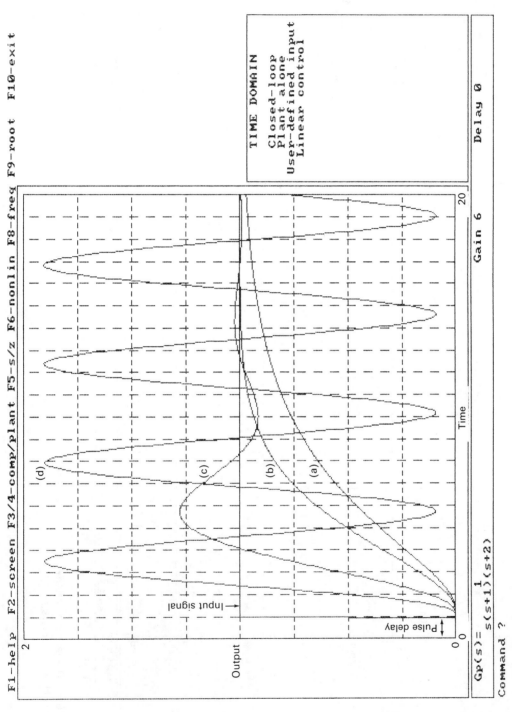

**Fig. 16.8** Examples of the closed-loop response to a step change in input signal: (a) overdamped, (b) critically damped, (c) underdamped, (d) conditionally stable

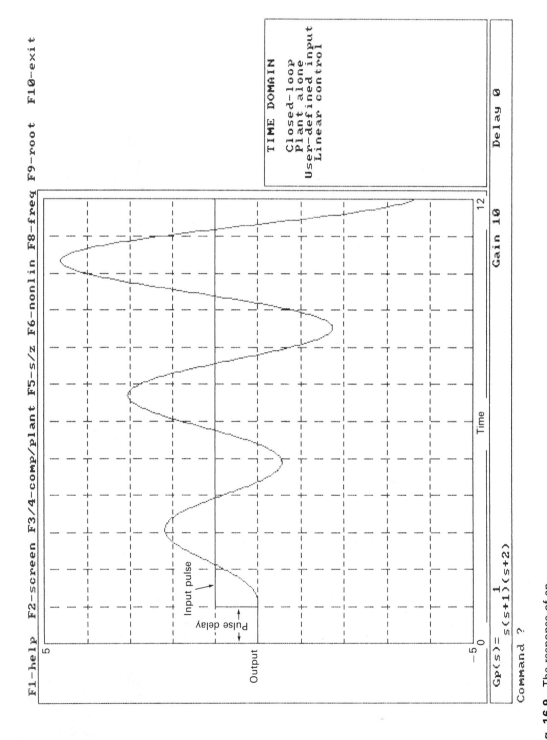

**Fig. 16.9** The response of an unstable closed-loop system

### Critically damped system

In this case [curve (b)] the damping is less than the overdamped case, the output from the system just reaches its steady-state value **without overshoot**. In this case it takes about 15 s to reach its final value.

### Underdamped system or damped system

Curve (c) shows a typical response; in this case the energy losses are even less than they are for the critically damped case. The system output oscillates with diminishing amplitude for a few cycles about the steady-state value and finally settles down (see also Fig. 16.7). This type of response is typical of the great majority of control systems, the amount of *first overshoot* often being a specified design figure for the system. Depending on the amount of first overshoot, this type of response often gives the most satisfactory performance.

### Conditionally stable system or undamped system

In this case [curve (d)], the system has no damping, the output continuously oscillating with a fixed amplitude about the desired steady-state value. This type of response is quite unsuitable for any practical control system, and special **compensating systems** are used to prevent it occurring. This type of response is also known as **marginal damping**.

### Unstable system

The amplitude of oscillations of the output from an unstable system (see Fig. 16.9) grow continuously with time until, finally, some part of the system breaks down! Once again, this type of system is totally unsuitable for a practical system. Again, compensating systems are used to overcome the problem.

### Summary of response curves

The response curves in Fig. 16.8 can be described in terms of the **damping factor** or **damping ratio** of the system, as follows:

*Overdamped system*: damping factor > 1.
*Critical damping*: damping factor = 1.
*Underdamped system*: damping factor lies in the range just over zero to just under unity.
*Undamped system*: damping factor = 0.

The damping factor is defined as

$$\frac{\text{The amount of damping in the system}}{\text{The amount of damping for critical damping}}$$

**16.11 Introduction to a process control system**

The control of flow of fluid to an industrial plant is typical of a process control system, a simple system and its block diagram being shown in Fig. 16.10.

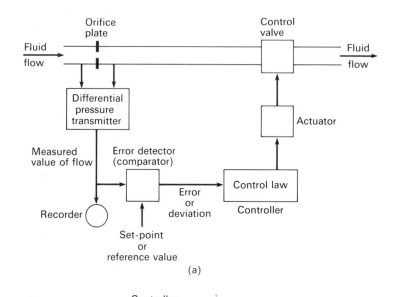

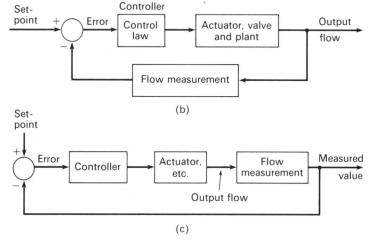

**Fig. 16.10** (a) The basis of a flow-rate process control system, (b) its non-unity feedback block diagram and (c) its unity feedback block diagram

The flow is measured by a simple orifice plate **transducer**, which produces a difference in pressure across an orifice plate when fluid flows in the pipe. The pressure difference is related to the fluid flow. The flow rate is converted to a suitable form of measured value by a **differential pressure transmitter**; the signal from the transmitter could be an air pressure, or an electric current, etc. We are not particularly concerned with the type of signal here, because transducers are available for converting a signal of any kind to a signal of any other kind.

It is usual (though not always necessary) in industry to continuously record the output of the process being controlled, as shown in Fig. 16.10. The *measured value* of flow is compared in an error detector with the *set-point* or *reference signal*; the resulting *error signal* is applied to a *controller*, which applies a particular form of *control law* to the error signal. We shall discuss control laws a little later. The error detector and controller are often, in practice, in an integral unit.

The output signal from the controller is applied to an *actuator* or *motor* which operates a *control valve* in the pipe, so controlling the flow of fluid.

The system in Fig. 16.10 is therefore a negative feedback control system. The set-point is usually fixed, so that the flow rate demand is assumed to be constant over a long period of time. Any variation in the plant, such as availability of the fluid whose flow is controlled, or the electrical supply voltage, etc. is accounted for by the closed-loop system, thereby maintaining a constant flow.

## 16.12  Exponential lag and distance/velocity lag

In electrical circuits we are familiar with the *exponential lag* in the form of a *time lag* in, say, $R$–$L$ and $R$–$C$ circuits (see also Chapter 13). This type of lag also occurs in process control systems as follows.

A *thermal exponential lag* occurs when a body or fluid is being heated. The body or fluid begins to heat up instantly, but it takes a finite time for it to reach its steady temperature. The time taken to reach the final temperature is controlled by the *thermal time constant* of the body or fluid. As explained in Chapter 13, it takes about five time constants to reach its final temperature.

A *flow exponential lag* occurs when fluid flows from one tank into another through a pipe. The time constant is fixed by the liquid capacity of the tank and the resistance of the pipe to fluid flow (analogous to the $R$–$C$ time constant of a resistance–capacitance circuit).

Clearly, exponential lags can have a destabilizing effect on a control system, as do exponential lags on an electrical system.

A **distance/velocity lag**, although not a purely mechanical feature, is present in practically all process control systems, and is the time interval between the alteration in the value of a signal and the time when its effect is measured. For example, suppose a control system regulates the flow of grain along a conveyor belt. If, due to the construction of the conveyor, the point at

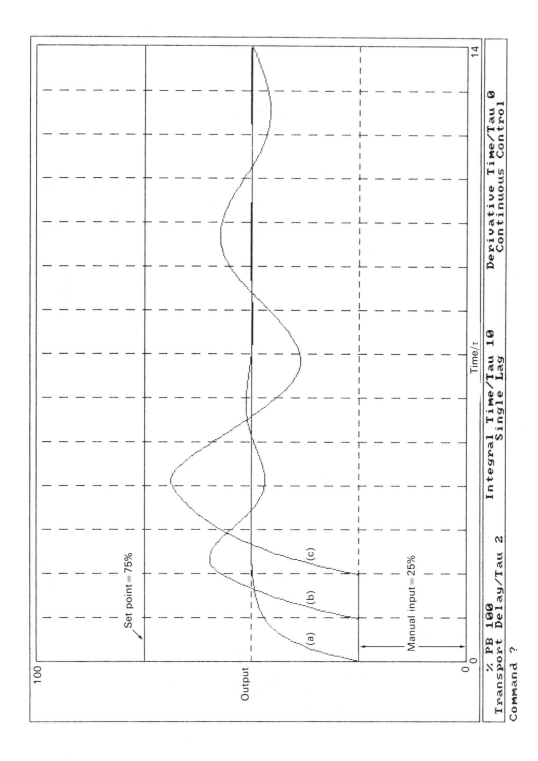

**Fig. 16.11** Response of a system with a time constant $\tau$ with a transport lag of (a) zero, (b) $T_d = \tau$ and (c) $T_d = 2\tau$

which the flow of grain is measured is distance $D$ away from the point where the grain is delivered on to the conveyor, then the distance/velocity lag, $T$, between the two points is

$$T = D/v$$

where $v$ is the velocity of the conveyor.

In high-frequency electrical systems, this type of delay also occurs in the propagation of signals in digital and analogue circuits, and also in transmission lines.

This type of delay is also known as a **transport delay**, or a **pure time delay**, or as a **dead time**. Both exponential and distance/velocity lags tend to destabilize systems.

Process control systems can be simulated by a number of computer packages, and here we will use the PCS package (Process Control Simulation) – see also Chapter 18. Figure 16.11 shows a typical PCS screen display for a process control system which has a single exponential time lag. To explain the graphs in Fig. 16.11, we must first look at the block diagram of the process control system in Fig. 16.12, which forms the basis of the PCS system.

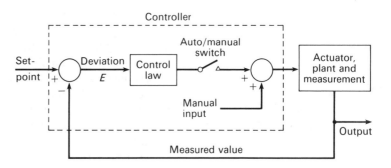

**Fig. 16.12** Block diagram of a process control system showing the controller

In a practical controller, the output of the system is initially controlled *manually*, the auto/manual switch being in the open position. When the output has reached the desired value, the auto/manual switch is closed so that the system operates in the *automatic* (or *auto*) mode. The effective output from the controller can be thought of as being

Manual output + function of the error

The 'function of the error' is the *control law*, and is described in section 16.13. Normally, the *manual input* signal remains constant, but the set-point value can be changed, depending on the demand placed on the system.

In Fig. 16.11, the manual input level is set at 25 per cent (%) of the maximum output, and the set-point is suddenly changed by 50% to 75%. We can, in fact, calculate the expected output value as follows. The closed-loop transfer function is [see also eqn [16.5] and Fig. 16.4(b)]

$$\frac{\text{Change in output}}{\text{Change in set-point}} = \frac{G}{1 + GH}$$

In this case $G = 1$ and $H = 1$, so that the value of the closed-loop transfer function is 0.5. Hence *the change in the value of the output variable* is

$$\text{Change in output} = 0.5 \times \text{change in set-point}$$
$$= 0.5 \times 50\% = 25\%$$

That is, the *steady-state output* from the system is

$$\text{Output due to manual control} + \text{change in output}$$
$$= 25\% + 25\% = 50\%$$

Curve (a) shows the response when the system does not have a transport delay (i.e. it only has a single exponential lag) and, quite clearly, it displays the characteristic exponential change in output from 25% (due to the manual input) to 50%.

The reader will note that the results from PCS are displayed in a 'non-dimensional' form. That is, the output amplitude is displayed in terms of the *percentage output change* that the plant can supply; clearly, this is in the range 0% (no output) to 100 (maximum output). Also the time scale is presented in terms of *the number of system time constants*. That is, if the system time constant is 1 s, then we are looking at time in seconds, and if the time constant is 1 h, then we are looking at time in hours!

If, in addition to the inherent exponential lag in the system, there is a distance/velocity lag equal to one time constant $(1 \times \text{tau})$, then the system response is as shown in curve (b) in Fig. 16.11. The reader will see that the response is oscillatory, but finally settles down to the same level of output as in case (a).

If the transport delay is $(2 \times \text{tau})$ the oscillations in the output level are not only larger, but are more prolonged [see curve (c)]. Quite clearly, an increase in transport delay makes the system more oscillatory (less stable).

**16.13 The control law of the controller**

The relationship between the error signal and the output of the controller is known as the **control law** of the controller (see also Fig. 16.10).

The simplest control law (**proportional control**) produces an output proportional to the error. The constant of proportionality is the **gain** of the controller, which is related to the **proportional band** of the controller; the latter is explained more fully in section 16.14.

Another form of control is known as **integral control**, and controllers which have **proportional plus integral control** are described as **PI controllers (two-term controllers)**.

**Derivative action control** is yet another form of control, and controllers which have **proportional plus integral plus derivative control** are described as **PID controllers** or **three-term controllers**.

Before going on to discuss the control law in detail, we need to look at the way quantities are described in process controllers. In a practical situation, the controller will only recognize variations of signal between the lowest possible level (0%) and the maximum level (100%). It is for this reason that process control engineers talk in terms of percentage values rather than actual values, i.e. flow rate, temperature, pressure. Typically, the set-point control on a controller is calibrated from 0 to 100%. If we consider a temperature control system, the error or deviation is in engineering units, i.e. °C, but in process control we can talk about the **per cent deviation** with respect to the **measurement span**. Suppose that the **measurement range** of our system is from 200 to 500 °C, then the measurement span is

$$500\,°C - 200\,°C = 300\,°C$$

That is, 200 °C corresponds to 0% of the measured value, and 500 °C corresponds to 100%.

If the set-point of the controller is 350 °C, and the value of the output temperature is 300 °C, the *error* or *absolute deviation* or *actual deviation* is 350 °C − 300 °C = 50 °C, and the per cent deviation is

$$\text{Per cent deviation} = \frac{\text{actual deviation}}{\text{measurement span}} \times 100$$

$$= \frac{50}{300} \times 100 = 16.67\%$$

Similarly, the controller output is expressed as a per cent of the controller output range (see Fig. 16.11).

**16.14 Proportional control**

While we have referred earlier to the 'gain' of an amplifier or controller, in process control we refer to the **per cent proportional band** (%PB) of the controller. This is best understood by means of the graphs in Fig. 16.13.

The proportional band of the controller is the per cent deviation which gives rise to 100% change in controller output. That is, a narrow proportional band implies that a small change in deviation produces a large change in controller output, or

> **a narrow proportional band implies that the controller has a high gain.**

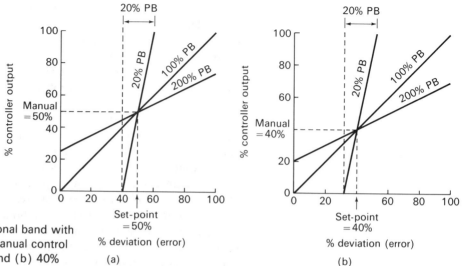

**Fig. 16.13** Proportional band with set-point (SP) and manual control (MAN) at (a) 50% and (b) 40%

A wide proportional band also implies a low value of controller gain.

The graphs in Fig. 16.13 have been restricted to limits of 100% for both input and output variation, and is common sense for a practical system. In the figure, a 20% PB means that a 20% change in controller input produces a 100% change in the controller output. It is unusual to have a proportional band less than about 10%, or greater than about 500%.

The relationship between the numerical gain of a controller and %PB is best obtained graphically as follows. Suppose a temperature controller has an input range of 100–200 °C, and its output is a current in the range 4–20 mA. The range of input and output values are shown on the axes of Fig. 16.14. The *numerical*

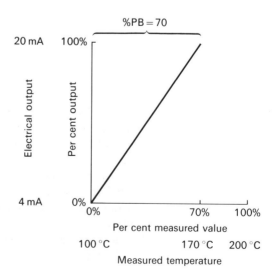

**Fig. 16.14** Characteristic of a controller with a 70% PB

*gain* of the controller is the numerical value of the slope of the output/input graph. If the controller has a 70% PB (as shown in the figure), then

Numerical gain of controller

$$= \frac{\text{fractional change in output}}{\text{fractional change in input}}$$

$$= \frac{100\% \text{ of } (20\,\text{mA} - 4\,\text{mA})}{70\% \text{ of } (200\,^\circ\text{C} - 100\,^\circ\text{C})} = \frac{16\,\text{mA}}{70\,^\circ\text{C}}$$

$$= 0.229 \text{ mA}/^\circ\text{C}$$

The %PB can also be calculated from

$$\%\text{PB} = \frac{100}{K} \times \frac{\text{output span}}{\text{input span}}$$

where $K$ is the controller gain. In some cases this can be simplified if output span = input span, when

$$\%\text{PB} = \frac{100}{K} \qquad\qquad [16.6]$$

Clearly, eqn [16.6] does not apply to the example in Fig. 16.14, because different engineering units are involved at the input and output of the controller. None the less, eqn [16.6] is a convenient tool to use in many cases.

Figure 16.15 shows a print-out from the PCS package of the response of a simple single-lag system in which the initial (manual) setting is 25%, and a sudden change of 50% is made to the set point. We see that with a 200% PB (a 'low' gain), the output from the process control system changes fairly slowly to a new value of about 42% [see graph (a)]. The new steady-state value can be calculated as follows. From eqn [16.6] we see that

$$K = 100/\%\text{PB} = 100/200 = 0.5$$

Since we are dealing with a unity feedback system, the loop gain is

$$\frac{\text{Change in output}}{\text{Change in input}} = \frac{K}{1 + K} = \frac{0.5}{1.5} = 0.333$$

hence

$$\text{Change in output} = 0.333 \times 50\% = 16.65\%$$

that is

New output = output due to manual input

+ change in output

$$= 25\% + 16.65\% = 41.65\%$$

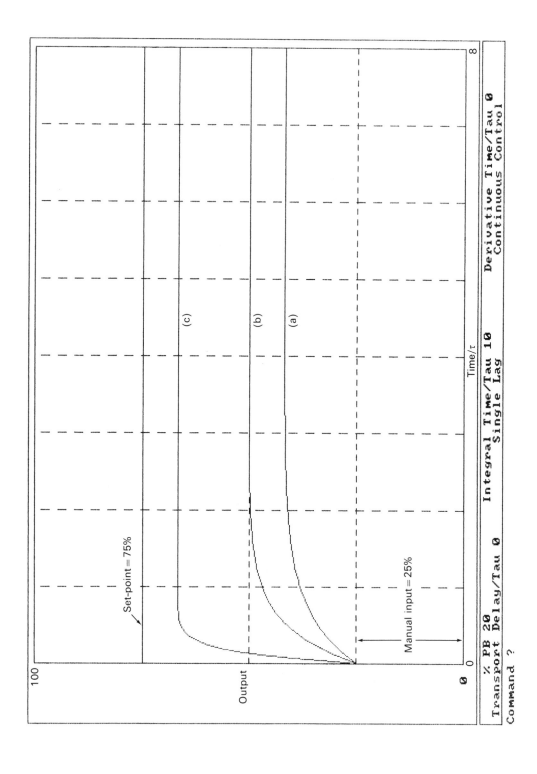

**Fig. 16.15** Response of a single-order control system having no transport lag and a proportional band of (a) 200%, (b) 100% and (c) 20%

If the proportional band is changed to 100% (corresponding to $K = 1$), the output responds rather more rapidly [see curve (b)], and the steady-state output is 50%. A further change in the proportional bandwidth to 20% produces an even more rapid response, and the steady-state output is 66.7% [see curve (c)].

An important point to note here is that, whatever the proportional band of the controller, **a steady-state error or offset always remains**. As explained earlier, the offset is necessary in order to 'drive' the system in order to produce the output. We will see how to reduce the offset in section 16.15.

Finally, we look at the response of the system when it contains a transport delay equal to the system time constant $\tau$. The reader will recall that a transport delay reduces the relative stability of the system, and the response of the system with the transport delay and a proportional band of (a) 100%, (b) 75% and (c) 50% is shown in Fig. 16.16. We see that while a proportional band of 100%, or even 75% is acceptable, a proportional band (in this case) of 50% gives an unacceptable response.

## 16.15  Proportional plus integral control

It was shown in section 16.14 (see also Fig. 16.15) that when a proportional controller is used, an offset remains in the steady-state no matter how high the gain of the controller. However, the fact that offset remains gives us a clue to a method of removing it.

What we need is a circuit within the controller which, so long as the offset exists, will cause the controller output to continue to increase. Only when the offset is zero will the controller output remain constant. Such an additional element or circuit will therefore produce a **reset action**, which resets the output to the desired value.

The reader will recall that mathematical integration is the process of adding the area under a graph. The controller therefore needs **integral action**, which will integrate the error signal and, so long as an error signal exists, will change the output from the controller.

We can think in electrical engineering of a capacitor as being an integrator, the 'input' or error signal being analogous to the charging current. So long as the charging current flows, the capacitor voltage increases; when the charging current stops, the capacitor voltage remains constant. So it is in controllers that we can use a capacitor (or its mechanical equivalent) as an integrator.

A block diagram of a controller which *incorporates proportional plus integral* (PI) control is shown in Fig. 16.17. The law of the controller is

$$\text{Controller output} = K\left[ E + \frac{1}{T_1} \int E \, dt \right]$$

$$= KE + \frac{K}{T_1} \int E \, dt$$

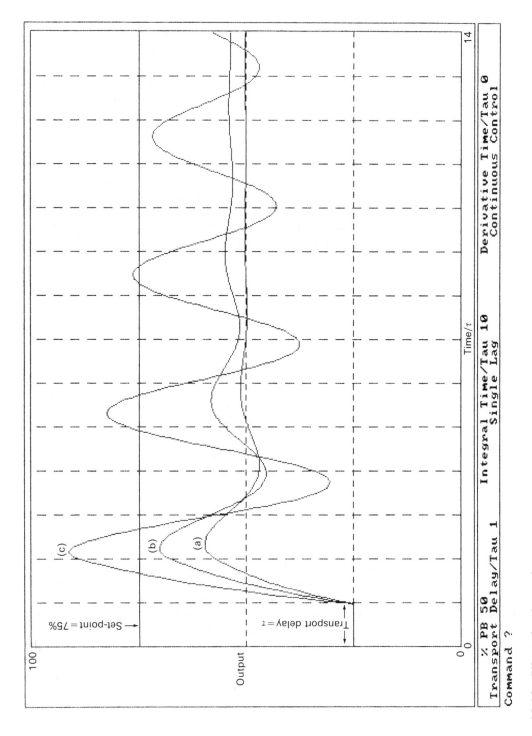

**Fig. 16.16** Effect of a change in proportional bandwidth (controller gain) on the system response: (a) PB = 100%, (b) PB = 75% and (c) PB = 50%

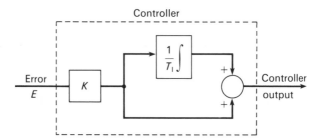

**Fig. 16.17** A proportional plus integral (PI) controller

where $KE$ is the 'proportional' contribution to the controller output, and $(K/T_I) \int E \, \mathrm{d}t$ is the 'integral' contribution.

The amount of integral contribution or *reset action* is weighted by the factor $1/T_I$, where $T_I$ is known as **integral action time**, or the **integral time**, or as **reset time**. Since $T_I$ is in the denominator of the equation, the larger its value, the less the amount of integral action. In the previous screen dumps from PCS, the integral action time was $10\tau$ which, in the case of PCS, can be considered to give $T_I = \infty$. That is, there was no integral action.

Figure 16.18 shows the effect of integral action on the system previously controlled by a proportional action controller [see Fig. 16.16, graph (a)]. Graph (a) in Fig. 16.18 shows the basic response *without integral action* when a 50% set-point change is made (the system has a transport delay of $\tau$ and %PB = 100). When the integral time is reduced to $4\tau$ (where $\tau$ is equal to the time constant of the system itself), we see that the output oscillations increase slightly in magnitude [curve (b)], but the integral action causes the deviation to reduce and will (eventually) reduce it to zero.

In graph (c) in Fig. 16.18, the integral action time is reduced to $1.5\tau$, and we see that the offset becomes zero after a few cycles. Clearly, *the smaller the integral time, the more rapid the reset action*.

Unfortunately, nothing in this world is perfect, and integral action has its drawbacks. We can see in Fig. 16.18 that, while reducing the reset time brings about a more rapid reset action, it also produces rather larger oscillations in the system output. Reducing the integral action time very much more can, in fact, significantly reduce the relative stability of the system.

**16.16  Proportional plus integral plus derivative action**

Let us consider the response of an otherwise 'ideal' system (see Fig. 16.19), and see if this can help us to improve the system response.

If we make a step change in the set-point, as shown in Fig. 16.19(a) and, assuming that the system does not have a transport delay, the output response should ideally have the general form in Fig. 16.19(b). The shape of the error waveform is the difference between graphs (a) and (b), and is shown in graph (c).

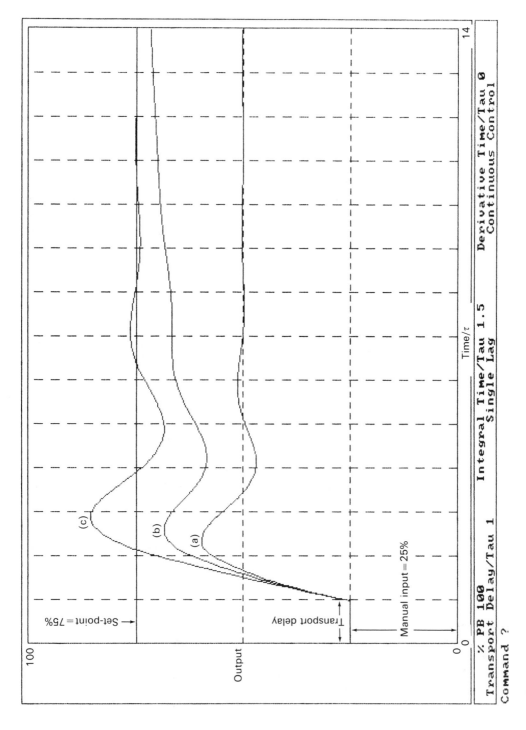

**Fig. 16.18** Effect of a change of integral action time (reset time) on system response

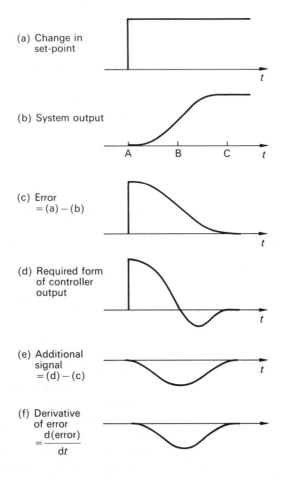

(a) Change in set-point

(b) System output

(c) Error = (a) − (b)

(d) Required form of controller output

(e) Additional signal = (d) − (c)

(f) Derivative of error = $\dfrac{d(\text{error})}{dt}$

**Fig. 16.19**  Response of an 'ideal' control system

However, the error waveform has the wrong 'shape' to produce the output waveform in graph (b), because the controller output signal must have a positive mathematical sign between A and B in order to cause the output to begin to respond rapidly, and it must have a negative mathematical sign between B and C in order to slow down the rate of change of output to cause it to reach the final value *without overshoot*.

Consequently, the required 'shape' of controller output should be as shown in curve (d). All we have to do is to find something which provides the correct waveshape within the controller. The additional signal 'shape' needed is therefore given by graph (d) − graph (c), and is shown in graph (e).

If we mathematically differentiate the error signal [curve (c)], that is we plot the rate of change of the error signal, we get the shape in curve (f). Clearly, this has the same general waveshape as that in graph (e). If, therefore, we differentiate the error signal and then add it to the error signal, we get a controller which gives a rapid response without overshoot. Quite clearly, the amount of derivative effect we add will affect the overall response.

The additional signal is known as **derivative action**, and is known by other names including *error-rate damping* or *error-derivative damping*.

When derivative action is combined with proportional plus integral control, we get **proportional plus integral plus derivative (PID) control** (*three-term control*). The control law of a PID controller is

$$\text{Controller output} = K\left[ E + \frac{1}{T_\text{I}} \int E \, \text{d}t + T_\text{D} \frac{\text{d}E}{\text{d}t} \right]$$

where $T_\text{D} \, \text{d}E/\text{d}t$ is the derivative contribution to the controller output, and $T_\text{D}$ is a weighting factor known as the **derivative action time** or **derivative time**; a value of $T_\text{D} = 0$ removes derivative action from the controller.

In the case of a controller providing *proportional plus derivative* (PD) control (another form of two-term control), the derivative element does not make any contribution to reducing the offset. As stated earlier, the function of derivative action is to reduce the overshoot and to make the system respond more rapidly.

Figure 16.20 illustrates the effect, using a PID controller, of applying various combinations of proportional, integral and derivative control. Graph (a) in Fig. 16.20 shows the basic output response of the system [also shown in graph (a) of Fig. 16.18] to a 50% change in set-point. The system has an exponential time constant of $\tau$, a transport delay of $\tau$, and a %PB of 100%. When integral action with $T_\text{I} = 1.5\tau$ is added, the response is as shown in graph (b) of Fig. 16.20 [see also graph (c) of Fig. 16.18], at which time the offset is reduced to zero after a few cycles, but the amount of first overshoot may be too large.

Finally, when derivative action with $T_\text{D} = 0.3\tau$ is introduced, the response is as shown in graph (c). In this case, the magnitude of the first overshoot is reduced, and the output oscillations rapidly die out.

**16.17 Selection of proportional, integral and derivative settings**

Process control engineers can, initially, set up the proportional band, integral and derivative times based on experience. However, there are tests which can be carried out to 'fine tune' the controller to give an optimum performance. One of the most popular is the *Ziegler–Nichols test*; this is beyond the scope of this book, but there are certain points to bear in mind as follows:

1. The more exponential lags there are in a system, the higher the probability of oscillation.
2. If the system contains one or more transport delays, the greater the probability of instability.
3. A low proportional bandwidth (a high gain) makes the system more responsive and reduces the offset. However, it does not eliminate the offset, and can reduce the stability.

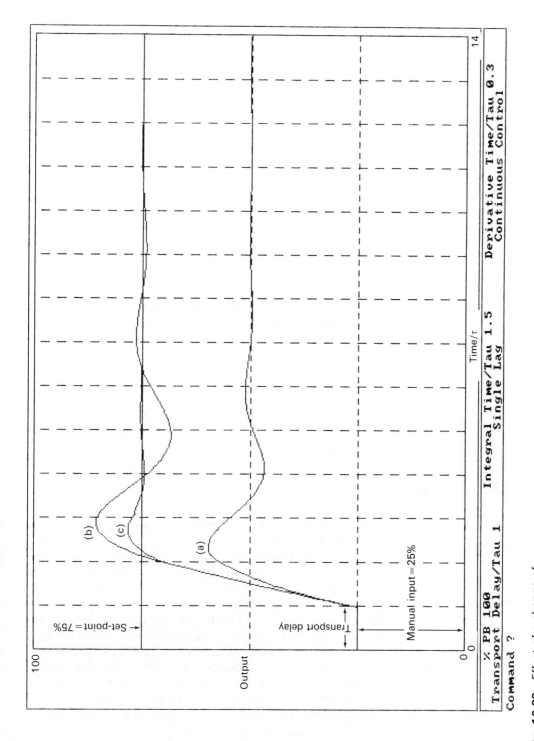

**Fig. 16.20**  Effect of a change of derivative time on system response

4. Integral action eliminates offset, but too rapid an integral action can reduce system stability.
5. Derivative action improves the damping of the system, so that it has a faster response and settles more quickly to its steady output.

The descriptions given above are in general terms, and it is not possible to alter the proportional band, integral and derivative control settings indiscriminately. In practice the proportional, integral and derivative settings are not completely independent of one another, and a change in one has some effect on the others.

CODAS and PCS simulation packages are particularly useful in studying the effects of controller settings.

**16.18   Introduction to operational amplifiers**

**Operational amplifiers**[†] (*op-amps*) are, strictly speaking, within the province of electronic engineering studies, but are so widely used in control engineering that we need to look at their basic operation.

A typical operational amplifier (see Fig. 16.21) has terminals for two separate input signals (both inputs being relative to earth), and an output terminal. It also requires two power supplies, one being positive with respect to earth, and the other negative.

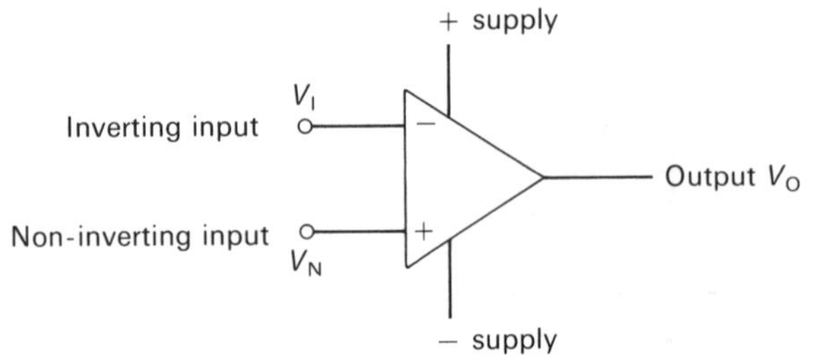

**Fig. 16.21**   An operational amplifier

When a signal is applied to the **non-inverting input**, $V_N$, the output voltage varies in sympathy with it. That is, if $V_N$ is positive, then $V_O$ is also positive; when $V_N$ is negative, then $V_O$ is negative. A signal applied to the **inverting input**, $V_I$, results in an output voltage of the opposite polarity. That is, if $V_I$ is positive, then $V_O$ is negative; when $V_I$ is negative, then $V_O$ is positive. The '+' sign inside the circuit symbol merely means that the input is the non-inverting input, and the '−' sign means that it is the inverting input. In the cases we consider, the '+' terminal is earthed.

The amplifier voltage gain between either input terminal and the output terminal is very high, typically 100 000 or more, and the impedance between the '−' and the '+' input terminals (known as the *differential input resistance*) is typically 1 MΩ.

---

[†] For more details see *Electronics III* by D. C. Green (Longman).

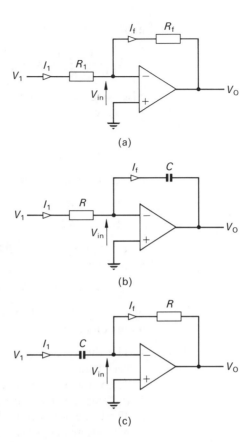

**Fig. 16.22** Operational amplifier circuits providing the basis of (a) proportional control, (b) integral control and (c) derivative action

The three circuits in Fig. 16.22 form the basis of proportional, integral and differential control, respectively, in process controllers. We will look at the three circuits and discuss their limitations.

**Proportional control [Fig. 16.22(a)]**

In this case, negative feedback is applied around the amplifier by resistor $R_f$. Let us assume for the moment that the two supply voltages to the amplifier are $\pm 10\,\text{V}$, so that the output voltage from the amplifier cannot exceed $\pm 10\,\text{V}$. Also if the gain of the amplifier is $10^6$, then the maximum voltage between the input terminals cannot be greater than $10/10^6 = 10\,\mu\text{V}$ if amplifier saturation is to be avoided. Since the '+' terminal is earthed then, for all practical purposes, the voltage at the '−' terminal is virtually zero; it is not zero, but it is virtually so. We therefore say that the phase-inverting input is a **virtual earth point**, that is $V_{in} \approx 0$. We will use this as the basis for the following discussion.

The current $I_1$ drawn from the input signal $V_1$ is

$$I_1 = \frac{V_1 - V_{in}}{R_1} \approx \frac{V_1 - 0}{R_1} = \frac{V_1}{R_1}$$

and

$$I_f = \frac{V_{in} - V_o}{R_f} \approx \frac{0 - V_o}{R_f} = -\frac{V_o}{R_f}$$

Also, since $V_{in} \approx 0$, no current flows into the amplifier, so that $I_f \approx I_1$, or

$$-\frac{V_o}{R_f} = \frac{V_1}{R_1}$$

hence the voltage gain of the amplifier is

$$\frac{V_o}{V_1} = -\frac{R_f}{R_1}$$

If, for example, $R_f = 1\,M\Omega$ and $R_1 = 100\,k\Omega$, then the gain of the amplifier is

$$V_o/V_1 = -1 \times 10^6/100 \times 10^3 = -10$$

or $V_o = -10V_1$, and the output voltage is 'proportional' to the input voltage. There are three important points to remember here:

1. The theory only works satisfactorily if the amplifier gain is very high.
2. The value of $R_1$ cannot be reduced to a 'low' value, otherwise the amplifier may be easily driven into saturation.
3. The value of $R_f$ cannot be too 'high', otherwise problems arise with electronic 'noise' in the system.

**Integral action control [Fig. 16.22(b)]**

Once again we will assume that the amplifier gain is very high, and that the phase-inverting terminal is a virtual earth point. The current $I_1$ flowing in $R$ is

$$I_1 = (V_1 - V_{in})/R \approx V_1/R$$

and

$$I_f = C \times \text{rate of change of capacitor voltage}$$

But, since $I_1 \approx I_f$, then

$$\frac{V_1}{R} = C \times \text{rate of change of capacitor voltage}$$

or

$$\text{Rate of change of capacitor voltage} = \frac{V_1}{RC}$$

Since the left-hand plate of the capacitor is earthed, and $I_f$ *flows out of the right-hand plate* then, for a positive input voltage, the output voltage is negative! That is

$$\text{Rate of change of } V_2 = -\frac{V_1}{RC} = -\frac{V_1}{T_I}$$

If, for example, $V_1$ is a constant (d.c.) voltage, then $V_2$ changes at a constant rate. That is, $V_2$ is the *integral* of $V_1$, and $RC$ is the integral time factor $T_I$ discussed earlier.

A limit to the use of an operational amplifier integrator circuit is that if a constant voltage is maintained at the input terminal, the amplifier will eventually saturate, and the circuit will no longer act as an integrator.

### Derivative action control [Fig. 16.22(c)]

As before, the amplifier is assumed to have a very high gain, and that the phase-inverting input terminal is a virtual earth point. In this case

$$I_1 = C \times \text{rate of change of capacitor voltage}$$

$$= C \times \frac{\text{d}}{\text{d}t} (V_1 - V_{\text{in}}) \approx C \frac{\text{d}V_1}{\text{d}t}$$

and

$$I_f = \frac{V_{\text{in}} - V_O}{R} \approx -\frac{V_O}{R}$$

Since $I_f \approx I_1$, then

$$-\frac{V_O}{R} = C \frac{\text{d}V_1}{\text{d}t}$$

or

$$V_O = -CR \frac{\text{d}V_1}{\text{d}t} = -T_D \frac{\text{d}V_1}{\text{d}t}$$

That is, the output voltage is the **derivative** of the input voltage, and is multiplied by $CR$, which is the *derivative time factor*, $T_D$, mentioned earlier.

Although the above argument is technically correct it has, unfortunately, a serious practical flaw. If we apply a sudden change of voltage to the input terminal of Fig. 16.22(c), the capacitor draws a large charging current from $V_1$. The value of the charging current is immaterial, because it has the effect of saturating the amplifier, so that the output voltage is not equal to the time differential of the input signal! We therefore have to revert to an *approximate differentiator circuit*, which involves connecting a resistor in series with the capacitor, as will be seen in section 16.19.

**16.19 PID controller using operational amplifiers**

A basic three-term PID analogue controller using an operational amplifier is shown in Fig. 16.23. Capacitor $C_1$ provides integral action and, when it is shorted out by $S_1$ (as shown), integral action is removed. Capacitor $C_D$ provides derivative action and, when $S_D$ is open-circuited (as shown), derivative action is removed.

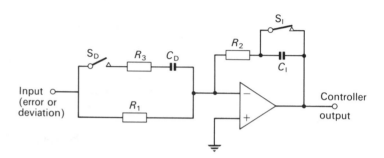

**Fig. 16.23** A basic electronic PID controller

Hence, with the switches in the position shown, the controller simply provides proportional control. Altering the value of $R_1$ (or $R_2$) modifies the gain of the amplifier, and therefore the proportional band of the controller.

When $S_1$ is opened, integral action is introduced; and when $S_D$ is closed, derivative action is introduced. In the latter case, the reader will note that $R_3$ has been included in series with $C_D$ to provide control over the problem of amplifier saturation mentioned in section 16.18.

**16.20 Digital process control**

Figure 16.24 shows a block diagram of a plant which is controlled by a digital controller. All the calculations relative to the control routine are carried out in the digital section of the system, the set-point being in digital (binary) form, and the measured value also being converted into digital form.

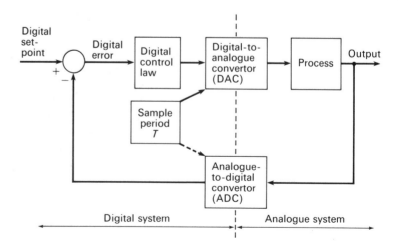

**Fig. 16.24** Basis of a digitally controlled process

The proportional band together with the integral and differential laws are controlled by the computer, so that the output signal from the controller is also in digital form. Later, we will look at the way in which the control law is implemented.

The digital output from the controller may be in the form of a series of pulses of varying height (representing the magnitude), as shown in Fig. 16.25(a), each being separated in time by the *sampling period, T.*

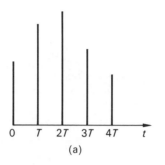 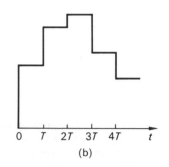

**Fig. 16.25** Simple digital-to-analogue conversion (DAC)

The digital output from the controller is converted to its analogue form by means of a **digital-to-analogue converter** (DAC) which, in turn, is controlled by the electronic sampling circuit. The DAC 'holds' the analogue output at the magnitude of the previous digital signal until the next pulse arrives at the DAC, giving an output of the form in Fig. 16.25(b). Devices which do this are known as **zero-order holds**. This signal is then applied to the process being controlled.

The process being controlled responds to this signal, and the measured analogue output from its output is applied to an **analogue-to-digital converter** (ADC) or **digitizer**. This device, under the control of the sampling pulses, samples the analogue signal at regular intervals of time, and digitizes it.

Providing that the sampling rate is sufficiently high, the controller works in much the same way as an analogue controller.

Let us take a look at a simplified flow chart (Fig. 16.26) for the control of such a system. After start-up, the controller would 'read' the set-point value, and would then 'read' the digital value from the ADC (the system output). These two values enable the error to be calculated, which is acted on by the control law of the controller. Once this has been done, the controller output is converted into analogue form and sent to the plant. This programming loop is then repeated time after time.

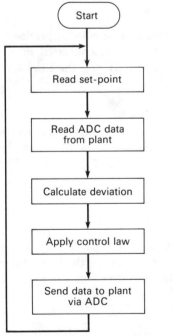

**Fig. 16.26** Simplified flow chart for a digitally controlled process

**16.21 Digital implementation of proportional action, integration and differentiation**

Computers are well suited to mathematical operations including addition, subtraction, multiplication, division, etc. It is simply a matter of combining these features in the software to provide a realization of proportional, integral and derivative action.

The relationship between the gain of the controller and its proportional band was discussed in section 16.14, and it is a fairly straightforward matter to obtain the proportional contribution of the output of the controller.

Assuming that the digital deviation signal has been converted into a series of pulses whose height represents the magnitude of the deviation (see Fig. 16.27), the software writer can use any one of several methods to produce *integral action*.

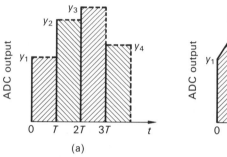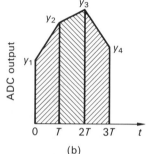

**Fig. 16.27** Digital approximation to integration using (a) rectangular approximation, (b) trapezoidal approximation

The simplest is the **rectangular approximation to integration** in Fig. 16.27(a). In this method, each pulse is converted into a rectangle of width $T$ (the sampling period), and the total area between $t = 0$ and $t = 3T$ is

$$T(y_1 + y_2 + y_3)$$

An alternative method is the **trapezoidal approximation to integration** in which each pulse is assumed to be linked to the following pulse by a straight line, as shown in Fig. 16.27(b). The total area between $t = 0$ and $t = 3T$ is

$$\frac{T}{2}(y_1 + y_4) + T(y_2 + y_3)$$

The *derivative* of the signal at any point in time between a pair of samples can be calculated from the slope of the line joining the two samples. For example, the derivative (the slope) of the graph between $t = T$ and $t = 2T$ is

$$(y_3 - y_2)/T$$

By combining the above it is possible, by software, to provide a good approximation to an analogue PID controller.

**16.22  Computer control of processes**

Increasingly, computer control is being used to replace analogue technology. A strategy known as **computer supervisory control** or **set-point control** uses microcomputers to control individual items of plant, the main computer or **supervisory computer** sending set-point information to **local computers** or **slave computers**. One advantage of this scheme over a completely computer-controlled system is that the slave computers continue to operate in the event of a breakdown either of the main computer or of the data link between them.

Another strategy known as **direct digital control** (DDC) uses a mainframe computer to control the complete system, as illustrated in Fig. 16.28.

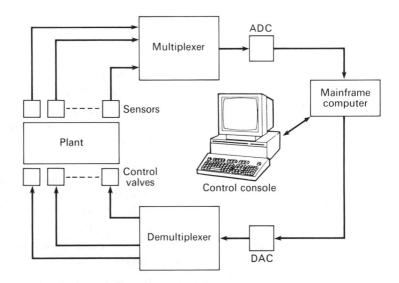

**Fig. 16.28** Direct digital control

In this method of control, data from all the sensors on the plant is sent to a **multiplexer**, which is an electronic version of a 'rotating contact' which samples the condition of each sensor for a period of time long enough to obtain a reliable reading. Each signal is digitized or quantized by an ADC, and sent on to the main computer for analysis. Finally, signals are sent from the software 'controllers' in the mainframe computer to a **demultiplexer**, which can also be thought of as another electronic 'rotating contact'. This device directs each signal to the correct control valve to govern the operation of the plant. This routine is repeated over and over again.

### Problems

**16.1** Explain the difference between an open-loop system and a closed-loop system. Describe the relative advantages and disadvantages of each.

**16.2** Draw a block diagram of a closed-loop control system, and describe the function of each element in the loop.

**16.3** Give four examples of closed-loop systems, explain the purpose of each system, and how each one functions.

**16.4**   Discuss the purpose of regulators and servosystems, and describe their similarities and differences.

**16.5**   If the amplifier gain in the control system in Fig. 16.5 is (a) halved, (b) doubled, calculate the value of the closed-loop transfer function in (rev/min)/V.

[(a) 50; (b) 80]

**16.6**   For Fig. 16.5 calculate, for a 10 V set-point voltage, (i) the steady-state output speed, and (ii) the steady-state error voltage.
[(a) (i) 500 rev/min, (ii) 800 rev/min; (b) (i) 5 V, (ii) 2 V]

**16.7**   What set-point voltage is required for the system in Fig. 16.5 to give an output speed of 500 rev/min?

[7.5 V]

**16.8**   Describe two non-engineering closed-loop systems other than the economic cycle in section 16.3.

**16.9**   What is meant by (a) a first-order control system and (b) a second-order control system. Sketch and describe the transient response of both types of system when a step change in input signal is applied.

**16.10**   In connection with control systems describe, with the aid of sketches of output response, what is meant by (a) an overdamped system, (b) a critically damped system, (c) an underdamped system, (d) an undamped system and (e) an unstable system.

**16.11**   With the aid of a block diagram, describe a process control system. Identify the transducers, controller, actuator and plant, and describe the function of each.

**16.12**   Explain the meaning of exponential lag and distance/velocity lag in association with a control system.

**16.13**   Describe what is meant by the *control law* of a controller. Explain the meaning of proportional band, integral control and differential control; what purpose does each perform in a PID controller?

**16.14**   Draw a circuit diagram of an operational amplifier circuit which can provide (a) proportional band control, (b) integral control and (c) differential control. Describe how each circuit carries out its function. A basic form of analogue PID controller is shown in Fig. 16.23; explain the purpose of each element in the circuit.

**16.15**   Describe how proportional band, integral control and differential control are implemented in a digital controller.

**16.16**   Outline the basis of direct digital control of a industrial process; explain how it is implemented.

# 17 Measuring instruments and measurements

## 17.1 Introduction

Instruments may be categorized not only by the quantity measured (i.e. voltage, current, power), but also by the way in which the measured quantity is displayed (i.e. in analogue or digital form). It is also possible to distinguish between instruments in terms of whether they are d.c. (mean) reading or r.m.s. reading.

Instruments are also categorized according to their application, i.e. as panel meters or as laboratory instruments, and are further subdivided by the accuracy of the displayed result.

An instrument is described as an **analogue instrument** when the indication is given by the position of a pointer on the face of the instrument scale. A **digital instrument** is one whose display is in the form of a series of decimal numbers. The choice between analogue and digital instruments involves many factors including the overall accuracy, the reliability of the instrument, its maintainability, etc.

## 17.2 Analogue instruments

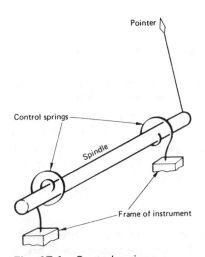

**Fig. 17.1** Control springs (deflecting system omitted for simplicity)

Most analogue instruments depend for their **deflecting torque** on the force acting on a current-carrying conductor in a magnetic field. Instruments in this category include *permanent-magnet moving-coil meters* (*galvanometers*) and *electrodynamic* (*dynamometer*) *instruments*.

Another popular range of instruments, known as *moving-iron instruments*, depend for their deflecting torque on the attraction or repulsion between pieces of magnetized iron.

Once an instrument has developed its deflecting torque, it is necessary to ensure that the pointer moves to a point on the scale which indicates the value of the measured quantity. That is, the result must be **repeatable** when the measured quantity next has the same value. Repeatability is ensured by the **controlling force** of the instrument. This is provided in many instruments by **spring control**; in the case where two control springs are used (see Fig. 17.1), movement of the instrument shaft increases the tension in one spring and reduces it in the other.

An additional (and essential) feature of practical measuring instruments is a means of **damping** the movement. This ensures that the pointer comes quickly to its final position without excessive oscillation. There are two principal forms of damping, namely:

(a) air vane damping; and
(b) eddy-current damping.

One form of **air vane damping** is shown in Fig. 17.2(a); in this case the movement is damped by the movement of a light vane or piston (usually made of aluminium) inside a cylinder. A gap is left between the vane and the cylinder to ensure that the damping is not excessive. This type of damping is widely used in moving-iron and electrodynamic instruments.

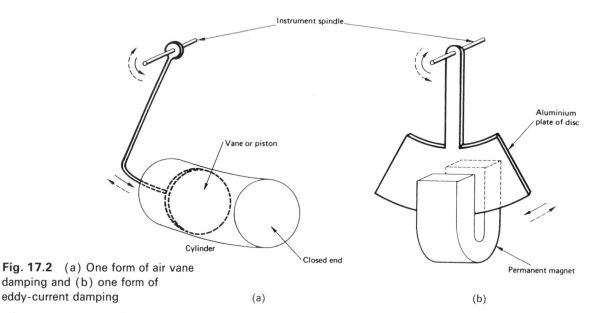

**Fig. 17.2**  (a) One form of air vane damping and (b) one form of eddy-current damping

(a)

(b)

In **eddy-current damping systems**, the movement of the instrument spindle causes a light conductor (usually aluminium) to move between the poles of a permanent magnet. The eddy current which is induced in the conductor dissipates energy, causing the instrument to be damped. One form of eddy-current damping is shown in Fig. 17.2(b). Eddy-current damping is used in moving-coil instruments (galvanometers), but in this case the eddy-current element is the aluminium former on which the coil of the instrument is wound (see Fig. 17.3).

**17.3  Permanent-magnet moving-coil instruments or galvanometers**

The permanent-magnet moving-coil instrument is a **direct current milliammeter** (or a microammeter) and, for reasons given below, requires additional electrical circuits to enable it to measure alternating current. Since the instrument can, in its normal form,

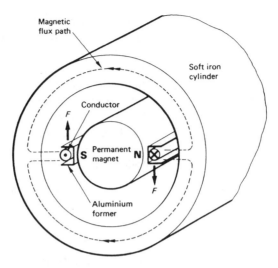

**Fig. 17.3** The basis of an 'internal' magnet moving-coil meter

only measure direct current and voltage, the manufacturer marks the polarity to be used at its terminals.

**In general, current must enter a terminal marked with a (+) sign, and leave a terminal marked with a (−) sign.**

The basis of one form of moving-coil instrument is shown in Fig. 17.3. The design is known as an **internal magnet instrument**, in which the coil (shown as a single conductor for simplicity) is wound on an aluminium former (for damping purposes) which can rotate around a cylindrical permanent magnet.

The movement is surrounded by a soft iron cylinder. The function of the cylinder is twofold; firstly, it ensures that the magnetic field between the magnet and the iron ring is radial and, secondly, it provides a magnetic screen around the instrument. The latter has the effect of reducing the possibility of errors arising from stray external magnetic fields. Since the magnetic field is radial at all points, the force on the conductors is uniform, allowing the instrument to have a linear scale calibration.

Some instruments use an **external magnet** construction, in which the positions of the magnet and the soft iron are interchanged.

In instruments in which the spindle of the instrument is supported by jewelled bearings, current is fed to both ends of the moving coil by two contra-wound springs (see also Fig. 17.1); these springs provide the controlling torque of the meter. In some instruments, the spindle (and coil) is supported by a taut beryllium–copper ribbon at either end of the spindle. This type is described as having a **pivotless suspension**, which virtually eliminates the frictional loss associated with pivoted supports. At the same time, the taut ribbon also provides not only a means of getting the current into the coil, but also the controlling torque of the meter.

When current flows in the direction shown in Fig. 17.3, the right-hand conductor experiences force $F$ in a downwards direction (reminder – use Fleming's left-hand rule), and the left-hand conductor experiences a force in the upwards direction. Since the magnetic field is radial, the torque on the coil causes it to rotate clockwise. When the torque produced by the coil just balances the torque in the controlling springs, the pointer reaches its steady deflection.

The equation of the moving-coil system can be deduced as follows. If

$B$ = flux density in the air gap;

$i$ = instantaneous current in the coil;

$N$ = number of *active conductors* on the coil;

$l$ = *active length* of one conductor;

$r$ = mean radius of the coil;

then the force $F_1$ acting on one conductor is

$$F_1 = Bil$$

and, since there are $N$ active conductors on the coil, the total force developed by the coil is

$$F_T = BilN$$

Hence, the gross torque, $T_G$, developed by the coil is

$$T_G = F_T \times \text{radius} = BilNr \qquad [17.1]$$

The controlling torque $T_C$ developed by the controlling springs is proportional to the angular deflection $\theta$ of the spring. That is

$$T_C = k\theta \qquad [17.2]$$

where $k$ is a deflecting constant of the spring, usually having the dimensions $N\,m/rad$ (or $N\,m/deg$). Under steady operating conditions, $T_C = T_G$, that is

$$k\theta = BilNr$$

hence

$$\theta = \frac{BlNr}{k} \times i \qquad [17.3]$$

In any given moving-coil instrument, the quantities $B$, $l$, $N$, $r$ and $k$ are constant, so that

$$\theta \propto i$$

That is, the deflection of the pointer is proportional to the current flowing through the meter. For this reason, **the scale of a moving-coil instrument is linearly calibrated**.

Although not apparent from the above discussion, the moving-coil instrument gives improved accuracy at high deflection. Thus, if a 2 A and a 1 A instrument are available, and we wish to measure 0.8 A, the 1 A meter should be selected.

The moving-coil instrument is unsuitable for the direct measurement of alternating current and voltage, because the alternating current produces a pulsating torque of the instrument movement. Above a frequency of about 10 Hz, the instrument gives zero deflection. However, if the meter is used in connection with a rectifier, it can be used to measure alternating voltage and current.

*Worked example 17.1*    The flux density in the air gap of a moving-coil instrument is 0.26 T, and the coil has $48\frac{1}{2}$[†] turns. The effective length of each conductor is 15 mm and the mean radius of the coil is 7 mm. If the constant of the control spring is $1.375 \times 10^{-6}$ N m/rad and the current in the coil is 0.8 mA, determine the steady angular deflection of the coil in degrees.

*Solution*    In this case $B = 0.26\,\mathrm{T}, l = 15 \times 10^{-3}\,\mathrm{m}, \mathrm{N} = 2 \times 48\frac{1}{2} = 97, r = 7 \times 10^{-3}\,\mathrm{m},$ $i = 0.8 \times 10^{-3}\,\mathrm{A}, k = 1.375 \times 10^{-6}$ N m/rad. From eqn [17.3]

$$\theta = \frac{BlNri}{k}$$

$$= \frac{0.26 \times (15 \times 10^{-3}) \times 97 \times (7 \times 10^{-3}) \times (0.8 \times 10^{-3})}{1.375 \times 10^{-6}}$$

$$= 1.54\,\mathrm{rad} \quad \mathrm{or} \quad 88.3°$$

## 17.4 Dynamometer (electrodynamic) instruments

A **dynamometer instrument** (see Fig. 17.4) is a form of moving-coil instrument which does not have a permanent magnet, the magnetic field being provided by means of a current $I_1$ which flows through a pair of coils which are fixed to the frame of the instrument (the **fixed coils**). Current $I_2$ flows through a third coil (the **moving coil**) which is pivoted about its centre, and the interaction between the two magnetic fields gives rise to a torque which results in the moving coil rotating about its axis.

The controlling torque is produced by springs (which are also the means of supplying current to the moving coil) attached to the spindle of the moving coil. Damping is provided by means of an air vane damper attached to the spindle of the instrument. It can be shown for this instrument that

Mean deflecting torque $\propto I_1 \times I_2$

---

[†]Since the current enters the coil via a spring at one end of the spindle, and leaves via a spring at the other end, the winding has an odd half-turn of wire.

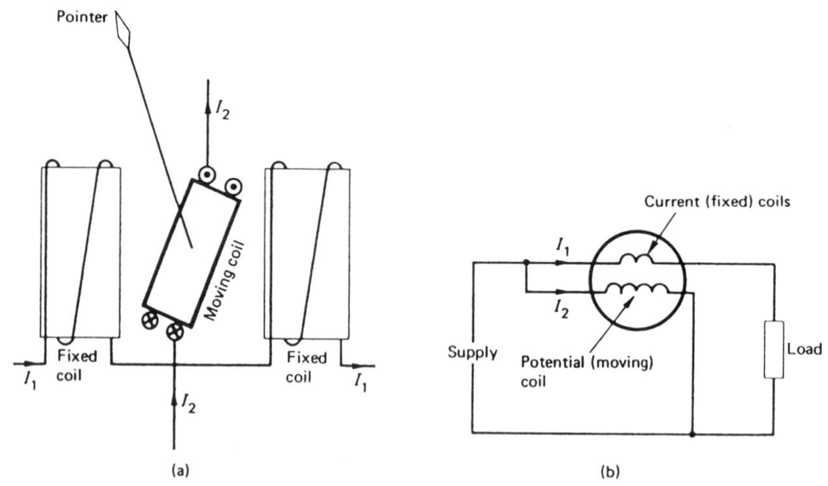

**Fig. 17.4** (a) The basis of a dynamometer instrument and (b) connection diagram for use as a wattmeter

In addition to being able to measure d.c. quantities, it can also be used to measure a.c. quantities for the following reason. When both coils are excited by alternating current, the magnetic flux produced by both coils reverses simultaneously. The reader will find it an interesting exercise to show that, under these circumstances, the direction of the torque produced remains unchanged. Consequently, the dynamometer instrument is an a.c./d.c. meter.

The most popular application for the dynamometer instrument is as a **wattmeter** [for connections see Fig. 17.4(b)]. In this case the fixed coils are wound with a few turns of large-section wire and carry the load current. The moving coil circuit has a high resistance, and the coil has many turns of fine wire. The supply voltage is connected to the moving coil, so that current $I_2$ is proportional to the voltage, $V_L$, across the instrument. Hence

$$\text{Deflecting torque} \propto I_1 \times I_2 \propto I \times V_L$$

$$\propto \text{load current} \times \text{load voltage}$$

$$\propto \text{power}$$

That is, the deflecting torque is proportional to the *average power* consumed by the load.

## 17.5 Moving-iron instruments

The majority of moving-iron instruments have a **repulsion-type movement** (see Fig. 17.5), in which the force of repulsion between two similarly magnetized iron rods or vanes causes the pointer to move. One of the vanes is fixed to the frame of the meter, and the other vane is attached to the spindle. When current passes through the coil, the two vanes are similarly magnetized and repel one another.

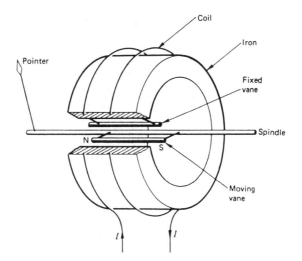

**Fig. 17.5** The basis of a repulsion-type moving-iron instrument

The controlling force of the meter is provided by a spring, and since current does not flow in the moving member, only one spring is necessary. Damping is usually provided by means of an air vane system.

In a simple instrument of this kind, it can be shown that the deflecting torque is proportional to the square of the current in the coil. As a result, the scale calibration is non-linear, being very cramped at the lower end of the scale. However, by using a wedge-shaped fixed vane, the scale calibration can be made almost linear over the majority of the scale.

The flux produced by the coil is related to the number of ampere-turns associated with the coil. Provided that the number of turns to give **full-scale deflection** (f.s.d.) is known for a particular design of meter, then it is possible to calculate the number of turns on the coil to give f.s.d. for a particular current.

*Worked example 17.2*  A moving-iron instrument requires an m.m.f. of 200 ampere-turns to give f.s.d. Determine the number of turns on the coil to give f.s.d. for a current of (a) 2 A, (b) 10 A.

*Solution*  In this case $NI = 200$ A or ampere-turns.
(a) $I = 2$ A, hence

$$N = 200/I = 200/2 = 100 \text{ turns}$$

(b) $I = 10$ A, hence

$$N = 200/10 = 20 \text{ turns}$$

**17.6  Extending the instrument range of a moving-coil meter**

A moving-coil meter is basically a low-current galvanometer, and only needs a low value of p.d. between its terminals to give f.s.d. To enable the instrument to *measure a large value of current* (say

10 A or 100 A), the meter must be shunted by a low resistance in order to shunt most of the current from the meter. Alternatively, if we need to *measure a high value of voltage* (say 100 V or 1 kV), it is necessary to connect a high value of resistance in series with the meter; this resistor 'drops' the majority of the voltage when current flows through the meter.

### 17.6.1 Extending the current range

The moving-coil meter is only capable of handling a small value of current, and when large currents are to be measured a low-resistance **shunt resistor**, $S$ (see Fig. 17.6), is connected in parallel with the instrument. The reader should note that the current enters the '+' terminal of the instrument. The following notation is adopted in Fig. 17.6:

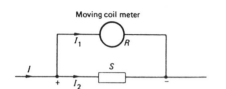

**Fig. 17.6** The moving-coil instrument as an ammeter

$I$ = current in the external circuit when the meter gives f.s.d.;

$I_1$ = current in the meter to give f.s.d.;

$I_2$ = current in the shunt when the meter gives f.s.d.;

$R$ = resistance of the meter;

$S$ = resistance of the shunt.

Applying KCL to any terminal of the meter gives

$$I = I_1 + I_2 \qquad [17.4]$$

Also, since the meter and the shunt are in parallel with one another, the same p.d. appears across both of them, hence

$$I_1 R = I_2 S \qquad [17.5]$$

The resistance of the shunt $S$ can be determined from eqns [17.4] and [17.5] as follows:

$S$ = p.d. across shunt/current in shunt

$$= \frac{I_1 R}{I_2} = \frac{I_1 R}{I - I_1} \qquad [17.6]$$

The shunt resistor must be manufactured to a very high degree of precision, and must ideally have a very low temperature coefficient of resistance; the latter ensures that the resistance of the shunt does not alter significantly with temperature change. However, the coil of the meter is wound with wire whose resistance does vary with temperature. The variation in meter resistance with temperature affects the 'current multiplying' ratio of the shunt, leading to a variation in the accuracy of the meter with temperature variation. To reduce error from this cause, a resistor known as a **swamp resistor**, $R_{\text{swamp}}$ (see Fig. 17.7) is connected in series with the meter; the resistance of the 'swamp' is typically three times the resistance of the meter movement.

*Worked example 17.3*　A moving-coil instrument gives f.s.d. for a current of 10 mA and has a resistance of 10 Ω. Calculate (a) the resistance of a shunt which enables the meter to be used to give an f.s.d. of 0.5 A, (b) the power consumed by the combination at f.s.d. and (c) the current in the main circuit when the p.d. across the instrument is 0.08 V.

*Solution*　The data given is $I_1 = 10 \times 10^{-3}$ A, $R = 10 \Omega$, $I = 0.5$ A.

(a) From eqn [17.6]

$$S = I_1 R/(I - I_1) = (10 \times 10^{-3}) \times 10/(0.5 - 10 \times 10^{-3})$$

$$= 0.2041 \ \Omega$$

(b) The power consumed can be calculated as follows. The equivalent resistance of the meter is

$$R_T = RS/(R + S) = 10 \times 0.2041/(10 + 0.2041)$$

$$= 0.2 \ \Omega$$

The total power consumed is

$$P_T = (\text{total current})^2 \times \text{equivalent resistance}$$

$$= 0.5^2 \times 0.2 = 0.05 \ \text{W}$$

(c) Since the p.d. across the instrument under the condition specified is 0.08 V, the current through the meter itself at this time is

$$I_m = 0.08 \ \text{V}/10 \ \Omega = 0.008 \ \text{A}$$

The meter current $I_1$ to give f.s.d. is 10 mA or 0.01 A and, since deflection $\propto$ current, then

$$\frac{\text{Deflection for } I_m}{\text{Deflection of } I_1 \text{ (f.s.d.)}} = \frac{I_m}{I_1}$$

or

$$\text{Deflection for } I_m = \frac{I_m}{I_1} \times \text{f.s.d.} = \frac{0.008}{0.01} \times \text{f.s.d.}$$

$$= 0.8 \times \text{f.s.d.} = 0.8 \times 0.5 = 0.4 \ \text{A}$$

### 17.6.2　The universal shunt

The use of a **universal shunt** (see Fig. 17.7) enables the user to measure any current within the range of an instrument. The reader is advised that, when measuring current with a **multi-range instrument**, it is advisable to *switch the meter to its highest current range* before connecting the instrument in the circuit, and then gradually reduce the current range until a satisfactory reading is obtained. The switching arrangement of the meter enables current ranges to be changed (a) without interrupting the current, and (b) without momentarily causing an excessive current to pass through the meter.

In the following we consider the design of a universal shunt

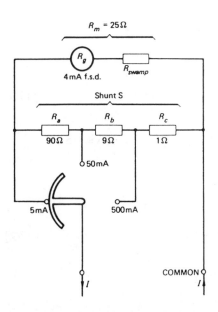

**Fig. 17.7**　A multi-range ammeter incorporating a universal shunt

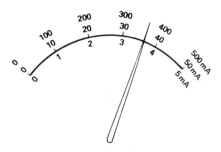

**Fig. 17.8** Meter scale for the multi-range ammeter in Fig. 17.7

suitable for a meter having an f.s.d. of 4 mA. The ranges on the resulting meter are to be 5, 50 and 500 mA respectively; the meter scales are shown in Fig. 17.8.

The resistance $R_m$ of the meter section in Fig. 17.7 is

$$R_m = R_g + R_{swamp}$$

where $R_g$ is the resistance of the galvanometer itself, and $R_{swamp}$ is the resistance of the swamp resistor.

The meter section is shunted by shunt $S$, which comprises resistors $R_a$, $R_b$ and $R_c$. The current needed to give f.s.d. on the meter is usually fairly close to the most sensitive current range on the multi-range instrument. For example, the galvanometer current to give f.s.d. in this case is 4 mA, while the most sensitive current range on our multi-range meter is 5 mA. If $R_m$ is 25 Ω, then the p.d. across the meter at f.s.d. is

$$4 \text{ mA} \times 25 \, \Omega = 0.1 \text{ V}$$

Under this condition the current in the main circuit is 5 mA, so that the current in the shunt is 1 mA. Hence

$$R_a + R_b + R_c = \frac{\text{p.d. across the shunt at f.s.d.}}{\text{shunt current at f.s.d.}}$$

$$= 0.1/1 \times 10^{-3} = 100 \, \Omega \qquad [17.7]$$

It can be shown for this type of shunt that

Most sensitive range $\times (R_a + R_b + R_c)$

$$= \text{next least sensitive range} \times (R_b + R_c)$$

$$= \text{least sensitive range} \times R_c$$

That is

$$5 \text{ mA} \times (R_a + R_b + R_c) = 50 \text{ mA} \times (R_b + R_c)$$

$$= 500 \text{ mA} \times R_c \qquad [17.8]$$

hence

$$5 \text{ mA} \times (R_a + R_b + R_c) = 5 \times 10^{-3} \times 100 = 0.5 \text{ V}$$

From eqn [17.8]

$$500 \text{ mA} \times R_c = 0.5$$

or

$$R_c = 0.5/500 \times 10^{-3} = 1 \, \Omega \qquad [17.9]$$

Also from eqn [17.8]

$$50 \text{ mA} \times (R_b + R_c) = 0.5$$

or

$$R_b + R_c = 0.5/50 \times 10^{-3} = 10 \, \Omega \qquad [17.10]$$

From eqns [17.9] and [17.10]

$$R_b = 10 - R_c = 10 - 1 = 9\ \Omega$$

and from eqns [17.7] and [17.10]

$$R_a = 100 - (R_b + R_c) = 100 - 10 = 90\ \Omega$$

### 17.6.3 Extending the voltage range

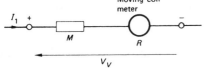

**Fig. 17.9** The moving-coil instrument as a voltmeter

At f.s.d., the p.d. across the movement of a moving-coil meter is only a fraction of a volt. In order to measure a high value of voltage, a resistor $M$ (see Fig. 17.9) known as a **voltage multiplier resistor**, is connected in series with the instrument. The following notation is adopted in Fig. 17.9:

$V_V$ = voltage applied to the terminals to produce f.s.d.;

$I_1$ = current to give f.s.d.;

$R$ = resistance of the galvanometer movement;

$M$ = resistance of the multiplier resistor.

Applying KVL to Fig. 17.9 gives

$$V_V = I_1(M + R) \qquad\qquad [17.11]$$

hence

$$M = \frac{V_V}{I_1} - R \qquad\qquad [17.12]$$

***Worked example 17.4*** A moving-coil instrument gives f.s.d. when a current of 1 mA flows through it. If the instrument resistance is 10 Ω, calculate the value of the multiplier resistor to enable it to have an f.s.d. of 300 V. Calculate the power consumed by the voltmeter when it indicates 240 V.

*Solution* From the data given $V_V = 300\ \text{V}$, $I_1 = 1 \times 10^{-3}\ \text{A}$, $R = 10\ \Omega$. From eqn [17.12]

$$M = \frac{V_V}{I_1} - R = \frac{300}{1 \times 10^{-3}} - 10 = 300\,000 - 10$$

$$= 299\,990\ \Omega$$

The total resistance of the voltmeter is

$$R_V = M + R = 300\,000\ \Omega$$

and the power consumed by the voltmeter when 240 V is applied to it is

$$P = V^2/R_V = 240^2/300\,000 = 0.192\ \text{W}$$

### 17.6.4  Voltmeter sensitivity

The sensitivity of a voltmeter is often quoted in **ohms per volt** (o.p.v.). This is defined as follows:

Sensitivity in ohms per volt

$$= \frac{\text{total resistance of the voltmeter}}{\text{voltage required to give f.s.d.}}$$

$$= \frac{M + R}{V_V} = \frac{R_V}{V_V}$$

The voltmeter in Worked Example 17.4 has a sensitivity of

$$300\,000 \ \Omega / 300 \ \text{V} = 1000 \ \Omega / \text{V}$$

The resistance of the voltmeter can also be calculated from the equation as follows:

$$R_V = \text{o.p.v.} \times \text{f.s.d. in volts}$$

From eqn [17.11] we have

$$\frac{1}{I_1} = \frac{1}{\text{current to give f.s.d.}} = \frac{M + R}{V_V}$$

$$= \text{voltmeter sensitivity in o.p.v.}$$

Applying this expression to the voltmeter in Worked Example 17.4 shows that its sensitivity is

$$1/1 \ \text{mA} = 1000 \ \Omega / \text{V}$$

The 'goodness' of a voltmeter is indicated by its sensitivity, and a 'good' voltmeter has a high o.p.v. sensitivity. A high-quality multi-range instrument may typically have a sensitivity of $20\,000 \ \Omega / \text{V}$ (corresponding to an f.s.d. of $50 \ \mu\text{A}$).

### 17.6.5  A multi-range voltmeter

**Fig. 17.10** (a) Circuit of a typical multi-range voltmeter and (b) instrument scales

A circuit of a typical three-range voltmeter is shown in Fig. 17.10(a). In this case a $50 \ \mu\text{A}$ movement is used, the resistance

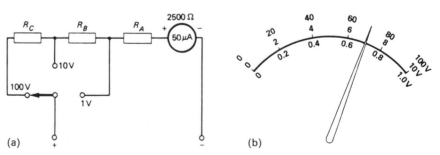

(a)          (b)

of the meter being 2500 Ω. Using the theory described earlier, it will be found that a multiplier of $R_A = 17\,500\,\Omega$ gives f.s.d. when 1 V is applied to the terminals of the meter, an additional multiplier of $R_B = 180\,000\,\Omega$ gives a 10 V range, and a further multiplier resistor of $R_C = 1.8\,M\Omega$ gives a 100 V range. The voltmeter has a sensitivity on all ranges of $20\,000\,\Omega/V$ (corresponding to an f.s.d. current of $1/20\,000\,\Omega/V = 50\,\mu A$).

## 17.7 Errors due to instrument connection in a circuit

All electrical instruments have some resistance. If it is a voltmeter, the resistance of the meter has some shunting effect on the circuit to which it is connected, with the result that the indicated voltage is lower than is the case without the voltmeter. If the instrument is an ammeter, the resistance of the meter increases the circuit resistance, and the measured value of current is less than is the case without the ammeter.

The reader should note that the phrase 'is less than' is a relative term, and can mean anything from a negligible change up to a large change. We look at this in the following sections.

### 17.7.1 The loading effect of a voltmeter

Consider the circuit in Fig. 17.11(a). Applying Ohm's law to the circuit, we see that the current is

$$I_1 = \frac{10\ V}{(10\,000 + 10\,000)\,\Omega} = 0.0005\ A \quad or \quad 0.5\ mA$$

and the voltage across $R_2$ is

$$V_2 = I_1 R_2 = 0.0005 \times 10\,000 = 5\ V$$

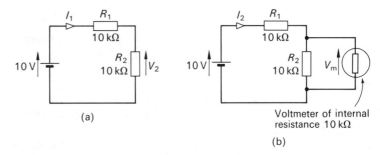

**Fig. 17.11** Resistive circuit (a) before a voltmeter is connected across $R_2$ and (b) with the voltmeter connected

(a)

Voltmeter of internal resistance 10 kΩ

(b)

If we attempt to measure $V_2$ using a voltmeter with an internal resistance of 10 kΩ [see Fig. 17.11(b)], we effectively shunt $R_2$ by 10 kΩ; the effective resistance of the parallel section of the circuit is 10 kΩ in parallel with 10 kΩ, or 5 kΩ. The new current in the circuit is therefore

$$I_1 = \frac{10\,V}{(10\,000 + 5\,000)\Omega} = 0.000667\,A \quad \text{or} \quad 0.667\,mA$$

That is, the voltmeter has **loaded** the circuit, causing it to draw more current from the supply. The voltage indicated by the voltmeter is equal to that across the parallel section of the circuit ($R_2$ in parallel with $R_V$), and is

$$V_m = I_2 \times \text{parallel circuit resistance}$$

$$= 0.000667 \times 5000 = 3.335\,V$$

and is 33.33 per cent low! Clearly, the connection of the voltmeter has resulted in the voltage across $R_2$ being reduced (and the subsidiary effect of increasing the voltage across $R_1$!).

As a general rule of thumb, **for a voltmeter to have little effect on the value of voltage it is used to read, the resistance of the voltmeter should be about 100 times greater than the resistance it shunts**.

Without going into the mathematics of the circuit, the correct voltage across $R_2$ can be calculated from the following formula:

$$\text{Correct voltage, } V_2 = V_m \left[ 1 + \frac{R_1 R_2}{R_V(R_1 + R_2)} \right] \qquad [17.13]$$

where

$$\frac{R_1 R_2}{R_V(R_1 + R_2)}$$

is a correction term. If we insert the values from Fig. 17.11(b) we get

$$V_2 = 3.335 \left[ 1 + \frac{10\,000 \times 10\,000}{10\,000(10\,000 + 10\,000)} \right]$$

$$= 3.335[1 + 0.5] = 5.0025\,V$$

The difference between this value and 5 V is due to rounding errors involved in the calculation.

### 17.7.2 Effect of connecting an ammeter in a circuit

The current in the simple circuit in Fig. 17.12(a) is

$$I_1 = 10\,V/0.4\,\Omega = 25\,A$$

If an ammeter of internal resistance $0.1\,\Omega$ is connected in the circuit [Fig. 17.12(b)], the current in the circuit is

$$I_2 = 10\,V/(R_1 + R_A) = 10/(0.4 + 0.1) = 20\,A$$

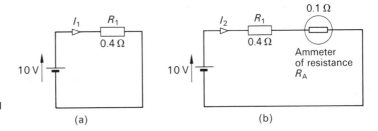

**Fig. 17.12** The effect of connecting an ammeter into a circuit

and this is the current which would be indicated by the meter! In this case, the ammeter reading is 20 per cent low!

As a general rule of thumb, **for an ammeter to have little effect on the current in the circuit to which it is connected, the resistance of the ammeter should be less than about 1 per cent of the resistance of the remainder of the circuit.**

Once again, without going into the mathematics of the matter, the current flowing in the circuit prior to the connection of the ammeter can be calculated from the following formula:

$$I_1 = I_2\left[1 + \frac{R_M}{R_1}\right] \qquad [17.14]$$

where

$$\frac{R_M}{R_1}$$

is a correction term. Inserting the values from Fig. 17.12(b) gives

$$I_1 = 20\left[1 + \frac{0.1}{0.4}\right] = 25 \text{ A}$$

**17.8 The moving-coil instrument as an ohmmeter**

The resistance of a resistor can be determined by measuring the current flowing in it when a voltage is applied to it. This is the basis of many ohmmeters. A simple ohmmeter circuit is shown in Fig. 17.13(a), and comprises a moving-coil meter in series with a battery (which is part of the ohmmeter) and a current-limiting resistor $R_L$ (which is the **set zero control** of the ammeter). The value of $R_L$ is manually adjusted to a value which allows the maximum current (i.e. f.s.d.) to flow through the meter when the test leads are short-circuited (1 mA in the case considered here).

It is worth pointing out that the polarity of the 1.5 V 'internal' battery results in the '+' terminal of the instrument being negative with respect to the '−' terminal. This fact is of importance when testing many types of semiconductor device such as diodes and transistors.

When the resistance of the 'unknown' resistor $R_U$ is zero (corresponding to the case when the test leads are short-circuited),

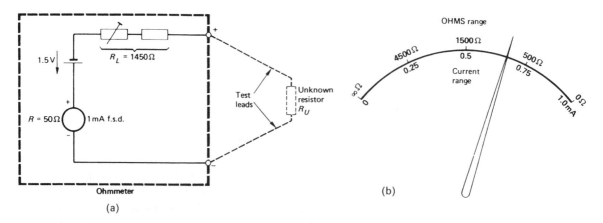

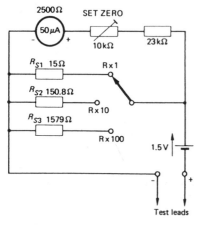

**Fig. 17.13** (a) Simple ohmmeter circuit and (b) its scale calibration

the meter current is a maximum. Thus *full-scale deflection corresponds to zero ohms*. When the value of $R_U$ is infinity (an open circuit), the meter current is zero. That is, *zero deflection corresponds to infinite resistance* [see Fig. 17.13(b)].

With the values in the circuit in Fig. 17.13(a), and $R_U = 1500\,\Omega$, the current in the circuit is

$$I = 1.5\,\text{V}/(R + R_L + R_U) = 1.5/(50 + 1450 + 1500)$$

$$= 0.5 \times 10^{-3}\,\text{A} \quad \text{or} \quad 0.5\,\text{mA}$$

That is, the deflection is one-half f.s.d. [see Fig. 17.13(b)]. It is the case in ohmmeters of this type that *the resistance indicated at one-half f.s.d. is equal to the total internal resistance of the meter*.

The reader should note from Fig. 17.13(b) that the resistance scale calibration is non-linear, being fairly open at the low-resistance (high current) end, and cramped at the high-resistance (low current) end.

Resistance values in both electrical and electronic engineering range from very low to very high values; practical ohmmeters must therefore be capable of measuring a wide range of resistance values. The basis of a practical multi-range ohmmeter is shown in Fig. 17.14. The instrument has a $10\,\text{k}\Omega$ SET ZERO control which must be manually adjusted when (a) the internal battery voltage changes with age, and (b) the ohmmeter is switched to another range (since this alters the resistance value in the circuit). The SET ZERO control must, of course, be initially adjusted with the test leads short-circuited.

**Fig. 17.14** Typical ohmmeter circuit

## 17.9 Extending the range of a moving-iron meter

Shunts are rarely used to alter the current range of a moving-iron ammeter because, to ensure accurate current division between the shunt and the meter, the shunt must have the same ratio of inductance to resistance ($L/R$) as that of the meter. One method of obtaining different ranges of current is to have a number of coils on the meter with a different number of turns on them (see also section 17.5).

Multi-range moving-iron ammeters have a number of coils which are connected in various series–parallel combinations to give the required ranges.

To measure a high value of alternating current, it is usual to use either a 0–1 A meter or a 0–5 A meter in conjunction with a current transformer.

Moving-iron voltmeters have a high resistance coil consisting of many turns of fine wire, the current required to give f.s.d. being in the range 0.05–0.1 A. In all but low-voltage meters, a voltage multiplier resistor is included in series with the instrument. To measure a high value of alternating voltage, a 0–110 V meter is used in conjunction with a potential transformer.

### 17.10 Extending the range of a dynamometer wattmeter

When a dynamometer wattmeter is used to measure power in a high-voltage high-current circuit, it is usual to use the instrument in association with a potential transformer (p.t.) and a current transformer (c.t.), as shown in Fig. 17.15. The ratings of the voltage and current coils are usually 0–110 V and 0–1 or 0–5 A respectively. The power consumed by the load is

Wattmeter reading × p.t. ratio × c.t. ratio

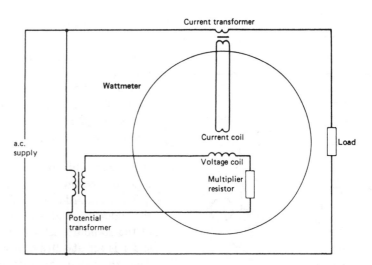

**Fig. 17.15** Extending the instrument range of a dynamometer wattmeter

### 17.11 The d.c. potentiometer

The **d.c. potentiometer** is an instrument primarily designed for the *comparison of direct voltages*. The usual measuring range is about 0–1.5 V, and can be adapted for higher voltages. It can also be used to measure other quantities such as current and resistance, and can be used for the calibration of ammeters and voltmeters.

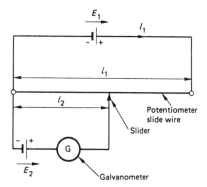

**Fig. 17.16** Basic circuit of a d.c. potentiometer

The basic circuit of a d.c. potentiometer is shown in Fig. 17.16; it consists of a length $l_1$ of resistance wire with a sliding contact on it, a battery of e.m.f. $E_1$, and a *centre-zero galvanometer*. The value e.m.f. $E_2$ is generally unknown.

The instrument is basically a *null balance* instrument; that is, when the circuit is 'balanced', no current is drawn from the sliding contact. Since this is the case, then *the p.d. per unit length along the wire is constant*. A grasp of this fact is critical to understanding the operation of the d.c. potentiometer.

The purpose of the centre-zero galvanometer is to detect the point at which no current is drawn from the slider. When the galvanometer reading is zero, the d.c. potentiometer is said to be **balanced**.

When the potentiometer is balanced, the p.d. per unit length on the slide wire is $E_1/l_1$, so that the p.d. across length $l_2$ is (see Fig. 17.16)

$$E_2 = \text{p.d. per unit length} \times l_2 = \frac{E_1}{l_1} \times l_2$$

or

$$\frac{E_1}{l_1} = \frac{E_2}{l_2} \qquad\qquad [17.15]$$

### 17.11.1  Measuring the e.m.f. of a cell using the d.c. potentiometer

A method which is used to determine accurately the e.m.f. of a cell is illustrated in Fig. 17.17. In this figure, $E_1$ supplies the slide wire, $E_2$ is the 'unknown' e.m.f., and $E_S$ is a cell with a 'standard' value of e.m.f. (and is usually described as a **standard cell**). Normally, the value of $E_1$ is also unknown, its value being unimportant so long as it is greater than either $E_2$ and $E_S$.

It is first necessary to calibrate the p.d. per unit length of the slide wire; this process is known as **standardization**, and is described below.

The standard cell is connected to the slider via resistor $R_2$, and the slider is set at a point corresponding to the known voltage of the standard cell. If the standard cell voltage is 1.018 V, the slider is set at 1.018 m from the left-hand end of the slide wire. The value of variable resistor $R_1$ is adjusted until the galvanometer gives zero deflection. At this time, resistor $R_2$ is connected in series with the galvanometer, its effect being to reduce the sensitivity of the instrument. To increase the overall sensitivity of the d.c. potentiometer, key K is pressed; this cuts $R_2$ out of the circuit, allowing a more precise null balance to be obtained. When balance is finally obtained, the p.d. per unit length of slide wire is 1 V/m.

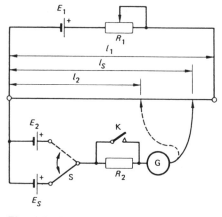

**Fig. 17.17** Comparison of e.m.f.s using a d.c. potentiometer

Once the slide wire has been standardized, any unknown e.m.f., $E_2$, can be determined. To measure $E_2$, the blade of switch S (see Fig. 17.17) is moved so that $E_2$ is connected to the slider via $R_2$. The position of the slider is then adjusted until balance is obtained on the galvanometer (*note*: the setting of $R_1$ *must not be altered*). The process is repeated once more with key K depressed to provide a more accurate balance condition. The value of $E_2$ is obtained from the equation

$$E_2 = \frac{E_S}{l_S} \times l_2 \qquad\qquad [17.16]$$

In our case $E_S/l_S = 1$ V/m, hence

$$E_2 \equiv l_2 \text{ volts}$$

When $E_S/l_S$ is not 1 V/m, $E_2$ can be calculated from eqn [17.16].

*Worked example 17.5*  A d.c. potentiometer is calibrated against a cell of e.m.f. 1.2 V, and balance is obtained against a length of 99 cm of slide wire. An unknown e.m.f. is then balanced by the potentiometer, and balance is obtained against 66 cm of slide wire. Determine the value of the unknown e.m.f.

*Solution*  The p.d. per unit length of the slide wire is

$$E_S/l_S = 1.2 \text{ V}/0.99 \text{ m} = 1.2121 \text{ V/m}$$

hence

$$\text{Unknown e.m.f.} = E_S l_2/l_S = 1.2121 \text{ V/m} \times 0.66 \text{ m}$$

$$\approx 0.8 \text{ V}$$

**17.12 Extending the range of the d.c. potentiometer**

In the following sections we consider various ways of extending the range of the d.c. potentiometer.

### 17.12.1 Measurement of a high value of direct voltage

To accommodate voltages higher than about 1.5 V, we make use of a **resistive potential divider** or **voltage ratio box** (also known as a *volt box*), shown in Fig. 17.18. The unknown voltage is connected to the terminals on the left-hand side of the potential divider, while the two lower terminals are connected to the potentiometer in the position where the unknown voltage is normally connected (see Fig. 17.17). The resistance values in the potential divider are arranged to give suitable multiplication factors such as 10, 100 and 200 at the 0–15, 0–150 and 0–300 V terminals, respectively.

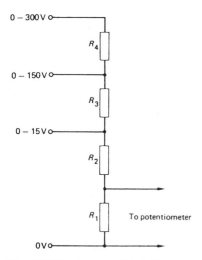

**Fig. 17.18** A potential divider or voltage ratio box

### 17.12.2 Measurement of direct current

The value of an unknown current $I$ can be determined by measuring the p.d. across a standard resistor $R$ (see Fig. 17.19). The value of the standard resistor used depends on the value of the current being measured; the p.d. developed by the current must be large enough to give a measurable voltage at the potentiometer terminals. The value of the current is calculated from the equation $I = E/R$, where $E$ is the p.d. determined by the d.c. potentiometer.

### 17.12.3 Measurement of resistance

The circuit in Fig. 17.19 can also be used to determine the value of resistance $R$. In this case, a known value of current, $I$, is passed through the resistor. The value of the resistance is determined by the equation $R = E/I$, where $E$ is the p.d. given by the potentiometer under balance conditions.

### 17.13 The Wheatstone bridge

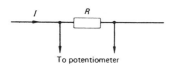

**Fig. 17.19** Basic circuit for current and resistance measurement with a d.c. potentiometer

The Wheatstone bridge is a set of resistors and a galvanometer connected as shown in Fig. 17.20, and is used to determine the value of one of the resistors.

A source of e.m.f., $E$ (whose value need not be known), is connected between points A and C, and a sensitive galvanometer G is connected between points B and D. In order to determine the value of *one of the resistors*, say $P$, we need to **balance** the bridge; the bridge is said to be balanced when the galvanometer deflection is zero.

Balance is obtained by adjusting one or more of the resistors $Q$, $R$ and $S$. When the bridge is balanced, the value of the 'unknown' resistor $P$ can either be calculated from the value of one resistor, say $Q$, and the ratio between the remaining two resistors (say the ratio $R/S$), or it can be calculated from a knowledge of the value of the other three resistors.

When we use the ratio of two resistors to calculate the unknown value, the two arms of the bridge whose ratio we use are known as the **ratio arms**.

When the bridge is balanced (see Fig. 17.20), the galvanometer current $I_G$ is zero; current $I_1$ flows in $P$ and $Q$, while $I_2$ flows in $R$ and $S$. Since no current flows in the galvanometer, the p.d. across it is zero, hence

p.d. across resistor $P$ = p.d. across resistor $R$

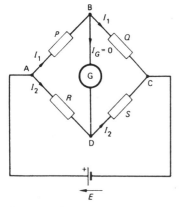

**Fig. 17.20** The Wheatstone bridge circuit at balance

that is

$$I_1 P = I_2 R \qquad [17.17]$$

Also

p.d. across resistor $Q$ = p.d. across resistor $S$

that is

$$I_1 Q = I_2 S \qquad\qquad [17.18]$$

Dividing eqn [17.17] by eqn [17.18] gives

$$\frac{P}{Q} = \frac{R}{S} \qquad\qquad [17.19]$$

The above equations can be rewritten in the form

$$PS = RQ \qquad\qquad [17.20]$$

Equation [17.20] is often a more convenient form to remember than eqn [17.19] since at balance, *the product of diagonally opposed pairs of resistors have equal values.*

*Worked example 17.6* A Wheatstone bridge of the type in Fig. 17.20 is supplied by a battery of e.m.f. 10 V and internal resistance $2\,\Omega$. An unknown resistor $P$ is connected between points A and B. At balance, the value of resistor $Q$ (between B and C) is $200\,\Omega$, resistance $S$ (between C and D) is $10\,\Omega$, and resistance $R$ (between D and A) is $20\,\Omega$. Determine the value of resistor $P$ and the current drawn from the battery at balance.

*Solution* The circuit conditions are $E = 10$ V of internal resistance $2\,\Omega$, $Q = 200\,\Omega$, $R = 20\,\Omega$ and $S = 10\,\Omega$. From eqn [17.20]

$$PS = RQ$$

or

$$P = RQ/S = 20 \times 200/10 = 400\,\Omega$$

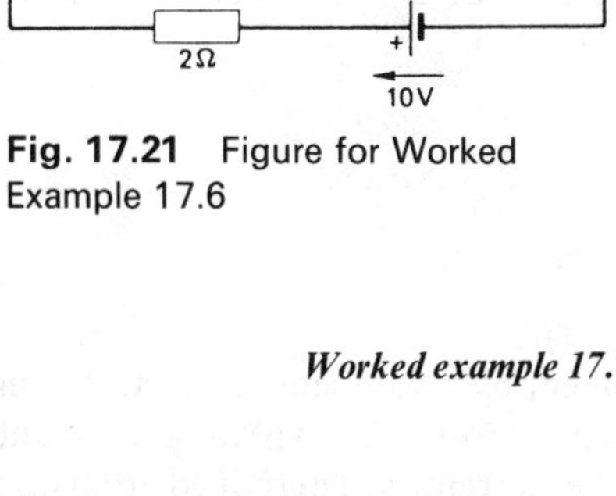

The equivalent electrical circuit of the bridge at balance is given in Fig. 17.21 (*note*: since no current flows through the galvanometer at balance, we can regard the path BD as an open circuit). The resistance of the circuit in Fig. 17.21 between A and C is

**Fig. 17.21** Figure for Worked Example 17.6

$$R_{\mathrm{AC}} = \frac{(400 + 200) \times (20 + 10)}{(400 + 200) + (20 + 10)} = 28.57\,\Omega$$

The current drawn from the battery is therefore

$$I = E/(R_{\mathrm{AC}} + 2) = 10/(28.57 + 2) = 0.327 \text{ A}$$

*Worked example 17.7* A resistance thermometer is in the form of the Wheatstone bridge in Fig. 17.22. The resistance thermometer element has a resistance of $65\,\Omega$ at a temperature of $25\,°\text{C}$. When the thermometer element is heated, the resistance of the variable resistor $R$ must be increased in value by $10\,\Omega$ in order to restore the bridge to balance. If the temperature coefficient of resistance of the thermometer element referred to $0\,°\text{C}$ is 0.001 per $°\text{C}$, calculate the temperature to which the element has been heated.

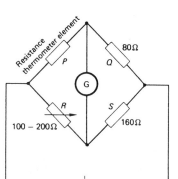

**Fig. 17.22** Figure for Worked Example 17.7

*Solution*  In this case $P = 65\,\Omega$ at $25\,°C$, $Q = 80\,\Omega$ and $S = 160\,\Omega$. At balance at $25\,°C$

$$PS = RQ$$

or

$$R = PS/Q = 65 \times 160/80 = 130\,\Omega$$

In order to maintain balance at the elevated temperature $\theta_2$, the value of $R$ is increased to $(130 + 10) = 140\,\Omega$. From eqn [1.14] we showed that

$$\frac{R_{P1}}{R_{P2}} = \frac{1 + \alpha_0\theta_1}{1 + \alpha_0\theta_2}$$

where $R_{P1}$ is the resistance of element P at $\theta_1$, $R_{P2}$ is the resistance of element P at $\theta_2$, and $\alpha_0$ is the temperature coefficient of resistance of the element referred to $0\,°C$. We solve for $\theta_2$ as follows. Rewriting the above equation gives

$$1 + \alpha_0\theta_2 = \frac{R_{P2}}{R_{P1}}(1 + \alpha_0\theta_1)$$

hence

$$\theta_2 = \left[\frac{R_{P2}}{R_{P1}}(1 + \alpha_0\theta_1) - 1\right]\bigg/\alpha_0$$

$$= \left[\frac{140}{130}(1 + [0.001 \times 25]) - 1\right]\bigg/0.001 = 103.8\,°C$$

## 17.14  Digital multimeters

Digital multimeters are *digital voltmeters* (DVMs) *with circuit modifications* which allow them to measure current and resistance. While the reader will appreciate that discussion of the operating principles of DVMs is beyond the scope of this book, we will explain in the following sections how a DVM can be modified to measure current and resistance.

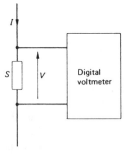

**Fig. 17.23** Measurement of current using a digital voltmeter and a shunt

### 17.14.1  Measurement of current

The basis of current measurement by a voltmeter is illustrated in Fig. 17.23, in which the voltage across a known value of shunt resistance, $S$, is measured. The current is determined from the equation $I = V/S$. In practice, resistance $S$ is incorporated in the multimeter, enabling the instrument to be calibrated in terms of current.

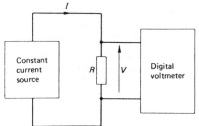

**Fig. 17.24** Measurement of resistance using a digital voltmeter

### 17.14.2 Measurement of resistance

When the digital multimeter is switched to an OHMS range, the internal circuitry is as shown in Fig. 17.24, where $R$ is the unknown resistance connected to the terminals of the instrument.

The constant current source is an electronic circuit which delivers a constant current $I$ to resistor $R$. The resistance is determined from the equation $R = V/I$, where $V$ is the reading of the DVM. The constant current source is incorporated in the multimeter, enabling the instrument to be directly calibrated in terms of resistance.

### 17.15 The decibel

When measuring the ratio of two similar quantities such as power at different points in a system, it is convenient to use the **bel** (B) or its subunit the **decibel** (dB). The power ratio in bels is defined as

$$N = \log_{10} \frac{P_2}{P_1} = \lg \frac{P_2}{P_1} \text{ bels}$$

where $P_1$ and $P_2$ are power values, $N$ the number of bels, and lg means 'the common logarithm' or logarithm to base 10. *One decibel is one-tenth of a bel*, hence the power ratio in decibels is

$$n = 10 \log_{10} \frac{P_2}{P_1} = 10 \lg \frac{P_2}{P_1} \text{ decibels} \qquad [17.21]$$

where $n$ is the number of dB.

The numerical power ratio can be determined from the decibel ratio as follows:

$$\frac{P_2}{P_1} = \text{antilog}_{10} \frac{n}{10} = 10^{(\text{dB value}/10)}$$

$$= 10^{(n/10)} \qquad [17.22]$$

Examples of power ratios in dB are given in Table 17.1. The reader will note that a power ratio of unity (that is, no change in power level) corresponds to a change of zero dB. If the power ratio is greater than unity (a power 'gain') the decibel ratio has a positive sign; if the power ratio is less than unity (an 'attenuation'), the dB ratio has a negative sign. It is of interest to note that a numerical power ratio of zero (i.e. $P_2 = 0$) corresponds to a power ratio of $-\infty$!

When electronic devices are cascaded, or connected in series, the overall numerical power gain is the product of the individual power 'gains'. That is, *the overall dB power ratio is the sum of the dB power gains of the system*.

**Table 17.1** Decibel power ratios

| Numerical power ratio | dB power ratio |
| --- | --- |
| 1000 | 30 |
| 20 | 13 |
| 2 | 3 |
| 1 | 0 |
| 0.5 | −3 |
| 0.05 | −13 |
| 0.001 | −30 |

*Worked example 17.8*   Calculate in decibels the ratio of output power to input power of a system containing two cascaded amplifiers having gains of 10 and 15 dB, respectively, together with a resistive network which introduces an attenuation of 5.6 dB. Determine the numerical value of the power gain or attenuation introduced by each section of the circuit, hence compute the overall numerical power gain or loss.

*Solution*   The overall power gain in dB is

$$10 + 15 - 5.6 = 19.4 \text{ dB}$$

The numerical power gain of the 10 dB section is

$$\text{Numerical ratio} = 10^{(\text{dB value}/10)} = 10^{(10/10)} = 10$$

For the 15 dB stage it is

$$\text{Numerical ratio} = 10^{(15/10)} = 31.62$$

and for the −5.6 dB stage it is

$$\text{Numerical ratio} = 10^{(-5.6/10)} = 0.275$$

that is, the −5.6 dB stage produces a reduction in power gain. The overall numerical power gain of the system is

$$10 \times 31.62 \times 0.275 = 87$$

This corresponds to an overall dB power gain of

$$10 \lg 87 = 19.4 \text{ dB}$$

## 17.16   dB reference levels

A datum power reference level frequently used in dB measurements is 1 mW (although any other level could be selected), and is given the symbol dBm. Thus a power level of $P$ mW is said to correspond to $10 \lg P$ decibel metres (dBm); a power level of 2 mW corresponds to $10 \lg 2 = 3$ dBm, and 0.5 mW corresponds to $10 \lg 0.5 = -3$ dBm, etc.

The dBm is widely used in telecommunications practice in conjunction with a 'standard' impedance of 600 Ω. It is the practice of many multi-range instrument manufacturers to include a dBm range on the scale of their meters, in which 0 dBm corresponds to the voltage at which 1 mW is developed in a 600 Ω resistor; this voltage is 0.775 V. It can be shown that if voltage $V_K$ develops $y$ decibel metres in a 600 Ω resistor, then

$$V_K = \sqrt{(0.6 \times 10(y/10))} \text{ volts}$$

Thus +2 dBm corresponds to 0.975 V and −5 dBm to 0.436 V. The relationship between the 1 V and the dBm scales on a meter is as shown in Fig. 17.25.

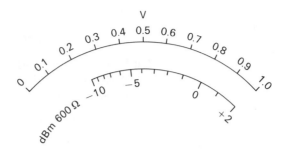

**Fig. 17.25**   Relationship between the voltage and dBm ranges

**17.17   The decibel as a voltage or current ratio**

If power $P_1$ is dissipated in resistance $R_1$, and $P_2$ in resistance $R_2$, then

$$n = 10 \lg\left[\frac{V_2^2/R_2}{V_1^2/R_1}\right]$$

$$= 10 \lg\left[\frac{V_2}{V_1}\right]^2 - 10 \lg\left[\frac{R_2}{R_1}\right]$$

$$= 20 \lg\left[\frac{V_2}{V_1}\right] - 10 \lg\left[\frac{R_2}{R_1}\right] \text{ decibels} \qquad [17.23]$$

In the case where $R_2 = R_1$, then

$$n = 20 \lg\left[\frac{V_2}{V_1}\right] \qquad [17.24]$$

Even though the value of $R_1$ and $R_2$ may not be equal in some cases, eqn [17.24] is frequently used for the dB voltage ratio in electronic calculations.

Also, since power $= I^2 R$, we may write (once again, assuming that $R_1$ and $R_2$ have the same value)

$$n = 20 \lg\left[\frac{I_2}{I_1}\right] \qquad [17.25]$$

*Worked example 17.9*   The output voltage from an amplifier is 2.5 V. If the voltage gain is 32 dB, determine the value of the input voltage. Assume that the amplifier input resistance and load resistance have equal values. What attenuation in dB must be provided by an attenuator connected to the output of the amplifier if the voltage from the attenuator is 0.3 V?

*Solution*   The voltage gain of the amplifier in dB is

$$32 = 20 \lg(2.5/V_1)$$

where $V_1$ is the input voltage to the amplifier. That is

$$2.5/V_1 = 10^{(32/20)} = 39.81$$

or

$$V_1 = 2.5/39.81 = 0.0628 \text{ V} \quad \text{or} \quad 62.8 \text{ mV}$$

If $V_3$ is the output voltage from the attenuator, and $V_2$ its input, then the voltage 'gain' of the attenuator is

$$n = 20 \lg(V_3/V_2) = 20 \lg(0.3/2.5) = -18.42 \text{ dB}$$

That is, 18.42 dB of attenuation must be provided by the attenuator.

## 17.18 Measurement of Q-factor

A simplified version of a circuit used to determine the $Q$-factor of a series-tuned circuit is shown in Fig. 17.26. The circuit is energized by an oscillator, which applies the voltage across resistor $r$ (which has a very low type, typically $0.025\,\Omega$ or less) to the series circuit. To measure the $Q$-factor of the circuit (that is of the coil and the capacitor), the frequency of the oscillator is adjusted until resonance occurs, at which time $V_r$ and $V_C$ are measured. Since $V_r$ acts as the supply source to the series-tuned circuit, then

$$Q\text{-factor} = \frac{V_C}{V_r} = \frac{V_C}{Ir} \qquad\qquad [17.26]$$

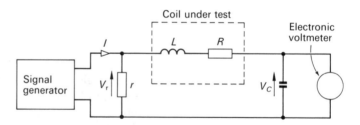

**Fig. 17.26** *Q*-factor measurement

## 17.19 The cathode-ray oscilloscope

The **cathode-ray oscilloscope** (CRO) provides the user with an 'electronic eye' with which waveshapes within circuits can be monitored.

A cathode-ray oscilloscope comprises a cathode-ray tube (CRT) together with power supplies, amplifiers and other components. A diagram of a basic CRO is illustrated in Fig. 17.27, together with the essential controls.

The **cathode-ray tube** consists of a glass envelope which encloses a number of electrodes which generate and control a beam of electrons inside the tube. The electron beam is produced by an **electron gun** as follows. A heater (H) is supplied with electrical energy, the heat produced causing a cloud of electrons to leave the surface of the metallic cathode (K).

The electrons are attracted to three anodes $A_1$, $A_2$ and $A_3$ in the **electron lens** system, the anodes being positive with respect

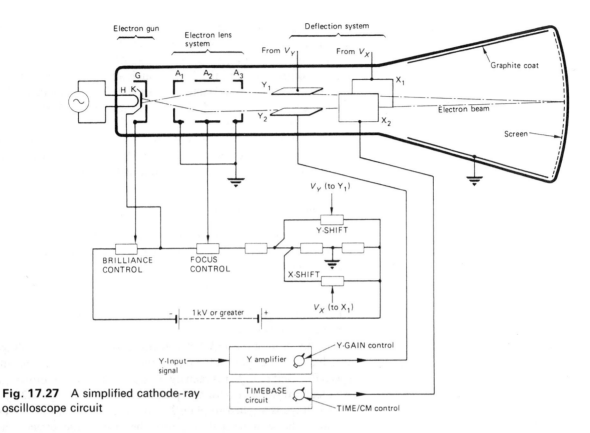

**Fig. 17.27** A simplified cathode-ray oscilloscope circuit

to the cathode (that is to say, the cathode is negative with respect to the anodes). Anodes $A_1$ and $A_3$ are discs having a hole in the centre, while $A_2$ is cylindrical. Due to this physical structure, the diverging beam of electrons which enters $A_1$ is converging when it leaves $A_3$. The operator of the CRO is provided with a **FOCUS CONTROL**, which enables the beam to be focused on the face of the tube.

After leaving the electron lens system, the beam enters the **deflection system**, whose function it is to cause the spot on the face of the tube to deflect both in the Y-direction (up and down) and in the X-direction (right and left). Each of the Y-deflection system and the X-deflection system has two deflecting plates, and the voltage between each pair is controlled by two independent voltages. For simplicity, each plate in Fig. 17.27 is shown as having its own voltage applied to it but, in practice, the voltage between a pair of plates is a composite voltage.

The controls marked **Y-SHIFT** and **X-SHIFT** are brought out on the control panel of the CRO for use by the operator. These controls allow the user to 'shift' the spot on the face of the tube in the Y- and X-directions, respectively. The 'shift' voltages are shown applied respectively to plates $Y_1$ and $X_1$.

The second signal applied to the Y-shift system is the **input signal** (which may have to be amplified if it is a small signal, or attenuated if it is a large signal). A **Y-GAIN** control is provided on the control panel of the CRO, which allows the user to make the trace large (or small) enough to fill the tube. The control is calibrated in VOLTS/CM, typical settings being 0.1, 0.2, 0.5, 1, 2, 5, 10, 20 and 50 V/cm.

A typical general-purpose CRT has a useful Y-deflection of up to 8 cm on the face of the tube, so that the smallest peak-to-peak alternating voltage giving a deflection of 1 cm on the face of the tube is

$$1 \text{ cm} \times 0.1 \text{ V/cm} = 0.1 \text{ V} \quad (\text{peak-to-peak})$$

which corresponds to a sine wave having an r.m.s. voltage of $0.1/(2\sqrt{2}) = 35.4$ mV. The largest peak-to-peak voltage which can be usefully viewed is

$$8 \text{ cm} \times 50 \text{ V/cm} = 400 \text{ V}$$

which corresponds to a sine wave with an r.m.s. voltage of $400/(2\sqrt{2}) = 141.4$ V.

The function of the **TIMEBASE** circuit is to produce a voltage which causes the spot on the face of the tube to scan at a constant speed from left to right (as seen by the operator) in the X-direction. An idealized waveform for this purpose is shown in Fig. 17.28(a). This voltage is applied to plate $X_2$, and the period of time taken to cause the spot to deflect across the face of the screen (the 'sweep' period in Fig. 17.28) is known as the **sweep time**.

Once the sweep is complete, the spot must return to the left-hand side of the screen. This is known as the **flyback** which, in the case of the idealized waveform in Fig. 17.28(a), takes zero time. In a

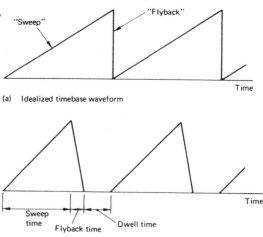

(a)  Idealized timebase waveform

(b)  Practical timebase waveform

**Fig. 17.28**  Timebase waveforms

practical circuit, it is not possible to reduce the voltage from its maximum voltage to zero in zero time, and it takes a time known as the **flyback time** [see Fig. 17.28(b)] for this to occur. Also, in order to prevent the 'sweep' commencing before the next cycle has begun, the X-deflection circuit contains a **synchronization circuit**, which delays the start of the sweep for a period of time known as the **dwell time**. When the synchronizing circuit detects the commencement of the next cycle, it allows the timebase to begin its sweep again. In this way, the waveform displayed on the face of the tube appears as a steady picture.

The timebase has a number of controls, the principal control being the 'speed' control (calibrated in TIME/CM) and a general-purpose CRO has settings of 1, 10 and 100 $\mu$s/cm, 1, 10 and 100 ms/cm. If the tube has a useful face width of 10 cm then, at the fastest sweep speed, the time taken to sweep across it is

$$10 \text{ cm} \times 1 \ \mu\text{s/cm} = 10 \ \mu\text{s}$$

corresponding to one complete cycle whose frequency is 100 kHz, or 10 cycles at 1 MHz. At the slowest sweep speed, the time taken is

$$10 \text{ cm} \times 100 \text{ ms/cm} = 1000 \text{ ms} = 1 \text{ s}$$

The electron beam finally strikes the inside face of the screen, which is coated with **phosphor**, which emits visible radiation when electrons strike it. The colour of the spot depends on the phosphor used, and is typically either green or blue. The brilliance of the spot depends on the magnitude of the beam current, which is controlled by the **BRILLIANCE** control (see Fig. 17.27). When the electrons have given up their energy to the phosphor, they make their way back to the power supply via the earthed graphite coat inside the flared neck of the CRT.

*Worked example 17.10*   A sinusoidal voltage is viewed on the screen of a CRT, the amplitude of the trace being 7.5 cm and one cycle is complete in 8.6 cm. If the Y-gain control is set at 5 V/cm, and the timebase control is set at 100 $\mu$s/cm, determine the r.m.s. value of the waveform and its frequency.

*Solution*   The peak-to-peak voltage of the waveform is

$$7.5 \text{ cm} \times 5 \text{ V/cm} = 37.5 \text{ V}$$

hence the r.m.s. voltage of the wave is

$$37.5/(2\sqrt{2}) = 13.26 \text{ V}$$

The periodic time of the wave is

$$T = 8.6 \text{ cm} \times 100 \ \mu\text{s/cm} = 860 \ \mu\text{s} = 0.86 \text{ ms}$$

and the frequency of the wave is

$$f = 1/T = 1/0.86 \times 10^{-3} = 1163 \text{ Hz}$$

**17.20 Measurement of voltage using a CRO**

A voltage can be measured by noting the Y-deflection produced by the voltage (see also Worked Example 17.10); this deflection is used in association with the Y-gain setting. However, **the reader is cautioned against using an oscilloscope for this purpose unless the CRO has been calibrated with a reliable voltage source.**

**17.21 Measurement of current and resistance using a CRO**

Using the general method shown in Fig. 17.23 and 17.24 (with the CRO replacing a voltmeter), a correctly calibrated CRO can be used for the measurement of either current or resistance. The relationship between the deflection of the spot on the face of the tube, the current and the resistance is

$$[\text{Deflection of the spot in cm}] \times [\text{volt/cm setting}]$$

$$= \text{current} \times \text{resistance}$$

Once the deflection of the spot, the volts/cm setting and either the current or the resistance are known, then the remaining unknown can be calculated.

**17.22 Measurement of frequency using the CRO**

A simple, though not very accurate, method of determining the frequency of a waveform is to estimate its periodic time on the screen of a CRT (see Worked Example 17.10).

A better method is to use a CRO which gives access either to the X-plates of the CRT or to the X-amplifier. The frequency of an unknown sinusoidal signal can then be determined by the method shown in Fig. 17.29. In this case, the signal $f_y$ of unknown frequency is connected to the Y-plates of the CRO, and a signal $f_x$ of variable but known frequency is connected either to the X-amplifier or the X-plates of the CRT. Initially, the trace on the screen will be a never-ending blur but, as the frequency of $f_x$ is altered, the trace will finally display a *single loop*, when $f_x = f_y$.

An alternative method using the same technique produces what are known as **Lissajous' figures** (named after the French physicist Jules Antoine Lissajous), typical displays for various values of

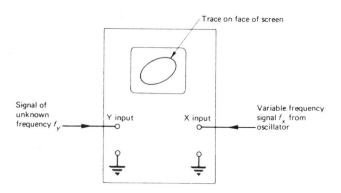

**Fig. 17.29** One method of determining an unknown frequency

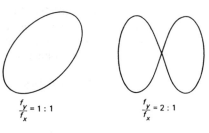 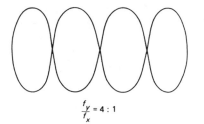 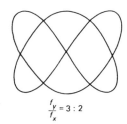

$\dfrac{f_y}{f_x} = 1 : 1$       $\dfrac{f_y}{f_x} = 2 : 1$       $\dfrac{f_y}{f_x} = 4 : 1$       $\dfrac{f_y}{f_x} = 3 : 2$

**Fig. 17.30** Lissajous' figures

the ratio $f_y/f_x$ being shown in Fig. 17.30. The value of the frequency ratio can be determined from the expression

$$\frac{f_y}{f_x} = \frac{\text{number of peaks along the top of the trace}}{\text{number of peaks along one edge of the trace}}$$

**Problems**

**17.1**    A moving-coil milliammeter is wound with $35\frac{1}{2}$ turns of wire. The effective radius of the coil is 1.0 cm and the effective length of each coil side is 2 cm. If the flux density in the air gap is 0.15 T, and the controlling torque produced by the hairsprings is $0.75 \times 10^{-6}$ N m/deg deflection, calculate the current needed to give a deflection of 90°.

[31.7 mA]

**17.2**    A moving-iron meter is wound with 50 turns and gives a full-scale deflection with a current of 5 A. If the instrument is rewound to give a full-scale deflection with a current of 12.5 A, how many turns are required?

[20]

**17.3**    A moving-coil instrument gives full-scale deflection with a current of 20 mA, the resistance of the meter being 10 Ω. Calculate (a) the resistance of the shunt required so that the meter may be used as a 5 A ammeter, and (b) the resistance of a voltage multiplier which allows the meter to be used as a 100 V meter.

[0.0402 Ω; 4990 Ω]

**17.4**    A moving-coil instrument gives full-scale deflection when the current is 10 mA, the terminal voltage of the meter being 70 mV. Calculate the value of a voltage multiplier resistor needed to allow the instrument to be used as a 50 V meter.

[4993 Ω]

**17.5**    A moving-coil meter has a maximum scale deflection of 100 divisions. If full-scale deflection occurs at a current of 20 mA, and the resistance of the meter is 10 Ω, calculate (a) the shunt resistor required to give a deflection of 0.2 A per scale deflection and (b) the series resistor needed to give 0.1 V per scale division.

[(a) 0.01 Ω; (b) 490 Ω]

**17.6**    Verify eqn [17.13] in section 17.7.1.

**17.7** A series circuit comprises a $50\,k\Omega$ resistor in series with a $10\,k\Omega$ resistor. If the circuit is supplied at 100 V d.c., determine the reading of a voltmeter of internal resistance $20\,k\Omega$ connected across the $10\,k\Omega$ resistor. What is the voltage across the $10\,k\Omega$ resistor when the voltmeter is disconnected?

[11.76 V; 16.67 V]

**17.8** Verify eqn [17.14] in section 17.7.2.

**17.9** A $5\,\Omega$ resistor is in series with a two-branch parallel circuit having $5\,\Omega$ in one branch and $8\,\Omega$ in the other, the complete circuit being supplied by a 10 V d.c. source. If the current drawn by the circuit is measured by an ammeter of resistance $1.0\,\Omega$, what current is indicated by the meter? What would be the current if the ammeter was short-circuited?

[1.1 A; 1.238 A]

**17.10** With the aid of a circuit diagram, describe how a d.c. potentiometer can be used to check the calibration of a direct current ammeter.

When a cell of e.m.f. 1.018 V is connected to a d.c. potentiometer, balance is obtained at 81 cm. A current is passed through an ammeter in series with a $0.2\,\Omega$ standard resistor, and the current is adjusted until the ammeter indicates 5 A. The p.d. across the standard resistor gives balance at 78 cm. Calculate the percentage error in the reading of the instrument.

[2.0 per cent high]

**17.11** A d.c. potentiometer is used to measure the e.m.f. of a supply source. Balance is obtained at 88.4 cm with the supply source in circuit, and at 45 cm with a standard cell of 1.018 V. When the terminal voltage of the supply source is measured by an accurate voltmeter of internal resistance $50\,\Omega$, the voltmeter indicates 1.96 V. Determine (a) the e.m.f. of the supply source and (b) its internal resistance.

[(a) 2 V; (b) 1.02 $\Omega$]

**17.12** An unknown resistance $R$ is measured using a Wheatstone bridge. At balance the ratio arms AD and DC have resistances of $1\,k\Omega$ and $10\,\Omega$, respectively, the resistance in arm BC being $5.6\,\Omega$. Determine the value of $R$, which is connected between A and B.

[560 $\Omega$]

**17.13** Four resistors are connected together in the arms of a Wheatstone bridge, the arms being between points A and B, B and C, C and D, and D and A. When the bridge is balanced, the value of the resistors between AB, BC and CD are 10.6, 10 and $21\,\Omega$, respectively. Determine the value of the other resistor.

A cell of e.m.f. 3.0 V and internal resistance $1.04\,\Omega$ is connected between points A and C. Calculate, at balance, the p.d. across the unknown resistor and the power consumed by the $10\,\Omega$ resistor.

[22.26 $\Omega$; 1.44 V; 0.184 W]

**17.14** Derive the conditions of balance for a Wheatstone bridge. A Wheatstone bridge is used to measure the resistance of a 10 V moving-coil voltmeter. At balance the resistances in the arms of the bridge are as follows: AB = $160\,\Omega$, BC = voltmeter, CD = $1\,k\Omega$ and DA = $20\,\Omega$. Calculate the resistance of the voltmeter.

If a 10 V d.c. supply is connected between points A and C, estimate the reading on the voltmeter when the bridge is at balance.

[8 kΩ; 9.8 V]

17.15 The power consumed by an a.c. load is measured by a dynamometer wattmeter in conjunction with instrument transformers. The wattmeter indicates 250 W. If the voltage coil is connected through a potential transformer with a step-down ratio of 1:10, and the current coil is connected through a current transformer with a step-up ratio of 2:1, calculated the power dissipated by the load. Neglect instrument and transformer losses.

[1250 W]

17.16 Express the following numerical power gains in decibels: (a) 0.01, (b) 0.3, (c) 1.0, (d) 3, (e) 15, (f) 600.

[(a) −20; (b) −5.23; (c) 0; (d) 4.77; (e) 11.76; (f) 27.78]

17.17 Some electronic circuits have the following respective power gains: 5, 15, −4 and −12 dB. If they are cascaded, calculate the overall power gain (a) in dB and (b) numerically.

[(a) 4 dB; (b) 2.51]

17.18 If only 3 per cent of the power supplied to a circuit appears at the output terminals, what is the power loss in dB?

[15.23 dB]

17.19 The voltage gains in dB of a number of circuits are 22, −6.2, −10.2 and 4.5. If the circuits are cascaded, calculate the overall power gain in dB. If 10 mV is applied to the input, calculate the output voltage.

[10.2 dB; 32.36 V]

17.20 An electronic voltmeter is to have a decibel range added to its scale, the reference level of 0 dB corresponding to the voltage at which a power of 10 mW is dissipated in a 1000 Ω resistor. Determine (a) the voltage corresponding to zero dB and (b) the voltage corresponding to power levels of (i) 2 dB, (ii) −2 dB and (iii) −10 dB.

[(a) 3.16 V; (b) (i) 3.98 V; (ii) 2.51 V; (iii) 1 V]

17.21 Describe how a Q-meter may be used to determine (a) the Q-factor of a coil, (b) the inductance of a coil, (c) the resistance of a coil at high frequency and (d) the capacitance of a capacitor.

17.22 Distinguish between an analogue measuring instrument and a digital instrument. Discuss the relative advantages and disadvantages of each.

17.23 Draw a block diagram of a cathode-ray oscilloscope and explain the purpose of each section. Describe the operation of the timebase circuit, and state why the timebase waveform has a sawtooth shape.

17.24 Describe how a CRO can be calibrated and used to measure (a) a direct voltage, (b) the r.m.s. value of a sinusoidal voltage, (c) the frequency of an alternating signal and (d) the phase angle difference between two sinusoidal signals.

# 18 Computer software for electrical and electronic principles

## 18.1 Introduction

The subject of electrical and electronic principles is wide ranging, and covers topics ranging from elementary circuits up to transients in electrical circuits and control systems. Although computer software is often thought of as something of an 'extra' it is, in fact, so useful that it should be thought of as vital not only in extending our problem-solving ability, but also of laboratory and project work. In fact, the software can be used to solve problems and support project work which are difficult in the laboratory unless specialized equipment and highly-trained staff are available.

It may also be thought that special software of this kind is very expensive, but this is not always the case. There are many low-cost 'student' versions available which are more than adequate to deal with problems at all levels. In most cases, documentation is available on the disc provided by the supplier. Students are advised to study the pages of magazines dealing with *computer shareware*.

We shall be looking at three items of software in this chapter. The first is a version of SPICE (Simulation Program with Integrated Circuit Emphasis) known as PSpice, a student or evaluation version being available either from the main supplier[†] or a good shareware supplier.

The remaining two items of software have already been introduced in Chapter 16, namely CODAS[‡] (COntrol system Design And Simulation) and PCS (Process Control Simulation) and are intended for the simulation of control systems and process control systems, respectively.

## 18.2 SPICE

SPICE was developed for integrated circuit simulation by the University of California, Berkeley. It is so versatile that even simple d.c. and a.c. circuits can be simulated using a very easy programming language. SPICE is perhaps the leading design software for electrical and electronic circuits, and enables engineers and technicians to develop a circuit from a basic design

---

[†] Available from ARS Microsystems, Herriard Business Centre, Alton Road, Basingstoke RG25 2PN.
[‡] CODAS and PCS are available from Golten and Verwer Partners, 33 Moseley Road, Cheadle Hulme, Cheshire SK8 5HJ.

concept to a working version with a minimum of delay. We shall be using PSpice, which is a version of SPICE; it uses the SPICE language, and is well-suited to the training and education of engineers and technicians.

SPICE enables circuits containing elements such as resistors, inductors, capacitors, transformers, semiconductors, voltage and current sources, transmission lines, operational amplifiers, etc. to be described to the computer in very simple terms. The circuit to be analysed is defined in an **input file** (one program line per component), which is read by a **software analyser**; the analyser checks the input file for errors and, if none exist, it performs the required analysis and supplies the results. Depending on the type of analysis needed, the results may either be in numerical form, or in graphical form, or both.

## 18.3 SPICE circuit description

Each terminal of every element is connected to a *node* in the circuit, and each node is given an identification number. Node zero is always the *reference node* or *zero voltage node* (take great care to use zero and not capital O).

The value of an element, such as the resistance of a resistor, can be entered in any one of many forms. For example, the resistance of a 1.8 kΩ resistor can be entered as 1800, or as 1.8K, or 1.8E3, etc. A list of SPICE symbolic values are given in Table 18.1. The reader should note carefully that M means **milli-**, and MEG means **mega-**; thus, in SPICE, a 10M resistor is a 10 milliohm resistor.

Two-terminal elements such as resistors, inductors and capacitors are described, for example, as follows:

**Table 18.1** SPICE and metric values

| SPICE suffix | Scale | Metric prefix |
|---|---|---|
| F | $10^{-15}$ | femto- |
| P | $10^{-12}$ | pico- |
| N | $10^{-9}$ | nano- |
| U | $10^{-6}$ | micro- |
| M | $10^{-3}$ | milli- |
| K | $10^{3}$ | kilo- |
| MEG | $10^{6}$ | mega- |
| G | $10^{9}$ | giga- |
| T | $10^{12}$ | tera- |

```
R1   1    2     10; 10 ohm resistor
L    0    4      5; 5 H inductor
C3   15   21    2.6U; 2.6 microfarad capacitor
```

The letter at the beginning of the line describes the type of element (R for resistor, L for inductor, etc.), and the number which follows it, i.e. R1, C3, is optional, and is merely the number we have given to it in the circuit. Alternatively, we can give the element a name such as Rbias, or Cblocking, or Lchoke, etc. The next two values in the element line are the nodes to which the element is connected; R1 is connected between nodes 1 and 2, L is connected between nodes 0 and 4, C3 is connected between nodes 15 and 21.

The final number in the line is the numerical value of the element. Unless a multiplying factor is involved, SPICE assumes that it is the basic value of the element. That is, it assumes that R1 has a value of 10 Ω and L has a value of 5 H. Since C3 has a multiplying factor associated with it, SPICE takes its value to be 2.6 $\mu$F.

The reader should note carefully that if we write the value of a capacitor as, say, 3F, SPICE assumes that the value is 3 femtofarads ($3 \times 10^{-15}$ F), *so do take care!* A value of 3 farads should simply be written down as 3; a 3 microfarad capacitor is written down as 3u or 3U (or as 3uF), and one of value 3 femtofarads as 3f or 3fF.

Each of the three element lines above have a semicolon (;) in it, followed by a statement. When SPICE finds a semicolon, it assumes that a **comment** follows the semicolon, which is ignored. A comment can be allocated a complete line to itself if the line commences with a '*' as follows:

```
* The following line defines a resistor
RS   9   10   2K
* This is another comment line
```

An **independent voltage source**, such as either of the batteries in Fig. 18.1, is defined as a 'V' source as follows:

```
V1   1   0   DC   10
V2   0   3   5.6
```

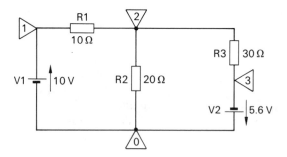

**Fig. 18.1**  A simple circuit for SPICE analysis

V1 is defined as being connected between node 1 and node 0, the first node (node 1) being positive with respect to the second node (node 0). Also V1 is specified as a DC source; this is optional and can be omitted, as it is in the case of V2. If it is an AC source, as we shall see later, the type of source (i.e. AC) **must be included** (see Table 18.14), otherwise SPICE assumes that it is a d.c. source. The final value in each line is the magnitude of the voltage.

**18.4  Analysis of a simple circuit**

When preparing a circuit for SPICE analysis, certain rules must be followed:

1. Draw the circuit diagram and give it a NAME which can be used in the title line [see 5(i) below].
2. Label EVERY element on the circuit with the element name you will use in the input file, and write its value on the circuit.

3. Clearly mark every node number on the circuit, the reference node or zero voltage node being node 0 (take care not to use capital O when typing the file). Each node must be numbered with a positive value, i.e. 0, 1, 2,..., 25,..., 64, etc.
4. Decide what type of analysis (if any) is to be performed on the circuit (we look at this later in the chapter).
5. Create a SPICE input file as follows:
   (i) the first line either contains a title line or is left blank, i.e. there is nothing in it; and
   (ii) write down a series of *element lines*, *comment lines*, *blank lines* and *control lines*. The order in which they appear is immaterial.
6. Terminate the input file with a '.END' line (the '.' before END is important).

The input file in Table 18.2 illustrates some of these points for the circuit in Fig. 18.1. The first line in the file is the text 'Solution to Fig. 18.1', which satisfies rule 5(i) in the above list. Alternatively the line may be left blank but, since this does not help to identify the purpose of the file, it pays to enter a sensible comment.

**Table 18.2**  Input file for the circuit in Fig. 18.1

```
Solution to figure 18.1


* Elements can be entered in any order

V1    1    0    10V; 'V' is optional

V2    0    3    5.6

R1    1    2    10ohms; 'ohms' is optional

R3    3    2    30

R2    2    0    20



* The following line saves paper when printing out

.OPTIONS NOPAGE

.END
```

The second line is left blank; this is optional and simply helps to make the file readable. The next line is the comment 'Elements can be entered in any order'; once again, this assists the reader to understand that we could easily alter the order in which the file is written. The next five lines contain the circuit elements in the manner described earlier.

Next, we say that the value of V1 is 10 V; the letter 'V' is optional and, since this letter does not appear in Table 18.1, we can use it as a unit symbol (which SPICE ignores). The value of

V2 is 5.6, and SPICE is told in this line that node 0 is positive with respect to node 3. Also, in the case of R1, 'ohms' is optional information, and since the letter 'o' in ohms does not appear in Table 18.1, SPICE ignores it.

After the elements have been defined, we leave a blank line once more.

Any line commencing with a '.' is known as a **control line**, which is used to control the operation of SPICE. The '.OPTIONS' control line contains statements which are used to transfer certain options to SPICE, one of which is the NOPAGE option. This is a routine which reduces the number of sheets of paper produced by SPICE at print-out time. In the absence of the NOPAGE command, the file in Table 18.2 produces two sheets of paper, one containing the input file or circuit description in Table 18.2, and the other containing the results.

When the circuit has been analysed, the output file is as given in Table 18.3. Unless otherwise told, SPICE carries out all analyses at its default temperature of 27 °C; we will see how this temperature can be altered in section 18.5. Since we have not specified a particular form of analysis in the input file (we will see how this is done in section 18.6), SPICE performs a 'SMALL SIGNAL BIAS SOLUTION', in which it prints out the voltage at each node, together with the current in each voltage source, i.e. the current in V1 and V2. However, the reader should carefully note that the current printed is that flowing *into the positive node* of the voltage source; this is the opposite way to the usual electrical circuit convention. The numerical values of current in Table 18.3 are correct, but should be interpreted as 0.5564 A *flowing out* of

**Table 18.3**   Output file for the circuit in Fig. 18.1

```
****     SMALL SIGNAL BIAS SOLUTION        TEMPERATURE =   27.000 DEG C

NODE   VOLTAGE      NODE   VOLTAGE      NODE   VOLTAGE

(   1)  10.0000  (    2)   4.4364  (    3)  -5.6000


      VOLTAGE SOURCE CURRENTS

      NAME          CURRENT

      V1            -5.564E-01

      V2            -3.345E-01

      TOTAL POWER DISSIPATION   7.44E+00  WATTS

            JOB CONCLUDED

            TOTAL JOB TIME         4.39
```

the positive terminal of V1, and 0.3345 A *flowing out* of the positive terminal of V2. We will see later how to 'measure' the conventional current in the circuit.

The reader will find it an interesting exercise to verify the results in Table 18.3 by conventional means.

**18.5 Temperature coefficient of resistance**

It was mentioned above that the default temperature at which SPICE assumes all values are given is 27 °C. Here we look at an input file in which not only do we alter the nominal temperature (TNOM) at which the circuit operates, but also introduce a method of dealing with the resistance–temperature coefficient of a resistor.

In this case the circuit comprises a voltage source, Vs, of 10 V d.c. connected to a resistor of resistance $100 \Omega$ at a temperature of 0 °C. The input file in Table 18.4 causes SPICE to calculate the current in the resistor at 0 and 100 °C. As with normal electrical circuit calculations, SPICE uses the equation

$$R_\theta = R_0(1 + \alpha_0 \theta)$$

where $R_\theta$ is the resistance at $\theta$ °C, $R_0$ is the resistance at 0 °C, and $\alpha_0$ is the linear resistance–temperature coefficient referred to 0 °C. SPICE can also deal with both linear and quadratic temperature coefficients.

**Table 18.4**   Effect of temperature coefficient of resistance

```
Vs   1    0    10V; supply voltage


*                    Linear coeff.
                          ¦
R    1    0    100ohms    TC = 0.00427; 100 ohms at TNOM


.TEMP        0C    100C; 'C' optional
.OPTIONS     NOPAGE    TNOM = 0C
.END
```

Once again, the first line in the file describes its purpose, and the second line is blank. Next we define the supply voltage Vs (node 1 is +10 V with respect to node 0), followed by another blank line.

Then comes the definition of the resistance, which differs slightly from the resistance values in Table 18.2 because, after the value of the resistance we specify the resistance–temperature coefficient (TC) of $\alpha = 0.00427$.

The value of the resistance is specified at some nominal temperature TNOM (usually 27 °C); however, in our case we want it to be at 0 °C. We therefore, in the '.OPTIONS' line, specify the value of TNOM to be 0 °C (the 'C' is optional in the input file, and can be omitted). Since we wish to perform the calculations both at 0 and 100 °C, we include a '.TEMP' control line in the file, which runs the analysis at the two temperatures. We could include several temperatures in this line. The print-out is listed in Table 18.5.

**Table 18.5**  Result of temperature coefficient calculation

```
****    SMALL SIGNAL BIAS SOLUTION        TEMPERATURE =    0.000 DEG C

  NODE   VOLTAGE

(    1)   10.0000

     VOLTAGE SOURCE CURRENTS

     NAME        CURRENT

     Vs         -1.000E-01

     TOTAL POWER DISSIPATION   1.00E+00  WATTS


****    TEMPERATURE-ADJUSTED VALUES      TEMPERATURE =  100.000 DEG C

**** RESISTORS

  NAME        VALUE

  R          1.427E+02

****    SMALL SIGNAL BIAS SOLUTION        TEMPERATURE =  100.000 DEG C

  NODE   VOLTAGE

(    1)   10.0000

     VOLTAGE SOURCE CURRENTS

     NAME        CURRENT

     Vs         -7.008E-02

     TOTAL POWER DISSIPATION   7.01E-01  WATTS
```

We see that, at 0 °C, the current drawn from the positive terminal of the battery is 0.1 A, i.e. the resistance is 100 Ω. At 100 °C, SPICE quotes the resistance as 142.7 Ω (the reader should verify this value), and the current drawn from the battery is 0.070 08 A. Clearly, it would not be a simple matter to solve the circuit in Fig. 18.1 for different temperature coefficients for each resistor over a range of temperatures.

**18.6   DC analysis**

A **DC analysis** produces a *sweep of a specified variable* such as a voltage, a current, a resistance or a temperature, and is set up by means of a '.DC' control line. In the following we will analyse the circuit in Fig. 18.1 with V1 having the values of 9, 10 and 11 V. That is V1 is *swept* or *stepped* from 9 to 11 V in 1 V steps. One advantage of using a 'sweep' analysis is that we can make SPICE print the current in each branch of the circuit as outlined below. The input file is given in Table 18.6.

**Table 18.6**   Input file containing a DC sweep analysis

```
Solution of figure 18.1 for several values of V1

* The value of V1 (below) must be given, but is
* overridden in a '.DC' analysis (see later)

V1     1    0    10V

*         Swept variable name
*         ¦ Start value   End value   Increment
*         ¦ ¦                 ¦            ¦
.DC    V1   9V               11V          1V

V2     0    3    5.6V
R1     1    2    10ohms
R2     2    0    20ohms
R3     2    3    30ohms

.OPTIONS NOPAGE
*PRINT command
* ¦   Analysis type      Variables to be printed
* ¦        ¦              ¦      ¦       ¦     ¦
.PRINT    DC          I(R1)   I(R2)   I(R3)  V(1,2)
.END
```

The first element line in the file specifies the value of V1 as 10 V. This value is, in fact, immaterial, because the value of V1 is overridden by the value given in the '.DC' line, namely that V1 should have a value commencing at 9 V, and should rise to 11 V in 1 V steps.

The circuit components are specified in much the same way as in Table 18.2, but with one very subtle and important difference.

That is, *the nodes of the resistors are specified in the order in which current is expected to flow through them*. This is particularly important in this input file, as we shall see.

Also included in the input file is a '.PRINT' control line, which commands SPICE to print (on the computer screen in the first instance) the values specified in the '.PRINT' line. There is one important thing to note, however, and that is whenever we have a '.PRINT' line (or a '.PLOT' line – see later), *we must specify the type of analysis which produces the results* (in this case a .DC analysis). Here we ask for the following to be printed:

$I(R1)$ – the current flowing in R1 from node 1 to node 2;
$I(R2)$ – the current flowing in R2 from node 2 to node 0;
$I(R3)$ – the current flowing in R3 from node 2 to node 3;
$V(1,2)$ – the voltage of node 1 with respect to node 2, i.e. $V_{12}$.

The ability of being able to ask for the current in an element is not available in all versions of SPICE, and PSpice allows us to do this. We will see in Table 18.10 how other versions of SPICE deal with this problem.

With the '.PRINT' line used here, PSpice prints the current flowing through the circuit element from the first node in the element line to the second node. That is, the current in R1 is assumed to flow from node 1 to node 2, the current in R2 is assumed to flow from node 2 to node 0, etc. Also we can ask for the voltage at a particular node, or for the voltage between a pair of nodes to be printed. In our case, we have asked for the voltage of node 1 with respect to node 2, i.e. $V_{12}$, to be printed. The results produced by PSpice are given in Table 18.7, and those for $V1 = 10\,V$ should be compared with those in Table 18.3.

**Table 18.7** Result of DC sweep analysis

| V1 | I(R1) | I(R2) | I(R3) | V(1,2) |
|---|---|---|---|---|
| 9.000E+00 | 5.109E-01 | 1.945E-01 | 3.164E-01 | 5.109E+00 |
| 1.000E+01 | 5.564E-01 | 2.218E-01 | 3.345E-01 | 5.564E+00 |
| 1.100E+01 | 6.018E-01 | 2.491E-01 | 3.527E-01 | 6.018E+00 |

## 18.7 Dependent or controlled sources

In studies of electrical engineering, we use voltage and current sources, but in electronics we meet with other types of source known as **dependent sources** or **controlled sources**. In these sources, a current (for example) in one part of a circuit may be dependent on the current in another part of the circuit. An example of this is found in the transistor, in which the collector current is controlled by the current in the base circuit. Yet another example is the operational amplifier, in which the output voltage is controlled by the voltage between the input terminals. SPICE has four types of controlled sources, namely:

1. Type E – voltage-controlled voltage source;
2. Type F – current-controlled current source;
3. Type G – voltage-controlled current source;
4. Type H – current-controlled voltage source.

Typical applications of these include:

(a) Type E – operational amplifier;
(b) Type F – bipolar junction transistor;
(c) Type G – field-effect transistor;
(d) Type H – separately excited generator.

We will consider the use of an E-element in the operational amplifier circuit in Fig. 18.2. This type of polarity-inverting amplifier was discussed in Chapter 16, where it was shown that the voltage gain of the circuit is

$$\frac{V_3}{V_{in}} = -\frac{R_f}{R_1}$$

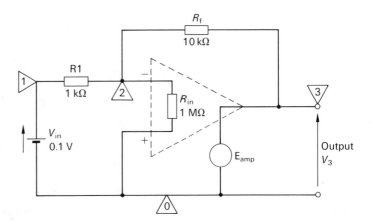

**Fig. 18.2** Simulation of an operational amplifier circuit

The input file for the circuit in Fig. 18.2 is given in Table 18.8, and the reader should note the order in which the nodes for the op-amp (Eamp) are written. The gain is given as a phase-inverting gain of one million, and the differential resistance between the input terminals, Rin is 1 MΩ.

It should be noted from the output file (Table 18.9) that the voltage at node 2 (which is the virtual earth point) is 1 μV, and the overall voltage gain is

$$\frac{\text{Output (node 3) voltage}}{\text{Input (node 1) voltage}} = \frac{-1}{0.1} = -10$$

**Table 18.8** The input file for the op-amp circuit in Fig. 18.2

```
Polarity-inverting op-amp circuit


Vin    1   0    DC    0.1V

R1     1   2    1Kohm;    input resistor

Rf     2   3    10Kohm;   feedback resistor

Rin    2   0    1MEGohm; differential input resistance


*Source name

*|      (+) output terminal

*|      |   (-) output terminal

*|      |    |   (+) input (controlling) node

*|      |    |    |   (-) input (controlling) node

*|      |    |    |    |    gain (volts/volt)

*|      |    |    |    |    |

Eamp   3   0    0    2    1MEG


.OPTIONS NOPAGE

.END
```

**Table 18.9** Output file for the circuit in Fig. 18.2

```
NODE    VOLTAGE     NODE    VOLTAGE      NODE    VOLTAGE

(    1)    .1000  (    2) 1.000E-06  (    3)   -1.0000


VOLTAGE SOURCE CURRENTS

NAME        CURRENT

Vin        -1.000E-04
```

**18.8 Use of expressions in SPICE**

A scientific or mathematical **expression** is a way of representing the value of a variable in words or symbols. For example, the resistance of a conductor can be represented by the equation

$$R = \rho l / a$$

where $\rho$ is the resistivity of the conductor, $l$ its length and $a$ its cross-sectional area. In this case, the *expression* for the resistance is $\rho l / a$.

SPICE allows us to specify a value in the form of an expression.

That is, instead of writing down the numerical value of the resistance, we can write it down in the form of an expression as follows:

R1  2  0  {rho * len/area}

where rho, len and area have been defined as parameters in a '.PARAM' statement in the input file; an example is given in Table 18.10, the output file being in Table 18.11.

**Table 18.10**  The use of expressions in a SPICE input file

```
Power consumed in a resistor



Vs    1    0    10V



* Vam is an 'ammeter' in the circuit

Vam    1    2    0V



* Define equation PARAMeters

.PARAM  rho = 17.3n len = 92.49,area = 0.8u

* NOTE: a space or a comma can be used to separate parameters



*        Expression for resistance

*                   ¦

R1    2    0    {rho*len/area}



*Source name

*¦        (+) output node

*¦        ¦  (-) output node

*¦        ¦  ¦            Expression

*¦        ¦  ¦                  ¦

Epower    10    0    VALUE = (I(Vam)*V(2))

* Every SPICE circuit node must have a 'd.c.' connection to node 0

Rpower    10    0    1ohm

.OPTIONS  NOPAGE

.END
```

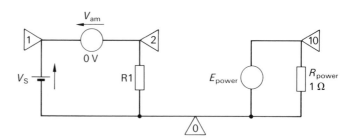

**Fig. 18.3**  Figure for the input file in Table 18.10

As stated in the title line of the file, the main purpose is to calculate the power consumed in a resistor but, as an aside, it is to demonstrate the power of expressions in SPICE. Moreover, the example enables us to demonstrate the use of a very popular form of SPICE 'ammeter'. The circuit is shown in Fig. 18.3 and, as we have already noted, SPICE can determine the current which flows through a voltage source (a 'V' source), and if the value of the source is zero volts, it can be used as an 'ammeter'. We use Vam in Fig. 18.3 (see also Table 18.10) in this way. However, we must take great care to ensure that *the anticipated direction of current flow is into the positive node of Vam*. Since it is reasonable to assume that current *flows out* of the positive terminal of $V_S$, then *node 1 must be the positive node of Vam, and node 2 its negative node*.

Next we define the PARAMeters in the expression for the resistance in a '.PARAM' line; here we use rho = $17.3 \times 10^{-9}\,\Omega\,m$ (typical of annealed copper), len (length) = 92.49 m, and area = $0.8 \times 10^{-6}\,m^2$; these values give a resistance of about $2\Omega$ (to within about 0.5 per cent). When 10 V is applied to the resistor, a current of 5 A (or thereabouts) flows in it, and the power consumed is $10\,V \times 5\,A = 50\,W$.

A feature of voltage-controlled sources (see also section 18.7) such as E-sources and G-sources, is that their value can also be represented in the form of an expression. In this example we determine the power consumed in resistor R1 by using an E-source (Epower) whose VALUE is given by

$$\text{VALUE} = \{\text{circuit current} \times \text{voltage across R1}\}$$

$$= \{I(Vam) * V(2)\}$$

The reader will see that, since the voltage of node 0 is zero volts, then the voltage at node 2 [i.e. V(2)] is the voltage across R1. Additionally, since SPICE 'knows' that (voltage × current = watts), the final line in the results table, Table 18.11, gives the power consumed as 50 W! Moreover, the voltage at node 10 is the power VALUE.

It will also be seen that we have separated Epower from the main circuit, with the exception that node 0 is common. Also, SPICE insists that every node should have a 'd.c. path' to node

**Table 18.11** Results produced by the input file in Table 18.10

| NODE | VOLTAGE | NODE | VOLTAGE | NODE | VOLTAGE |
|------|---------|------|---------|------|---------|
| ( 1 ) | 10.0000 | ( 2 ) | 10.0000 | ( 10 ) | 49.9980 |

```
VOLTAGE SOURCE CURRENTS

NAME            CURRENT

Vs            -5.000E+00

Vam            5.000E+00

TOTAL POWER DISSIPATION   5.00E+01   WATTS
```

0, and it will not allow the input file in Table 18.10 to run unless node 10 has a d.c. link to node 0. To satisfy this requirement, we have connected a $1\,\Omega$ resistor (Rpower) between node 10 and node 0; strictly speaking, this resistor could have almost any value (except zero).

## 18.9 Using SPICE to plot transients

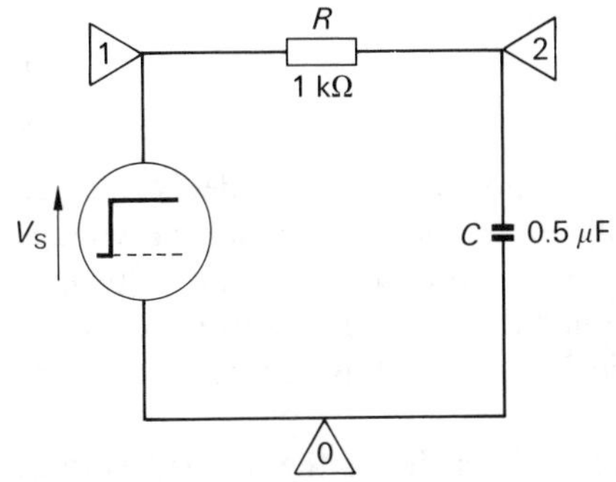

**Fig. 18.4** Circuit for transient analysis in section 18.9

SPICE can not only calculate and print results, it can also PLOT them. There are two separate and quite distinct ways it can do this.

The first is to plot the results as standard text symbols on a printer, i.e. in the form of a series of '*' or '+' characters). The advantage of this arrangement is that it is simple and can be used with any printer; the disadvantage is that it gives a crude print-out, and the plotted points do not always lie in smooth curves.

The second method is to use a plotter or a printer which can operate in a graphics mode. PSpice has an optional graphics post-processor called PROBE, which effectively allows the computer and monitor to be used as a *software oscilloscope*.

We now look at the way in which SPICE handles transients in the circuit in Fig. 18.4. Since we are dealing with transients in a circuit, we must specify the 'waveform' applied to the circuit as a time-varying quantity. We do this in the input file which describes the circuit (see Table 18.12) by means of a PWL wave (Piece-Wise Linear wave).[†] What we do here is to join a number of specified points on a graph by a series of straight lines. This is satisfactory in our case because we are applying a step change or sudden change in voltage to the circuit.

Thus, in the Vs line in Table 18.12, we say that Vs is zero from $t = 0$ to $t = 0.4999$ ms (see also below) and, at that time, the

---

[†] SPICE can deal with the following waveforms: PWL – Piece-Wise Linear (containing up to 3995 pairs of points), PULSE (a repetitive pulse with controllable rise-time, fall-time, pulse-width, etc.), EXP (EXPonential rise and fall wave), and SFFM (Single-Frequency Frequency Modulated wave).

**Table 18.12** Transient response of the circuit in Fig. 18.4

```
Transients in an R-C circuit


*Name of source

*¦            Type of source

*¦                   ¦  T1,V1      T2,V2        T3,V3

*¦                   ¦    ¦          ¦            ¦

Vs    1    0    PWL( 0s,0V   0.4999ms,0V   0.5ms,20V )


R    1    2    1kohm

C    2    0    0.5ufarad


*Analysis type

*¦  Step time     End of TRANsient analysis

*¦         ¦           ¦

.TRAN    0.25ms    3.5ms


*PLOT the result on a printer

*¦ Type of analysis to be PLOTted

*¦          ¦   Variables to be PLOTted

*¦          ¦      ¦          ¦

.PLOT    TRAN    V( c )    V( 1,2 )

.OPTIONS   NOPAGE

.END
```

voltage rises to 20 V; that is the supply voltage is suddenly changed at $t = 0.4999$ ms. SPICE only allows us to use 'practical' voltage sources, and insists that it takes a finite time for the voltage to 'rise' to 20 V. We have assumed that it takes 0.0001 ms or 0.1 $\mu$s to reach 20 V; for all practical purposes, this can be taken to be zero time!

We must specify the points which define the applied voltage wave by means of pairs of values inside brackets on the PWL line, as follows:

0s,0V means that when $t = 0$, Vs $= 0$;
0.4999ms,0V means that when $t = 0.4999$ ms, Vs $= 0$ V;
0.5ms,20V means that when $t = 0.5$ ms, Vs $= 20$ V.

Since no further values are given, SPICE assumes that Vs remains at 20 V thereafter. The pairs of points which specify the waveform are in the order TIME, VOLTAGE, and can be separated either by a comma or a space.

Finally, the analysis type is specified in a '.TRAN' line, in which we give the step interval between the points on the transient analysis, and the length of time over which the analysis is to be performed.

The time constant of the circuit is

$$\tau = RC = 1000 \times 0.5 \times 10^{-6} = 0.5 \, \text{ms}$$

We can therefore expect the transients in the circuit to be completed by $5 \times 0.5 = 2.5$ ms after the voltage change is applied. In our case the voltage change occurs at $t = 0.5$ ms, so that the transients will have decayed after 3 ms. For this reason, we have restricted the length of the transient analysis to 3.5 ms. For comparison, the reader may like to know that this problem was solved earlier in Worked Example 13.1.

In the simple text-type print-out in Table 18.13, the following characters are used:

'X' – both voltages have the same value;
'*' – the capacitor voltage, V(C);
'+' – the resistor voltage, V(1,2).

**Table 18.13**   Text PLOT from the input file in Table 18.12

```
LEGEND:
*: V(C)
+: V(1,2)
  TIME        V(C)
(*+)--------   0.0000E+00   5.0000E+00   1.0000E+01   1.5000E+01   2.0000E+01
              - - - - - - - - - - - - - - - - - - - - - - - - - - - - -
  0.000E+00  0.000E+00 X          .            .            .            .
  2.500E-04  0.000E+00 X        . /            .            .            .
  5.000E-04  2.001E-03 *          .            .            .            +
  7.500E-04  7.851E+00 .          .       *    .       +    .            .
  1.000E-03  1.266E+01 .          .          + .            *    .       .
  1.250E-03  1.554E+01 .          .   +.      .            .  .*         .
  1.500E-03  1.730E+01 .       +     .        .            .      *      .
  1.750E-03  1.836E+01 .   +        .         .            .         *   .
  2.000E-03  1.901E+01 . +          .         .            .          * .
  2.250E-03  1.940E+01 . +          .         .            .           * .
  2.500E-03  1.964E+01 .+           .         .            .            *.
  2.750E-03  1.978E+01 .+           .         .            .            *.
  3.000E-03  1.987E+01 +            .         .            .            *
  3.250E-03  1.992E+01 +            .         .            .            *
  3.500E-03  1.995E+01 +            .         .            .            *
              - - - - - - - - - - - - - - - - - - - - - - - - - - - - -
```

The reader will recall that the applied voltage must have a finite rise-time, so that 20 V is applied to the circuit marginally before $t = 0.5$ ms. This allows us to see the change in the voltage both across the capacitor [V(C)] and across the resistor [V(1,2)].

The two sets of values printed to the left of Table 18.13 are: firstly, the time and, secondly, the value of the first-named variable in the '.PLOT' line, namely the capacitor voltage V(C). The reader will find it of interest to verify that the capacitor voltage, at any given point in time, is in very close agreement with the theoretical value (see Chapter 13). A table of $V_R$ and $V_C$ can also be obtained by including a '.PRINT' line in the file.

When we look at the PROBE print-out of the capacitor voltage [V(2)] and the resistor voltage [V(1,2)] in Fig. 18.15 we see that, even on a dot matrix printer, the curves are much better than those in Table 18.13. The curve traced out by the black squares is $v_R$, and that traced out by the 'white' squares is $v_C$.

**Fig. 18.5** PROBE print-out from a 24-pin dot matrix printer for $V_C$ [i.e. v(2)] and $V_R$ [i.e. v(1,2)] for the circuit in Table 18.12

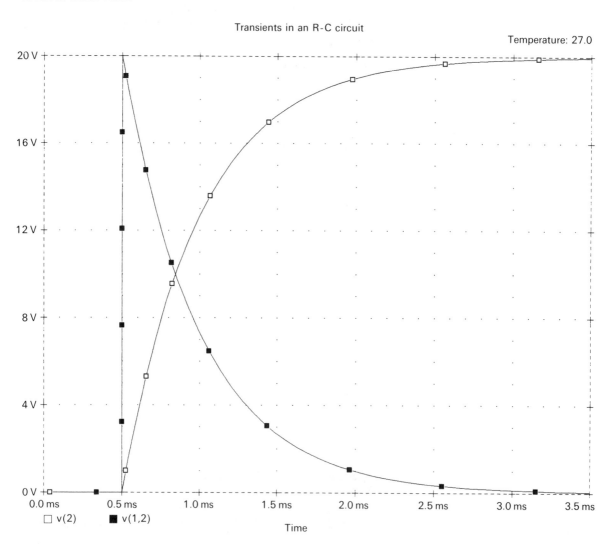

Transients in an R–C circuit

Temperature: 27.0

## 18.10 A.c. circuit analysis

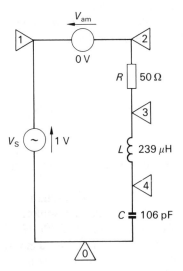

**Fig. 18.6** Resonance in a series R–L–C circuit

In this case we will look at the series *R–L–C* circuit in Fig. 18.6, whose supply frequency is 'swept' through its resonant frequency. It is left as an exercise for the reader to verify that the resonant frequency is 1 MHz, and that the *Q*-factor at resonance is 30.

In this case we use a 1.0 V a.c. source; the fact that it is 'a.c.' must be included in the 'Vs' line in the input file in Table 18.14. As with one of the earlier circuits, we will use a null voltage source, Vam, as an 'ammeter'.

**Table 18.14** AC analysis of a series R–L–C circuit

```
Series RLC resonance

Vs      1    0    AC     1V; 'AC' MUST be stated

Vam     1    2          0V; Vam is a SPICE ammeter

R       2    3    50ohm

L       3    4    239uH

C       4    0    106pF

.OPTIONS NOPAGE


*AC analysis

*¦  LINear sweep

*¦      ¦   No. of points in sweep

*¦      ¦      ¦ Start frequency (Hz)

*¦      ¦      ¦      ¦ End frequency (Hz)

*¦      ¦      ¦      ¦              ¦

.AC   LIN    9    0.9MEGhz     1.1MEGhz


.PRINT    AC    IM(Vam)    IP(Vam)    VM(3,4)

.END
```

When using an '.AC' line, SPICE assumes that the frequency is 'swept' from a 'start' frequency to an 'end' frequency, and we need to state the following in the '.a.c.' line:

1. the way in which the frequency is swept;
2. the number of frequency values used during the sweep;
3. the value of the 'start' and 'end' frequencies.

There are various ways in which we can 'sweep' through the frequency values, and we use a LINear sweep in which each frequency is equidistant from the next. We have chosen the 'start' frequency to be 0.9 MHz, and the 'end' frequency to be 1.1 MHz and, in selecting an odd number of frequency points, we ensure that the resonant frequency is in the centre of the range.

The '.PRINT' line must be specified as dealing with an 'AC' analysis, and the quantities to be printed differ from those used hitherto. In AC analysis, 'I' and 'V' can be modified by a suffix, and we use the following:

IM(Vam) – Magnitude of I in Vam;
IP(Vam) – Phase of I in Vam;
VM(3,4) – Magnitude of V of node 3 relative to node 4 (i.e. the voltage across L).

The reason that we have chosen the above values is that we are particularly interested in (i) the current at resonance, and (ii) the $Q$-factor at resonance.

For the circuit in Fig. 18.6, the current at resonance is $I_0 = V_S/R = 1/50 = 0.02\,\text{A}$, and the phase angle at resonance is $0°$ ($I$ is in phase with $V_S$). We see from the results in Table 18.15, that SPICE predicts a current of 0.02 A at 1 MHz and a phase angle of $-0.25°$ (which is practically zero). Moreover, the $Q$-factor at resonance is

$$Q = \frac{\text{voltage across L at resonance}}{\text{supply voltage}} = \frac{\text{VM}(3,4)}{\text{Vs}}$$

$$= \frac{30.03}{1} = 30.03$$

which is practically equal to the theoretical value!

**Table 18.15** AC frequency 'sweep' of the circuit in Fig. 18.6

| FREQ | IM( Vam ) | IP( Vam ) | VM( 3,4 ) |
|---|---|---|---|
| 9.000E+05 | 3.118E-03 | 8.103E+01 | 4.214E+00 |
| 9.250E+05 | 4.177E-03 | 7.795E+01 | 5.802E+00 |
| 9.500E+05 | 6.180E-03 | 7.200E+01 | 8.817E+00 |
| 9.750E+05 | 1.101E-02 | 5.660E+01 | 1.612E+01 |
| 1.000E+06 | 2.000E-02 | -2.516E-01 | 3.003E+01 |
| 1.025E+06 | 1.116E-02 | -5.609E+01 | 1.717E+01 |
| 1.050E+06 | 6.448E-03 | -7.119E+01 | 1.017E+01 |
| 1.075E+06 | 4.479E-03 | -7.706E+01 | 7.230E+00 |
| 1.100E+06 | 3.434E-03 | -8.011E+01 | 5.672E+00 |

## 18.11 CODAS and PCS software

CODAS software (COntrol system Design And Simulation) is an integrated graphics-based package for designing and simulating feedback control systems. It is a powerful tool not only for the practising control engineer, but also for the technician.

The package not only deals with the transient response of systems (see typical screen dumps in Figs 16.6–16.9, inclusive), but also with the frequency response of systems and the response of non-linear systems (such as those which include saturation effects, backlash in gears, etc.). The systems can be excited with a wide range of input signals, including step functions, ramp functions, sine waves and special user-defined signals.

Although the systems are defined mathematically (see, for example, the edit box at the foot of any one of the above figures), the software is particularly useful as a demonstration aid for almost any type of control system.

PCS software (Process Control Simulation) provides a very easy method of simulating process control plants using one-, two- or three-term controllers. Features relating to proportional, integral and derivative control can be studied (see Chapter 16), together with the effects of exponential and distance/velocity lags.

CODAS is more mathematically based, and is the more general and powerful form of the two.

# Index